D1739110

Nanoscience and Nanotechnology for Chemical and Biological Defense

ACS SYMPOSIUM SERIES **1016**

Nanoscience and Nanotechnology for Chemical and Biological Defense

R. Nagarajan, Editor
Natick Soldier Research, Development & Engineering Center

Walter Zukas, Editor
Natick Soldier Research, Development & Engineering Center

T. Alan Hatton, Editor
Massachusetts Institute of Technology

Stephen Lee, Editor
U.S. Army Research Office

Sponsored by
The ACS Division of Colloid and Surface Chemistry

American Chemical Society, Washington DC

Library of Congress Cataloging-in-Publication Data

Nanoscience and nanotechnology for chemical and biological defense / R. Nagarajan ... [et al.], editor ; sponsored by the ACS Division of Colloid and Surface Chemistry.

p. cm. -- (ACS symposium series ; 1016)

Proceedings of a symposium held at the 234th ACS National Meeting in Boston, MA on August 19-23, 2007.

Includes bibliographical references and index.

ISBN 978-0-8412-6981-1 (alk. paper)

1. Nanotechnology--Congresses. 2. Biotechnology--Congresses. 3. Chemical detectors--Congresses. 4. Chemical warfare--United States--Safety measures. 5. Biological warfare--United States--Safety measures. I. Nagarajan, R. (Ramanathan) II. American Chemical Society. Division of Colloid and Surface Chemistry. III. American Chemical Society. Meeting (234th : 2008 : New Orleans, La.)

TP248.25.N35N365 2009

623.4'59--dc22

2009024531

ISBN 978-0-8412-6981-1

The paper used in this publication meets the minimum requirements of American National Standard for Information Sciences—Permanence of Paper for Printed Library Materials, ANSI Z39.48–1984.

Distributed by Oxford University Press

PRINTED IN THE UNITED STATES OF AMERICA

Foreword

The ACS Symposium Series was first published in 1974 to provide a mechanism for publishing symposia quickly in book form. The purpose of the series is to publish timely, comprehensive books developed from the ACS sponsored symposia based on current scientific research. Occasionally, books are developed from symposia sponsored by other organizations when the topic is of keen interest to the chemistry audience.

Before agreeing to publish a book, the proposed table of contents is reviewed for appropriate and comprehensive coverage and for interest to the audience. Some papers may be excluded to better focus the book; others may be added to provide comprehensiveness. When appropriate, overview or introductory chapters are added. Drafts of chapters are peer-reviewed prior to final acceptance or rejection, and manuscripts are prepared in camera-ready format.

As a rule, only original research papers and original review papers are included in the volumes. Verbatim reproductions of previous published papers are not accepted.

ACS Books Department

Contents

Introduction

Detection of Chemical and Biological Agents

Protection from Chemical and Biological Agents

Indexes

Preface

This book results from the symposium "Nanoscience and Nanotechnology for Chemical and Biological Defense" held at the 234th ACS National Meeting in Boston, MA on August 19-23, 2007. The symposium was sponsored by the ACS Division of Colloid and Surface Chemistry as part of its continuing symposia series on nanoscience and nanotechnology. Approximately 80 papers were presented at this symposium and the book chapters represent a sampling of problems addressed. This is the first time this critical research area of chemical and biological defense has been addressed comprehensively at any ACS National Meeting and this is the first book on this topic to appear as part of the ACS Symposium Series.

The first two chapters provide a broad view of the field. Chapters 3 to 17 describe a number of detection methods applicable to chemical and biological threats. Chapters 18 to 24 describe a number of approaches to provide protection from chemical and biological agents. Many of the chapters describe not only outcomes of fundamental research but also how this research can be transitioned to produce practical detection devices and novel protective materials.

The first chapter is an invited commentary on the perspective of the Defense Threat Reduction Agency (DTRA), the federal organization that is vested with the responsibility to sponsor all research connected to chemical and biological defense. This chapter by Charles Bass Jr. clearly outlines DTRA's programmatic goals and the priority research needs. The second chapter is an invited assessment of nanotechnology research dedicated to chemical and biological defense. Zukas et al have provided a comprehensive view of all nanotechnology research efforts with particular emphasis on the projected needs and technology gaps for protection, detection and decontamination.

The remaining chapters are devoted to specific research problems concerning the detection of chemical and biological threats and protection against such threats. In Chapter 3, Mutharasan et al report on their recent work with piezoelectric-excited millimeter-sized cantilever (PEMC) sensor for detecting biological threat agents. PEMC sensors immobilized with an antibody specific to the target agent are shown to be highly sensitive for detecting 10 -100 Bacillus anthracis spores or the pathogen, E. coli O157:H7, in buffers and also in complex food matrices. In Chapter 4, Dai et al report on recent progress on electrospun polymers and polymer/CNT composite nanofibers, with an emphasis on those based on conducting polymers. The electrospun nanofibers based on polymer/carbon nanotube (CNT) composites are attractive multifunctional nanomaterials for many potential applications. In Chapter 5, Marcus Lay and coworkers describe the use of two dimensional networks of SWNTs for gas sensing application. They show how the networks can be grown by high temperature techniques, or room-temperature liquid deposition techniques to generate highly aligned SWNT networks on a wide variety of substrates.

Sooter et al report on their development of microfluidic sorting chips to rapidly screen for molecular recognition elements that can be used in immunochromatographic detection of biowarfare agents, in Chapter 6. The result is a convenient handheld device for the assay of biological agents. Another portable analytical system is described in Chapter 7. There Lin et al describe on-site detection of exposure to organophosphate (OP) pesticides and chemical nerve agents, using nanomaterials functioning as transducers. In Chapter 8, Driskell et al. investigate the use of aligned Ag nanorod arrays, as surface-enhanced Raman scattering (SERS) substrates for the identification and quantitation of viral pathogens. Paresh Ray and collaborators report in Chapter 9 on the development of a miniaturized, and battery operated ultra-sensitive gold nanoparticle-based surface energy transfer probe for screening of the bioagents DNA with excellent sensitivity and selectivity. They demonstrate multiplexed detection of two target sequences, oligonucleotide sequence associated with the anthrax lethal factor and sequence related to positions of the E.coli DNA which codes for the 23S rRNA, based on this approach.

In Chapter 10, Samuel Hernandez-Rivera and coworkers show that Raman detection of trace amounts of explosives and other hazardous materials on surfaces can be improved by ten to one thousand times with the addition of colloidal metallic nanoparticles to contaminated areas. Banahalli Ratna and colleagues demonstrate that organized spatial distribution of fluorescent reporter molecules on a virus capsid eliminates the commonly encountered problem of fluorescence quenching. Using such viral nanoparticles, they show in Chapter 11 that enhanced fluorescence for the detection of protein toxins is possible.

Kyung Son and coworkers have developed micro chemical sensor nodes consisting of GaN HEMT (High Electron Mobility Transistor) sensors and a RF communication link for long range chemical threat detection and early warning. They report in Chapter 12, results on high selectivity detection of chemical agents (stimulants) using the GaN HEMT sensors and optimal operating parameters for such high sensitivity detection. Reppy and Pinzola describe their work with polydiacetylene as a sensing material for point-detection of biological agents. In Chapter 13, they show that polydiacetylene with switchable fluorescence can be used either as coatings on nanoporous membranes or as liposomes attached to fiber supports, for the detection of biological targets. Weiss and Rong describe their development of a porous silicon resonant waveguide biosensor in Chapter 14 for the selective detection of DNA and other small molecules. Experimental demonstrations of the selective detection of complimentary DNA oligos, with negligible response for both non-complementary DNA and buffer controls, are reported.

Fan et al report on a method for detecting toxins that inhibit protein synthesis in Chapter 15. To detect a toxin, a group of proteins are simultaneously synthesized in a microfluidic array device. The production yields of these proteins are inhibited differentially by the toxin. The toxin can thus be identified based on the unique response pattern (or signature) of the array. The limit of detection for ricin, a bioterrorism agent, is determined at 10 pM. In

Chapter 16, Samuel Hernandez-Rivera and coworkers focus on a new class of SERS materials. They use crystalline nanoparticles of TiO_2, SnO_2 and Sc_2O_3 and show that polymorphism plays an important role in the Raman signal enhancement when using metal oxides. Nanoparticle structure and electronic properties of these systems are expected to allow for applications in sensors development for use in nitroaromatic explosives detection at sub-picogram level. Samuel Hernandez-Rivera et al also show in Chapter 17 that bimetallic Au/Ag nanoparticles can be used for detecting TNT in solution with high sensitivity and molecular specificity. The detection was an indirect method via alkaline hydrolysis of TNT using a strong base and the SERS signals can identify TNT nitro degradation products at 10^{-15} g.

Chapters 18 to 24 describe nanotechnology based methods to provide protection against chemical and biological threat agents. Peterson et al describe in Chapter 18, a method to increase the mesoporosity of zeolites by treating with sodium hydroxide. They show that an increase in mesoporosity, coupled with optimizing the acidity while maintaining the hydrophobicity of the zeolite resulted in the largest increase in the ethylene oxide (a toxic industrial chemical) breakthrough time. In Chapter 19, Grassian and Larsen demonstrate that nanocrystalline zeolites with crystal sizes less than 100 nm, possess enhanced absorptive and catalytic properties and can very effectively deactivate chemical warfare agent simulants. Kaiser and coworkers present their results on the adsorption capacity and off gassing properties of activated carbon based fabrics for the chemical warfare agent HD, a vesicant, and for one of its simulants, in Chapter 20 and argue that the average pore size of the carbon is a critical variable controlling adsorption capacity.

An electrostatic filter containing powdered activated carbon is described in Chapter 21. The filter is shown to retain high levels of microbes while the activated carbon within the structure is shown to display a high dynamic response as a chemical sorbent. In Chapter 22, Luzinov et al describe the surface modification of poly(ethylene terephthalate) (PET) membranes by grafting various polymers using the "grafting to" technique in order to reduce the permeability of the membrane to chemical agent vapors. Elabd and coworkers report on the synthesis of polymer-polymer nanocomposites membranes in Chapter 23. The synthesis is done either by filling nanoporous polymer membranes with polyelectrolyte gels or by encapsulating electrospun polyelectrolyte nanofiber meshes within polymer membranes. These unique membrane systems were shown to be have high permeability for water but low for chemical agents and responsive to an electrical stimulus. They can be candidates for protective clothing, electrochemical devices (e.g., fuel cells, batteries, actuators, sensors) and other perm-selective membrane applications. In the last chapter, Seong Kim et al report on the development of a scanning atmospheric radio-frequency (rf) plasma source for surface decontamination and protective coating deposition. This system was used for decontamination of organophosphorus nerve agents and deposition of hydrophobic and superhydrophobic coatings on various substrates including ceramics, metals, fabrics, and nanofibers.

The editors are thankful to the numerous reviewers. The reviewers of individual chapters contributed prompt critical reviews that have helped improve the quality of the manuscripts. The Editors acknowledge the support from their institutions, Natick Soldier Research, Development & Engineering Center, Massachusetts Institute of Technology and the US Army Research Office, that allowed them to organize the symposium and develop this book based on selected contributions.

R. Nagarajan
Molecular Sciences and Engineering Team
Natick Soldier Research, Development & Engineering Center
Natick, MA

Walter Zukas
Chemical Technology Team
Natick Soldier Research, Development & Engineering Center
Natick, MA

T. Alan Hatton
Department of Chemical Engineering
Massachusetts Institute of Technology
Cambridge, MA

Stephen Lee
U.S. Army Research Office
Research Triangle Park, NC

Introduction

Chapter 1

Challenges for Chemical and Biological Protection

Charles A. Bass, Jr.

Defense Threat Reduction Agency, Fort Belvoir, VA 22060

The Joint Science and Technology Office for Chemical and Biological Defense (JSTO-CBD), which is operated within the Defense Threat Reduction Agency's (DTRA) Chemical and Biological Technologies Directorate, has the responsibility to develop and manage the technology base for chemical and biological passive defense capabilities. The challenge of novel protective systems focus on factors such as burden, costs, duration of performance and effectiveness against a full range of agents. These challenges require the development of materials and systems that capture, block or destroy agents more effectively than the current systems. These technology solutions must also decrease material weight and costs, while reducing burden and lessening the impact on mission performance. Specific challenges in different technology areas and some approaches to addressing them are briefly described.

Introduction

The fall of the Berlin Wall two decades ago signaled the close of the Cold War and yet the approach to current capabilities for individual and collective chemical and biological protection has not sufficiently considered the evolution of the threat. Like Improvised Explosive Devices (IEDs) currently used against U.S. forces in Iraq and Afghanistan, an innovative threat force will be able to compensate for inadequacies and synthesize new ways to deliver Chemical and Biological (CB) agents on target. So, future use of CB weapons may be characterized as immediate, intense, and local, rather than the massive barrages and tons per hectare scenarios characteristic of cold war threat analysis. Threats may include classical CB agents, as well as non-traditional agents and emerging biological threats. It also can include Toxic Industrial Chemicals/Toxic Industrial Materials (TIC/TIM) that could threaten operational forces through accidental or malicious release. As a consequence, the risk may be rising. Although, protective equipment may not have to defend against the same quantities previously established as the standard, it may have to address a broader spectrum of agents. Further, the nature of the threat may require greater availability of protection that is built-in, rather than applied on the top of vehicles, vessels, platforms, structures and standard battledress. Protection that is always present will be available when needed. Providing this continuous, built-in protection without increasing the burden on the warfighter is where the challenge lies. Meeting this challenge is the focus of science and technology efforts within the protection capability area.

Challenges

Factors such as burden, costs, duration of performance, and effectiveness against the full spectrum of potential agents must be addressed. These operational challenges require the development of new materials and systems that capture, block or destroy agents more effectively than the current systems. Technology solutions must also decrease material weight and costs, while reducing burden and lessening the impact on mission performance. Specific challenges in material science can be subdivided into the technology theme areas of air purification and protective barriers.

Removal of hazardous high-volatility, low-molecular weight chemical vapors is the limiting performance characteristic of adsorption-based air purification technologies. The primary challenge is to develop broad spectrum treatments, while reducing size, flow resistance, and power demand. Particulates and aerosols have increasingly become important considerations for both biological and chemical threats, but current High Efficiency Particulate Air (HEPA) technologies create significant pressure barriers (pressure drops) to airflow and have diminished performance for the smallest particles. These traditional approaches also place a significant logistical burden that limits the wide-spread application to building protection. New technologies must address the high costs as well as limited shelf and service life of filters.

The development of protective barrier materials used for protective clothing addresses the spectrum of percutaneous threats, but must achieve protection with a relatively thin material over a large area. The balance of maximizing protective performance while minimizing thermal burden has been the essence of the challenge for years, and the increasing need to protect against particulates further complicates the issue. The goal of incorporating these materials into full-time wear uniforms is especially challenging because increased durability is needed to meet service life and the threshold of acceptable thermal burden is much lower. Self-detoxifying materials are seen as a means of increasing service-life and performance, while simultaneously decreasing thermal burden, but these materials need to be safe, shelf-stable and effective.

Novel, high-performance materials often cost more, so a better understanding of human performance and the relative system science will facilitate the analysis of trade-offs to apply these materials to the greatest effect. The burden and degradation of mission performance are warfighters' primary issues with currently fielded individual protection systems. Burden and mission degradation are based on a multitude of both cognitive and physiological factors. Some factors, such as heat burden, are well quantified, but others, such as cognitive effects of encapsulation, are not. Currently fielded and developmental individual and collective protection still use a variation of a basic design derived many decades ago. Whole system analysis is needed to determine the value of new and emerging technologies that may allow a revolutionary systems approach to individual and collective protection.

Approaches

Present investments have focused on the development of the materials that will feed into a complete integrated protective ensemble. Traditionally, protective mask and protective clothing have been developed as separate programs and little has been done to co-develop CB protection with ballistic protection and on-board optical and communication sub-systems that are now part of a modern fighting ensemble. A holistic approach can balance various elements, such as power demand, mission interface and thermal burden. As a concept refinement strategy, thermal performance can be managed as an independent variable by setting the desired thermal load equal to that of a standard battledress uniform, then determine the best achievable chemical/ biological performance under that constraint. Such an approach does a better job at spurring technological innovation and will provide the user with a better understanding of the tradeoffs between performance and physiological burden.

Much of the cognitive burden and mission degradation lies in the interface between and the chemical/biological protective ensemble and the balance of mission equipment. The chemical/biological protective ensemble must integrate with ballistic protection, optics, and communication. The interface of the protective mask with the helmet accounts for a number of issues. Problems include disrupting the helmet suspension system, potentially compromising the seal of the mask, and the creation of irritating "hot spots." Early integration between these systems must occur during concept refinement and technology development. An integrated chemical/biological ensemble that is a component or accessory to an integrated warfighter system, like the Army's Ground Soldier System, may be the desired outcome. Achieving low-burden necessitates the innovative application of new technologies.

A number of new technologies offer considerable opportunities in achieving integrated low-burden protection within a broadening threat spectrum without compromising needed performance. The growing field of nanostructured materials provides an opportunity to move beyond traditional materials such as activated carbon and zeolites. One of the most exciting areas is reticular chemistry, which is described as the "linking of molecular building blocks of synthetic and biological origin into predetermined structure using strong bonds" (UCLA Center for Reticular Chemistry, http://crc.chem.ucla.edu/). The most well known class of these materials is Metal Organic Frameworks (MOFs) which have already exhibited absorbency potentials that far exceed activated carbon, and are currently being manufactured in commercial quantities. These compounds can be tailored to target specific classes of chemicals that include performance-limiting, high-volatility TICs. MOFs and similar compounds could be used to design smaller and lower-profile filter beds that protect against the expanding spectrum of threats. Smaller and lower profile filters decrease weight and reduce interference of the respirator with other mission systems. Reticular chemistry has the potential to go further. Reticular structures could be designed to contain an internal catalyst that detoxified the agent, or structures could be designed to adsorb as well as detect the compound or agent adsorbed.

Another promising area has been the development of manufacturing techniques for nanofibers. These fibers have diameters on the scale of the mean-free-path of an air molecule. When nanofibers are used as a filter media the aerodynamics are governed by the "slip flow" or Knudsen regime with much lower drag than conventional flows. This implies that it may be possible to produce particulate filters with order-of-magnitude lower pressure drops, and high efficiency particulate filtration capabilities built into the clothing. In recent years, a number of innovations have just about made the production of these materials commercially viable. Additional developing technologies will make it possible to assemble these fibers into nano-composites that will include built-in adsorption, reactive, anti-microbial and sensing capabilities into a thin coating. This could revolutionize protective clothing and produce unconventional and extremely low burden approaches to respiratory protection.

Incorporation of novel materials may allow the reinvention of collective protection systems. Novel, low-cost and scalable approaches that allow seamless incorporation into all building and vehicle designs and support rapid (field) conversion of fixed facilities are being sought. Novel system approaches may allow less dependence on overpressure techniques and rely more heavily on network integration and rapid response. Technologies that include strippable coatings, self-detoxifying surfaces and responsive (switchable) surfaces could support such novel configurations that address protection against external, as well as internal fugitive sources of contamination. Perhaps, more importantly, intrinsic and universal collective protection may be more constrained by our thinking. A more flexible approach that identifies areas and points of specific vulnerability may be in order. Settling for less protection in lower threat areas can facilitate investments in lower-cost approaches. Lower costs allow for universal design of protection. This will enable a sufficient degree of protection to be always available and will likely reduce the overall risk.

Conclusion

Solutions to the challenge of low-burden protection depend on discovery, development and implementation of novel technologies. Of particular concern are technical challenges and expense associated with the development of new materials. Only by leveraging other funded areas of research can success be achieved. This requires partnering with industry through Cooperative Research and Development Agreements (CRADA) and developing and making use of cooperative agreements with allied countries. Despite the difficulty and expense of developing new materials, the technical opportunities present today assure a good chance of success. The risks imposed by the evolving nature of the threat demand the effort.

Chapter 2

Assessment of Nanotechnology for Chemical Biological Defense

Walter Zukas[1], Catherine Cabrera[2], James Harper[2], Roderick Kunz[2], Theodore Lyszczarz[2], Lalitha Parameswaran[2], Mordechai Rothschild[2], Michael Sennett[1], Michael Switkes[2], and Hema Viswanath[2]

[1]U.S. Army Natick Soldier Research, Development and Engineering Center, Natick, MA 01760
[2]Massachusetts Institute of Technology Lincoln Laboratory, Lexington, MA 02420

Nanotechnology encompasses a broad and rapidly growing set of projects in basic research, applied research, early attempts at commercialization, and even mature technologies. A multidisciplinary team from the Massachusetts Institute of Technology Lincoln Laboratory (MIT-LL) and the U.S. Army Natick Soldier Research, Development, and Engineering Center (NSRDEC) was tasked by the Special Projects Office at the Joint Science and Technology Office for Chemical and Biological Defense (JSTO-CBD) to assess ongoing nanotechnology research efforts and commercial off-the-shelf nanotechnologies relevant to CBD, with particular emphasis on projected needs and technology gaps in the Protection, Detection, Decontamination, and Medical CBD capability areas. This broad-based assessment resulted in a list of recommended technologies to address these needs. This paper addresses the process utilized for the assessment and provides the technical detail for the recommended technologies. Early access to advanced CBD capabilities based on the unique properties of nanotechnology has high potential to provide significantly better performance than incremental improvements in current technology.

Background

The term nanotechnology generally refers to the evolving body of tools and knowledge that allow manipulation of material structures at the scale of approximately 10-100 nanometers and to understand the relationship between nanometer scale features and the macroscopic properties of materials. Rapid progress in the development of analytical tools to probe the nanometer scale and to manipulate materials at this scale has led to a dramatic increase in the number and diversity of research programs focused on nanoscience and nanotechnology. There are now a significant number of federally funded nanotechnology programs being conducted by a variety of government agencies, as part of The National Nanotechnology Initiative (NNI), established by Congress in FY01. The latest (December 2007) NNI Strategic Plan *(1)* is available at "www.nano.gov". It lists the NNI participating agencies for FY08, which include Department of Defense, Department of Homeland Security, National Science Foundation, National Institutes of Health, and Department of Justice.

The essentially new phenomena encountered at the nanoscale, and which are not easily accessible at the macro level, present new opportunities. First and foremost, the deeply scaled dimensions of nanoparticles enhance the surface-area-to-volume ratio, and suitable surface chemistry can then lead to highly efficient sensing schemes or catalytic reactions.

In the sensing area we note that mechanisms for sensing chemicals can, in the most general terms, be considered to be based either on ionization, spectroscopic signatures, or sorption. Of these three sensing modalities, sorption-based detection is the one most suited for applications of nanotechnology. This is not surprising because, on the nanoscale of living systems, all sensing and molecular recognition functions are based on sorption of some sort, suggesting that man-made sensing systems aimed at mimicking biological systems will most likely also use this mechanism. When a molecule adsorbs to a surface it not only changes the mass at the surface, but it can also impart changes in the electronic structure of the adsorbate-substrate interface which in turn can result in changes in electrical, optical, and/or luminescent properties, all of which have been explored as mechanisms for chemical sensing. High surface-area-to-volume ratio nanostructures generally exhibit amplified responses to these properties, leading to sensor demonstrations with unprecedented sensitivity.

Nanotechnology may also play a role in development of non-caustic decontamination treatments. Most non-caustic decontamination chemicals exhibit slower reaction rates with agents than the caustic chemicals such as bleach or sodium hydroxide. To make up for this difference, liquid decontaminants will need to be finely and uniformly atomized to ensure efficient conversion to vapor which will facilitate penetration into all contaminated areas; whereas solid, particulate decontaminants will need to be very small to allow for a large effective surface area. This might lead to high-volume liquid dispensing systems that create micrometer- or even nanometer-sized aerosols that can be used to create vaporous hydrogen peroxide, or special preparation methods to produce nanoscale neutralizing powders such as titanium dioxide. Furthermore, the desire to create self-cleaning surfaces may require the development of nano-composites or nano-porous surfaces that offer a large functional surface area upon which sorbed threat chemicals can react with impregnated catalysts.

A unique property of nano-structured materials is the ability to achieve unprecedented levels of various properties and performance (notably permeation barrier properties, chemical selectivity and reactivity) in known materials. The ability to process known materials to create controlled nano-structures, or to combine known materials to create these structures, is a significant advantage over traditional material development programs based on iterative programs of synthesis, purification and characterization in that the constituent materials in the new nanomaterials tend to be commercially available in large scale. This advantage can facilitate development and has led in some cases to significantly faster transition from laboratory discovery to fielded application, as in the case of nanocomposite nylon materials which were first reported in the scientific literature in the early 1990's and were in mass production for commercial application in less than 10 years.

Besides the high surface-area-to-volume ratio afforded by nanostructures, unique aspects of nanotechnology have been demonstrated by the application of deeply scaled microfabrication technologies, which enable the preparation of miniaturized yet complex mechanical, electrical, electro-mechanical, or opto-electro-mechanical devices. Nanometer-scale resonant cantilevers have been demonstrated as sensors with mass sensitivity to zeptogram (roughly 1000 molecules) levels *(2)*, and some groups have even demonstrated selective responses by proper engineering of the surface coating. Because of their dimensions, nanoscale devices can be envisioned as components of arrays of massively parallel and/or redundant sensors or data processors. Along these directions, research is being conducted on "smart dust" sensors *(3)* comprised of micrometer-sized particles possessing nanometer-scale structures that undergo changes in reflectivity or luminescence upon adsorption of an analyte molecule and that can be probed remotely.

A range of ongoing work in nanotechnology is directed towards improving diagnostic and therapeutic capabilities. This work includes efforts to miniaturize sensors for evaluating the normal functioning of the body (e.g., glucose monitors), the development of nano-encapsulation technologies and targeting mechanisms to deliver drugs to specific cell types within the body, and the development of molecular machines and motors that can be operated within the body. Progress in these fields has direct application to the CB arena, as the success of any of these approaches makes the development of the equivalent device for biological or chemical agents significantly more feasible, whether it be blood-based organism detection or nerve-cell specific nerve agent antidote dispensing.

Given the promise of nanotechnology on the one hand and its relative immaturity on the other hand, the DoD has a unique opportunity to harness nanotechnology to solve CBD problems by steering, guiding, and accelerating nanotechnology development in directions most suited for its needs. Indeed, the DoD is actively pursuing some of the available nanotechnology approaches for these applications at this time, though on a limited scale. Examples include the development of nanocomposite elastomers with enhanced barrier properties and the use of reactive metal oxide nanoparticles as decontaminants for CW agents. Nevertheless, while some work is underway, there are numerous additional opportunities for nanotechnology to be applied to the CBD mission. An early step in this direction is a critical assessment of the ever-growing number of nanotechnology-based projects related to CBD.

Relevance

CB agents pose extreme challenges for detection, protection, and decontamination. Their characteristic feature is their high lethality, so that even minute amounts (micrograms to milligrams) can constitute a lethal dose. Therefore, the fundamental challenge of CBD is to develop products which are highly sensitive, selective, and efficient. Sensors must detect agents at levels well below LD50, and still having extremely low levels of false alarms. Protective gear or filters must be highly efficient in neutralizing or blocking the penetration of CB agents, while being specific enough so as not to impede the flow of harmless materials. Decontamination techniques must be equally efficient and selective, in order to enable rapid restoration of functionality to areas that have been attacked. In all these applications, nanotechnology holds the promise of enabling novel approaches and eventually novel solutions that would also be affordable and easily deployable. For instance, properly nanostructured materials may be engineered into highly efficient filters for building protection or into uniforms and masks for individual protection of the warfighter. The warfighter would also greatly benefit from the availability of distributed arrays of nanosensors that detected CB agents rapidly and with high specificity. These considerations indicate that the potential of nanotechnology should be exploited as quickly as possible, and be readied for transitioning to the warfighter in a streamlined process.

However, the risk exists that not every nanotechnology-based approach can meet its stated goal. There may be difficulties in meeting performance targets because of fundamental scientific limitations. More probably, the difficulties may arise when specific concepts of operation are examined more closely, and are found to limit the full utilization of the proposed device or process. Finally, scale-up in production of nanostructured materials or nano-devices may encounter unexpected challenges in quality control, health-related effects, and cost. Since nanotechnology is such a new field, there is little experience in turning a promising concept into a useful product. Therefore, the benefit to the warfighter would be greatly enhanced by a systematic and critical evaluation of the various nanotechnology-based projects with applications to CBD. Such an evaluative process will set its goal to accelerate and streamline the technology transfer of the most promising concepts to their intended users.

Assessment Structure

The nanotechnology assessment was conducted by a multidisciplinary team from the MIT-LL and NSRDEC on the basis of literature and patent searches, attendance of conferences, site visits, and direct contacts with nanotechnology researchers. The assessment included projects funded by a broad range of government agencies, as well as those funded by industry. The projects are carried out in various organizations, including academia, government laboratories, industry, and consortia. The evaluative process emphasized an integrated systems-oriented approach, anchored in basic science and phenomenology, but focused on impact on CBD needs, feasibility of implementation, and estimated time to market.

Team members attended the American Chemical Society Fall 2006 National Meeting, the 3rd International Conference on Bioengineering and Nanotechnology, Nanotech 2006, the 7th International Conference on MEMS and BioMEMS, and the 3rd International Conference on Photonic Nanosystems. Team members, together with JSTO personnel, visited the nanotechnolgy laboratories at Lawrence Livermore National Laboratory, Sandia National Laboratory, Los Alamos National Laboratory, Oak Ridge National Laboratory, the MIT Institute for Soldier Nanotechnologies, and the University of Massachusetts Center for Hierarchical Manufacturing. The MIT-LL/NSRDEC team also met with the capability area managers of JSTO-CBD, and obtained information from them on their research priorities and future plans. These were factored into the team's assessment and recommendations.

Recommendations

The following recommendations were ultimately guided by the potential applications to CBD, specifically, applications to the detection of chemical and biological warfare agents, protection from attack by such agents, post-attack decontamination, and medical countermeasures including diagnostics, prophylaxis, and therapeutics.

Sensitive Detection of Molecule-Receptor Binding Events without Chemical Modification of the Receptor

The ability to detect molecule-receptor binding events is a critical aspect for chemical detection methods based on molecular recognition. Currently, fluorescence is a primary means by which these events are detected. *(4-6)* However, this often requires modification of the receptor to incorporate the fluorescent center and/or immobilization of the receptor molecule onto a surface. Far more desirable would be a means to directly detect the molecule-receptor interaction in solution without the need to modify the naturally occurring receptor. In the absence of any receptor modification, however, there are no signatures stemming from the molecule-receptor interaction that are measurable using currently available means. One possible means to detect these binding events would be to sense the heat of reaction from these small ensembles of molecule-receptor reaction pairs. *(7)* Calorimetric means to detect this heat would obviate the need for engineering fluorescent centers into the receptor, and could result for a whole new class of sensors, but other detection methods may also be feasible. The recommendation is to further explore the feasibility of such alternative sensing technologies.

Ultrahigh-Gain Plasmonic Surfaces

Plasmonic phenomena relying on locally enhanced electric fields have been the subject of numerous projects aimed at ultra-sensitive detection, especially as surface-enhanced Raman spectroscopy (SERS) and plasmon-assisted fluorescence *(8-12)*. However, the elegant demonstrations of single-molecule spectroscopy in localized "hot spots" *(13)* do not constitute a practical field sensing strategy. Practical use of this phenomenon requires substrates that can be inexpensively fabricated that possess a very high density of "hot spots", such that the average enhancement is high enough to result in real performance benefits.

The recommendation is to specifically focus on efforts to achieve a high density ($>10^4/mm^2$) of adsorption sites producing the highest plasmon-assisted electric field enhancements. The surfaces must exhibit these structures over large areas ($>mm^2$) and be fabricated using low-cost means. Localized nanometer-scale structures that produce high field enhancements that enable single-molecule detection are not of interest. The surfaces can include either two or three dimesional arrangements of these high-field structures. The expected use of these surfaces is as substrates for surface-enhanced Raman spectroscopy or plasmon-assisted fluorescence. The efforts would also include modeling that includes entire nanostructure ensembles to get ensemble-averaged, as opposed to single-molecule, sensitivity enhancements

Nanotechnology-Based Sensor Components

The vast majority of nanotechnology-based CB sensor research has focused on ultra-sensitive transducers such as nanowires, nanotubes, and cantilevers *(14-16)*. However, sensing elements are only useful if particles of interest are present in the sample volume being interrogated; as that volume decreases, the effective concentration in the sample must increase *(17)*. A simple reduction in volume while retaining the materials in the initial sample may not be sufficient if the initial sample contains materials that could cause either false positives or false negatives. Therefore, both sample purification and sample concentration are necessary steps for sample preparation. While sample purification and concentration methodologies are readily available for benchtop applications *(18)*, methodologies and devices that readily interface with nanoscale detectors are still in the early stages of development *(19-21)*. Furthermore, nanotechnology has the potential to further improve the performance of the subsystems of sample collection *(22)*, preconcentration *(23-25)*, transducer functionalization *(26)*, and so on. The recommendation is to focus on the integration of several of these technologies into a single sensor system, so as to bring nanotechnology closer to deployment, and allow direct comparison with more conventional detection technology.

Sensors Using Single-Walled Carbon Nanotubes (SWNTs) as Transducers

Changes in the electrical properties of CNTs, induced by adsorbates, can provide a platform for the development of highly sensitive chemical and gas sensors *(27-29)*. For such an application, one needs to fabricate arrays of CNT-based field-effect transistors (FETs), and here one of the critical issues of using arrays of CNTs as sensors is in achieving individual addressability of each of the nanotubes *(30-32)*. In addition, optical interrogation may be a viable alternative transduction modality, since SWNTs are highly fluorescent in the range of 900 to 1600 nm, a region that is transparent to biological materials such as flesh and blood, and they also exhibit a strong Raman signal in that range *(33-38)*. Like many fluorescent particles, their optical properties can change with the local environment *(36)*. While much effort has been expended on improving the sensitivity of CNT sensors, a continuing problem is improving target-sensor interaction, especially in liquid samples. Developing methods to rapidly bring the target in proximity to the nanotube sensing element is critical in improving response time of nanotube sensors. The electrokinetic behavior of SWNTs in solution is also an area of great potential *(21)*, but also great uncertainty, and would benefit from additional basic research.

The recommendation is to focus on the following aspects of SWNT transducers: development of structure and fabrication methods for large arrays of SWNT that are addressable either individually or in controlled groups; development of high-performance SWNT-based FET arrays, and development of controllable contacts to SWNTs; conductivity modulation of SWNT by modifying structure through introduction of defects/dopants *(39,40)*; investigation of the optical properties of SWNTs in the context of their use as chemical and gas sensors.

Nanotechnology-Based Methods for Protein Stabilization

Many proposals for applying nanotechnology to CBD involve the use of proteins (i.e., enzymes, antibodies). Antibodies provide the specificity, and to some extent sensitivity, for many proposed nano CB sensors (e.g., nanowires *(41,42)*, cantilevers *(43)*, SPR *(44)*, etc.), while others rely on amperometric detection of enzyme catalysis *(45)*. In addition, enzymes that specifically attack CB targets show some promise for relatively benign decontamination. However, one of the major obstacles to the adoption of these promising protein-based nanotechnologies is the fragility of the proteins under normal environmental conditions. Even in solution, protein half-lives at room temperature can be as short as a few hours or days, and the problem is only compounded by dehydration or elevated temperatures.

Recently several nanotechnology-based techniques have emerged with the promise of extending protein lifetimes to months or even years under fairly rigorous conditions *(46-56)*. The nano size scale allows a much greater surface/volume ratio than in previous protein stabilization efforts, greatly improving the enzyme/substrate interaction rate. In addition, the generally gentle synthesis conditions allow a greater preservation of protein activity than previous techniques. The recommendation is to focus on improving these new techniques and demonstrate them specifically on enzymes, antibodies, and potentially other proteins or peptides of interest for CBD.

Controlled Synthesis of SWNTs or Other Organic Structures with Similar Functionality

Numerous reports have demonstrated properties of SWNTs that may enable improved materials, sensors, and devices with the potential to revolutionize CBD capabilities (see topic above). A primary obstacle in realizing this potential has been the lack of scalable synthesis methods that provide high purity, low defect rate, and defined molecular structure (e.g., chirality, diameter, and length) of as-grown SWNTs *(57,58)*. All of the existing methods are similar in that they require high temperatures in excess of 600°C and tight control over both the atmosphere in the reaction chamber and rate of delivery of vapor-phase carbon sources to support growth. The as-grown products of all of these methods are also characterized by the presence of numerous structural defects, a variety of different tube sizes and structures, and varying degrees of impurities, including metal catalyst particles, fullerenes, multiwall carbon nanotubes (MWNTs) and amorphous carbon that require multiple post-synthesis purification steps to produce clean nanotubes *(59,60)*. The recommendation is to focus on projects that combine development of highly controlled SWNT synthesis methods with demonstration of improved properties of the defined products for applications relevant to CBD.

The area of organic supramolecular assemblies is based on recent advances in synthetic organic, colloid, and polymer chemistries *(61-69)*. The broad field of organic supramolecular assemblies has largely avoided the label of "nanotechnology," but it certainly possesses the requisite properties to be recognized as such. Organic supramolecular assemblies may provide breadth of functionality greater than that of carbon nanotubes (CNTs), while enabling simpler and better paths to their synthesis. CNTs have shown promise as structural, barrier, sensing, and antimicrobial materials. However, as noted above, one of the biggest challenges is to separate and purify them *(70)*, as many of these applications require monodisperse nanotube compositions. Organic supramolecular assemblies have similar dimensionality as CNTs, but their synthesis and functionalization are carried out in solution, and therefore have the potential of larger diversity and much better control. The recommendation is to explore the inherent capabilities of this field, and engage a largely untapped resource of researchers to explore specific applications to CBD.

Nanocomposite Barrier Materials with Novel Functionalities

The field of nanplatelet polymeric additives is well established *(71-75)*. The recommendation here is to emphasize new functionalities, including catalytic reactivity and selective binding. There is potential for improved agent barrier performance from these materials for protection applications, as well as the addition of multifunctionality to include sensing and decontamination. Projects addressing the chemical modification of the nanoparticle reinforcements would advance the use of this class of materials for CBD applications.

"Bucky-paper" membranes *(76,77)* have not yet been explored for filtration applications, especially in the liquid state, and it is recommended to investigate it more closely. Permeants though bucky-paper are interacting with the outer walls of the nanotubes in these membranes, and pass though irregularly shaped pores with a range of size distributions. By functionalizing the surface of the nanotubes, it is expected that the transport properties and permselectivity of the bucky-paper membranes can be controlled.

Metal-Organic Frameworks as Novel Adsorbents/Reactants

Metal-organic frameworks (MOFs) are synthetic microporous materials with well defined and well–controlled pore size and the potential for a broad range of functionalities *(78-80)*. MOFs present a three-dimensional hybrid structure between inorganic zeolites and fully organic molecules, and therefore they offer the prospect for high degree of tailorability and chemical flexibility. MOFs have also been shown to have a very large degree of porosity, and therefore have been explored as platforms for effective gas storage. They may prove to be equally effective in CBD, but to date little research has been carried out in this direction. The recommendation is to explore the promise and applicability of MOFs to CBD. Specific topics could include: synthesizing/tailoring MOFs for the effective adsorption of chemical warfare agents; synthesizing/tailoring MOFs for subsequent reaction or catalysis of decontamination reactions with chemical warfare agents; showing that the physical properties of the MOFs are sufficient for envisioned CBD applications; and showing that commercial production of useful MOFs is feasible and cost effective when compared to currently used adsorbents.

Novel Developments in Electrospun Membranes

Electrospun polymeric membranes have been investigated for CBD applications for some time *(81-83)*. Early materials lacked the flexibility and durability required in many protective systems, and efforts did not sufficiently address the issue of production. Current advances in the area of modeling electrospun membranes *(84,85)*, elastomeric membranes *(86)*, and effective use of electrospun membranes as supports for reactive nanoparticles could assist in the understanding and application of this class of materials in chem./bio defense. The recommendation is to focus on projects that specifically address the membrane structural tailorability, the membrane flexibility and durability, the effective addition of agent reactivity into the membrane, and the commercial scale production of membranes.

Permselective Membranes Based on Carbon Nanotubes

Nano-permeable membranes (NPMs), especially those based on carbon nanotubes, have been the focus of extensive research. Recently, several groups have reported that the transport of water through the nanotube pores is orders of magnitude higher than predicted by classical hydrodynamic theories *(87-89)*. However, research on NPMs, has been thus far primarily confined to studies of small molecules, primarily in the gas phase, though the most recent publications cited also deal with liquid permeants. Little significant work has been done either theoretically or experimentally to explore the interaction of NPMs with chemical and biological warfare agents or simulants. The field appears to have excellent potential to yield substantial valuable results from an investment focused on projects specifically tailored to address chem/bio protection, and the long-standing need for permselective membranes with improved water transport and high selectivity.

Nanoscale Materials for Medical Countermeasures and Exposure Characterization

Nanoparticles and nanofibers have been reported to have utility for a variety of medical applications including imaging contrast agents, cancer therapeutics, targeted drug delivery, controlled-release vehicles for pharmaceuticals, vaccine adjuvants, and gene therapy vectors. The recommendation is to focus on projects that seek to extend these results to applications of direct relevance to CBD (e.g., chemical and biological agent prophylaxis, vaccines, diagnostics, and treatments) and that seek to demonstrate that the nanoscale features of these methods to offer revolutionary capability improvements when compared to traditional approaches.

Acknowledgements

The authors wish to acknowledge the support and assistance of Dr. Jerry C. Pate of the Joint Science & Technology Office, Defense Threat Reduction Agency (DTRA). The Lincoln Laboratory portion of this work was sponsored by the DTRA under Air Force Contract FA8721-05-C-0002. Opinions, interpretations, conclusions, and recommendations are those of the authors, and do not necessarily represent the view of the United States Government.

References

1. *The National Nanotechnology Initiative Strategic Plan*, National Science and Technology Council, Arlington, VA, December 2007.
2. Nishio, M.; Sawaya, S.; Akita, S.; Nakayama, Y. *Appl. Phys. Lett.* **2005**, *86*, 133111.
3. Warneke, B.; Last, M.; Liebowitz, B.; Pister, K. *IEEE Computer* 2001, *34(1)*, 44.
4. Zhang, S-H.; Swager, T. *J. Am. Chem. Soc.* **2003**, *125*, 3420.
5. Aslan, K.; Gryczynski, I.; Malicka, J.; Matveeva, E.; Lakowicz, J.; Geddes, C. *Current Opinion in Biotechnology* **2005**, *16*, 55.
6. Viveros, L.; Paliwal, S.; McCrae, D.; Wild, J.; Simonian, A. *Sensors and Actuators B: Chem.* **2006**, *115*, 150.
7. Fon, W.; Schwab, K.; Worlock, J.; Roukes, M. *Nano Lett.* **2005**, *5*, 1968.
8. Orendorff, C.; Gearheart, L.; Jana, N.; Murphy, C. *Phys. Chem.:Chem. Phys.* **2006**, *8*, 165.
9. Lakowicz, J. *Plasmonics* **2006**, *1*, 5.
10. Lakowicz, J.; Chowdury, M.; Ray, K.; Zhang, J.; Fu, Y.; Badugu, R.; Sabanayagam, C.; Nowaczyk, K.; Szmacinski, H.; Aslan, K.; Geddes, C. *Proc. SPIE* **2006**, *6099*, 34.
11. Yan, F.; Wabuyele, M.; Griffin, G.; Vo-Dinh, T. *Proc. SPIE* **2004**, *5327*, 53.
12. Szmacinski, H.; Pugh, V.; Moore, W.; Corrigan, T.; Guo, S.-H.; Phaneuf, R. *Proc. SPIE* **2005**, *5703*, 25.
13. Jiang, J.; Bosnick, K.; Maillard, M.; Brus, L. *J. Phys. Chem B* **2003**, *107*, 9964.
14. Hierold, C.; Jungen, A.; Stampfer, C.; Helbling, T. *Sensors and Actuators A: Phys*ical **2007**, *136*, 51.
15. Chopra, N.; Gavalas, V.; Hinds, B.; Bachas, L. *Anal. Lett.* **2007**, *40*, 2067.
16. Li, M.; Tang, H.; Roukes, M. *Nature Nano.* **2006**, *2*, 114.
17. Sheehan, P.; Whitman, L. *Nano Lett.* **2005**, *5*, 803.
18. Commercial biological-reagent kits are available.
19. Austin, R. *Nature Nanotechnology* **2007**, *2*, 79-80.
20. Yang, L. J.; Banada, P.; Chatni, M.; Lim, K.; Bhunia, A.; Ladisch, M.; Bashir, R. *Lab On A Chip* **2006**, *6*, 896.
21. Zhou, R.; Wang, P.; Chang, H. *Electrophoresis* **2006**, *27*, 1376.
22. Sigurdson, M.; Mezic, I.; Meinhart, C. in *BioMEMS and Biomedical Nanotechnology*, v. 4; Bashid, R.; Wereley S. Eds.; Springer; **2005**
23. MacDonald, M.; Spalding, G.; Dholakia, K. *Nature* **2003**, *426*, 421.
24. Wang, Y-C.; Stevens, A.; Han, *J. Anal. Chem.* **2005**, *77*, 4293.
25. Karnik, R.; Castelino, K.; Majumdar, A. *Appl. Phys. Lett.* **2006**, 88, 123114.
26. Bunimovich, Y.; Ge, G.; Beverly, K.; Ries, R.; Hood, L.; Heath, J. *Langmuir* **2004**, *20*, 10630.
27. Sood1, A.; Ghosh, A. *Phys. Rev. Lett.* **2004**, *93*, 086601.
28. Byon, H.; Choi, H. *J. Am. Chem. Soc.* **2006**, *128*, 2188.
29. Kim, S.; Rusling, J.; Papadimitrakopoulos, F. *Advanced Materials* **2007**, *19*, 3214.

30. Tseng, Y.; Xuan, P.; Javey, A.; Malloy, R.; Wang, Q.; Bokor, J.; Dai, H. *Nano Lett.* **2004**, *4*, 123.
31. Qi, P.; Vermesh, O.; Grecu, M.; Javey, A.; Wang, Q.; Dai, H. *Nano Lett.* **2003**, *3*, 347.
32. Durkop, T.; Kim, B.; Fuhrer, M. *J. Phys.: Condens. Matter* **2004**, *16*, R553.
33. Barone, P.; Parker, R.; Strano, M. *Anal. Chem.* **2005**, *77*, 7556.
34. Barone, P.; Baik, S.; Heller, D.; Strano, M. *Nature Materials* **2005**, *4*, 86.
35. Heller, D.; Baik, S.; Eurell, T.; Strano, M. *Advanced Materials* **2005**, *17*, 2793.
36. Bachilo, S.; Strano, M.; Kittrell, C.; Hauge, R.; Smalley, R.; Weisman, R. *Science* **2002**, *298*, 2361.
37. Heller, D. A.; Jeng, E. S.; Yeung, T-K; Martinex, B. M.; Moll, A. E.; Gastala, J. B.; Strano, M S.; Science 2006, 311, 508-511.
38. Rao, A.; Richter, E.; Bandow, S.; Chase, B.; Eklund, P.; Williams, K.; Fang, S.; Subbaswamy, K.; Menon, M.; Thess, A.; Smalley, R.; Dresselhaus, G.; Dresselhaus, M. *Science* **1997**, *275*, 187.
39. Sternberg, M.; Curtiss, L.; Gruen, D.; Kedziora, G.; Horner, D.; Redfern, P.; Zapol, P. *Phys. Rev. Lett.* **2006**, *96*, 075506.
40. Lee, Y-S.; Marzari, N. *Phys. Rev. Lett.* **2006**, *97*, 116801.
41. Patolsky, F.; Timko, B.; Zheng, G.; Lieber, C. *MRS Bulletin* **2007**, *32*, 142.
42. Zheng, G.; Patolsky, F.; Cui, Y.; Wang, W.; Lieber, C. *Nat. Biotechnol.* **2005**, *23*, 1294.
43. Wee, K.; Kang, G.; Park, J.; Kang, J.; Yoon, D.; Park, J.; Kim, T. *Biosens. Bioelectron.* **2005**, *20*, 1932
44. Yu, F.; Persson, B.; Löfås, S.; Knoll, W. *Anal. Chem.* **2004**, *76*, 6765.
45. Deo R.; Wang, J.; Block, I.; Mulchandani, A.; Joshi, K.; Trojanowicz, M.; Scholz, F.; Chen, W.; Lin Y-H. *Anal. Chim. Acta.* **2005**, *530*, 185.
46. Kim, J.; Grate, J.; Wang, P. *Chem. Eng. Sci.* **2006**, *61*, 1017.
47. Luckarift, H.; Spain, J.; Naik, R.; Stone, M. *Nature Biotech.* **2004**, *22*, 211.
48. Sheldon, R.; Schoevaart, R.; van Langen, L. *Biocatal. Biotrans.* **2005**, *23*, 141.
49. Kim, S-H. *Analytical Chem.* **2006**, *77*, 6828.
50. Mulchandani, A. *J. Ind. Microbiol. Biotechnol.* **2005**, *32*, 554.
51. Lejeune, K. *Nature* **1998**, *395*, 27.
52. Kim, P.; Kim, D.; Kim, B.; Choi, S.; Lee, S.; Khademhosseini, A.; Langer, R.; Suh, K. *Nanotechnology* **2005**, *16*, 2420.
53. Zong, S.; Cao, Y.; Ju, H. *Electroanalysis* **2007**, *19*, 841.
54. Wu, S.; Ju, H.; Liu, Y. *Adv. Funct. Materials* **2007**, *17*, 585.
55. Gribenko, A.; Makhatadze, G. *J. Molecular Biology* **2007**, *366*, 842.
56. Asuri, P.; Karajanagi, S.; Yang, H.; Yim, T.; Kane, R.; Dordick, J. *Langmuir* **2006**, *22*, 5833.
57. Sinnott, S.; Andrews, R. *Critical Reviews in Solid State and Materials Sciences* **2001**, *26*, 145.
58. Dai, H. *Acc. Chem. Res.* **2002**, *35*, 1035.
59. Park, T-J.; Banerjee, S.; Hemraj-Benny, T.; Wong, S. *Journal of Materials Chemistry* **2006**, *16*, 141.
60. Banerjee, S.; Hemraj-Benny, T.; Wong, S. *Journal of Nanoscience and Nanotechnology* **2005**, *5*, 841.

61. Kinbara, K.; Aida, T. *Chem Rev.* **2005**, *105*, 1377.
62. Yamamoto, Y.; Fukushima, T.; Saeki, A.; Seki, S.; Tagawa, S.; Ishii, N.; Aida, T. *J. Amer. Chem Soc.* **2007**, *129*, 9276.
63. Motoyanagi, J.; Fukushima, T.; Kosaka, A.; Ishii, N.; Aida, T. *J. Polym. Sci. A: Polym. Chem.* **2006**, *44*, 5120.
64. Wusong, J.; Fukushima, T.; Niki, M.; Kosaka, A.; Ishii, N.; Aida, T. *Proc. Nat. Acad. Sci.* **2005**, *102*, 10801.
65. Szymanski, C.; Wu, C.; Hooper, J.; Salazar, M.; Perdomo, A.; Dukes, A.; McNeill, J. *J. Phys. Chem. B* **2005**, *109*, 8543.
66. Wang, H.; Gu, L.; Lin, Y.; Lu, F.; Meziani, M.; Luo, P.; Wang, W.; Cao, L.; Sun, Y-P. *J. Am. Chem. Soc.* **2006**, *128*, 13364.
67. Rajangam, K.; Behanna, H.; Hui, M.; Han, X.; Hulvat, J.; Lomasney, J.; Stupp S. *Nano Lett.* **2006**, *6*, 2086.
68. Stone, M.; Moore J. *Org. Lett.* **2004**, *6*, 469.
69. Wolffs, M.; Hoeben, F.; Beckers, E.; Schenning, A.; Meijer, E. *J. Amer. Chem. Soc.* **2005**, *127*, 13484.
70. Kim, S.; Zachariah, M. *Nanotechnology* **2005**, *16*, 2149.
71. Giannelis, E. *Advanced Materials* **1996**, *8*, 29.
72. Messersmith, P.; Giannelis, E. *J. Polym. Sci. A* **1995**, *33*, 1047.
73. Lebaron, P.; Wang, Z.; Pinnavaia, T. *Applied Clay Science* **1999**, *15*, 11.
74. Gilman, J.; Kashiwagi, T.; Lichtenhan, J. *SAMPE Journal* **1997**, *33*, 40.
75. Lee, S; Cho, D.; Drzal, L. *J. Mater. Sci.* **2005**, *40*, 231.
76. Wang, Z.; Liang, Z.; Wang, B.; Zhang, C.; Kramer, L. *Composites, Part A: Applied Science and Manufacturing* **2004**, *35A(10)*, 1225.
77. Endo, M.; Muramatsu, H.; Hayashi, T.; Kim, Y.; Terrones, M.; Dresselhaus, M. *Nature* **2005**, *433*, 476.
78. Rowsell, J.; Yaghi, O. *Microporous and Mesoporous Mater.* **2004**, *73*, 3.
79. Reiter, W.; Taylor, K.; An, H.; Lin, W.; Lin, W. *J. Am. Chem. Soc (Comm.).* **2006**, *128*, 9024.
80. Snurr, R.; Hupp, J.; Nguyen, S. *AIChE J.* **2004**, *50*, 1090.
81. Reneker D.; Chun, I. *Nanotechnology* **1996**, *7*, 216.
82. Gibson, P.; Gibson, H.; Rivin, D. *AIChE J.* **1999**, *45*, 190.
83. Subbiah, T.; Bhat, G.; Tock, R.; Parameswaran, S.; Ramkumar, S. *J. Appl. Polym. Sci.* **2005**, *96*, 557.
84. Deitzel, J.; Kleinmeyer, J.; Harris, D.; Beck Tan, N. *Polymer* **2001**, *42*, 261.
85. Teo, W.; Ramakrishna, S. *Nanotechnology* **2006**, *17*, R89.
86. Threepopnatkul, P.; Murphy, D.; Mead, J.; Zukas, W. *Rubber Chem. Technol.* **2007**, *80*, 231.
87. Hummer, G.; Rasaiah, J.; Noworyta, J. *Nature* **2001**, *414*, 188.
88. Majumder, M.; Chopra, N.; Andrews, R.; Hinds, B. *Nature* **2005**, *438*, 44.
89. Holt, J.; Park, H.; Wang, Y.; Stadermann, M.; Artyukhin, A.; Grigoropoulos, C.; Noy, A.; Bakajin, O. *Science* **2006**, *312*, 1034.
90. Chen, J.; Saeki, F.; Wiley, B.; Cang, H.; Cobb, M.; Li, Z-Y.; Li, X.; Xia, Y. *Nano Lett.* **2005**, *5*, 473.
91. Kobayashi, H.; Brechbiel, M. *Advanced Drug Delivery Reviews* **2005**, *57*, 2271.
92. Rhyner, M.; Smith, A.; Gao, X.; Mao, H.; Yang, L.; Nie, S. *Nanomedicine* **2006**, *1*, 209.

93. Ferreri, M. *Nature Reviews* **2005**, *5*, 161.
94. Sinha, R.; Kim, G.; Nie, S.; Shin, D. *Mol. Cancer Ther.* **2006**, *5*, 1909.
95. Park, K. J. *Controlled Release* **2007**, *120*, 1.
96. Koo, O.; Rubinstein, I.; Onyuksel, H. *Nanomedicine: Nanotech. Bio. Med.* **2005**, *1*, 193.
97. Fifis, T.; Gamvrellis, A.; Crimeen-Irwin, B.; Pietersz, G.; Li, J.; Mottram, P.; McKenzie, I.; Plebanski, M. *J. Immuno.* **2004**, *173*, 3148.
98. Scheerlinck, J.; Gloster, S.; Gamvrellis, A.; Mottram, P.; Pleblanski, M. *Vaccine* **2006**, *24*, 1124.

Detection of Chemical and Biological Agents

Chapter 3

Piezoelectric-excited Millimeter-sized Cantilever (PEMC) Sensors for Detecting Bioterrorism Agents

Gossett Campbell, David Maraldo, and Raj Mutharasan

Department of Chemical and Biological Engineering
Drexel University
Philadelphia, PA 19104

We summarize our recent work with piezoelectric-excited millimeter-sized cantilever (PEMC) sensor for detecting biological agents that may be used as biothreat agents. PEMC sensors immobilized with an antibody specific to the target agent is shown to be very highly sensitive for detecting 10 - 100 biothreat agents (*Bacillus anthracis* spores, *E. coli* O157:H7), both in buffers and in food matrices. After a brief introduction to the principles of sensing and response characteristics of PEMC sensors, results from detection experiments are described.

Introduction

The area of biosensors has been an active area of development for detecting pathogenic bacteria and toxins in a variety of applications such as food analysis, clinical diagnostics, and for biothreat monitoring (*1-17*). Current methods rely on enrichment culture wherein the target bacteria is grown selectively, and then identified with labeled reagents, or from extracted DNA followed by polymerase chain reaction (PCR). If the pathogenic agent is present in copious amounts, one can use traditional methods such as enzyme-linked immunosorbent assays (ELISA) which has a high limit of detection - ~10^5 cells for pathogens and ~ng/mL for toxins. While such methods are well established and can accurately identify biothreat agents, they are laborious and time-consuming and often require 8-24 hours for detection and identification (*18*). Therefore, there is a great need for biosensors that accurately and rapidly detect a few biothreat agents in real matrices. In this regard, cantilever biosensors have attracted considerable interest for label-free detection of proteins and pathogens because of their promise of very high sensitivity (*19,20*). The binding of an antigenic target (biothreat agent) to an antibody-immobilized cantilever sensor causes a resonance frequency decrease due to increase in mass, and is measured electronically. Two excellent reviews on cantilever sensors are available and the reader is referred to them (*21,22*). The cantilever sensors reported in the literature, by and large, use optics or piezoresistive transduction mechanism for measuring resonance frequency. In this paper, we describe a new class of cantilever sensors developed in the author's laboratory over the past five years. They are self-excited and self-sensing and have been labeled as piezoelectric-excited millimeter-sized cantilever (PEMC) sensors. This new class of cantilever sensors have an attached piezoelectric layer which is used both for excitation and sensing. The authors have shown that these sensors exhibit sub-femtogram sensitivity and were successfully used to detect *Bacillus anthracis* (*6, 23-25*), *Eschericia coli* O157:H7 (*26-30*) and food toxins (*31,32*). More recently we showed the same sensor can be used to detect single stranded DNA (*33*). In this paper we describe the applications for the biothreat agents taken from our recently published work.

Physics of Millimeter-sized Cantilever Biosensors

The piezoelectric ceramic used in the construction of PEMC sensors is made from oxides of lead, zirconium, and titanium and is referred to as lead zirconate titanate (PZT). The PZT gives a sensitive response to weak stresses due to the direct piezoelectric effect and generates high strain via the converse piezoelectric phenomenon (*34*). PEMC sensors are fabricated to provide predominantly the bending mode vibration. An electric voltage applied across the thickness of the PZT film will lengthen or shorten the film depending on the polarity of the electric field. Such a change in dimension causes the cantilever to bend, twist or buckle. If the applied electric field is alternated periodically, the composite cantilever would vibrate, and would resonate when the electric excitation frequency coincides with the mechanical or natural resonance frequency. The natural frequency of the cantilever depends on the bending modulus and the mass density of the composite cantilever. At resonance the cantilever undergoes a significantly higher level of vibration and larger stresses. As a consequence, the PZT layer exhibits a sharp change in impedance, and is conveniently followed by measuring the phase angle between excitation voltage and the resulting current (*35*). Further details on construction and properties of PEMC sensors may be found in our earlier publications (*23,34*).

Antibody Immobilization

A PEMC sensor surface (glass) is immobilized with an antibody (Ab) to the bioterrorism agent of interest. Several functionalization methods have been reported in the literature (*36*). In the current work, two main functionalization schemes were used. These are (1) derivatize the glass surface with an amino silane, (2) gold coat the glass surface and then, immobilize protein G. The Ab-immobilization technique via amine-terminated silane involves three main steps: cleaning, silanization, and antibody immobilization. After cleaning, the glass surface is silanylated with 0.4% 3-aminopropyl-triethoxysilane (APTES; Sigma-Aldrich) in deionized water at pH 3.0 and 75°C for 2 hours. APTES reacts with glass leaving a free amine group for further reaction with carboxyl group to form a peptide bond. The carboxyl group present on the antibody is activated using the zero length cross linker 1-ethyl-3-(3-dimethylaminopropyl)-carbodiimide (EDC; Sigma-Aldrich) and promoted by sulfo-N-hydroxysuccinimide (Sigma-Aldrich). EDC converts carboxyl groups into reactive unstable intermediates susceptible to hydrolysis. Sulfo-NHS replaces the EDC producing a stable reactive intermediate that is susceptible to attack by amine group on sensor. Covalent coupling of the stable intermediate with the silanylated glass surface is carried out at room temperature for 2 hours. The glass surface with the immobilized antibody is used to detect the antigen of interest. In the case of Au-coated sensor, it is exposed to 10 μg/mL Protein G (Pierce) for 2 h followed by dipping in Ab solution of interest, also at 10 μg/mL for 1 hour at room temperature. For all BA experiments we used antibody against spore surface antigents obtained from Chemicon International (Temecula, CA) or a rabbit polyclonal antibody provided kindly by Professor Rest (Drexel University College of Medicine). Anti-E. coli O157:H7 was obtained from KPL (Gaithersburg, MD)

Resonance Characteristics of PEMC Sensors

PEMC sensors exhibited dominant modes in the region 0 – 200 kHz and 700 kHz – 1.0 MHz with Q-values ranging from 20 – 110. In Figure 1 the resonance spectrum (phase angle versus excitation frequency) in air of a typical sensor and its geometric design are shown.

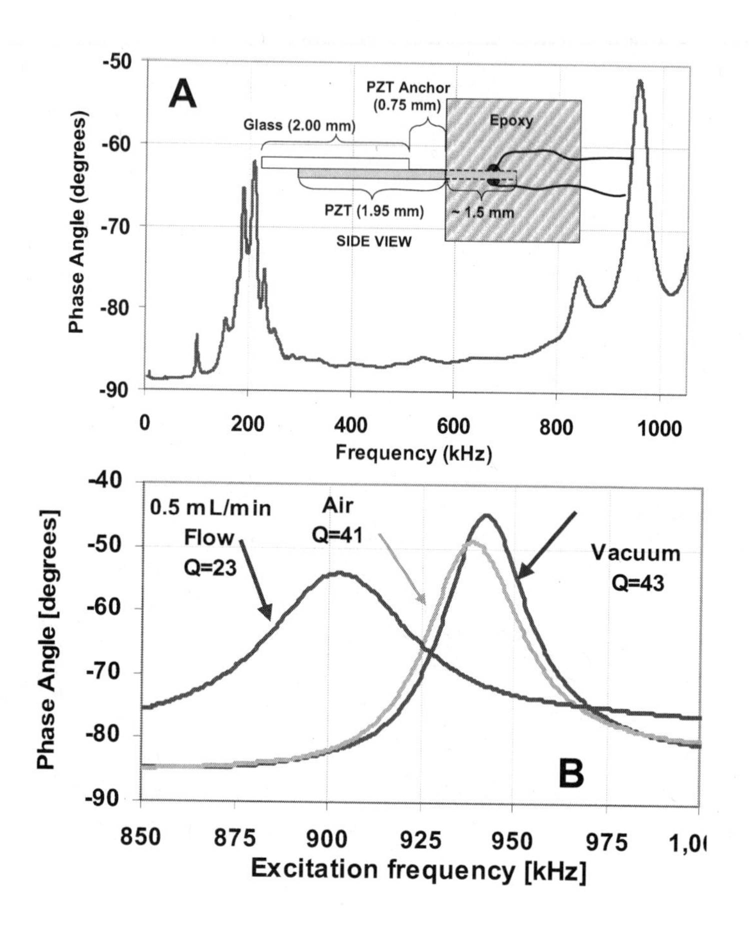

Figure 1. Resonance spectra of a PEMC sensor A: Resonance spectrum exhibiting resonant modes near 100 kHz, 200 kHz, 850 kHz, and 944 kHz. B. The resonant frequency increased from 941.5 ± 0.05 kHz to 944.50 ± 0.05 kHz in vacuum and when submerged in PBS the resonant frequency decreased to 902.5 ± 0.05 kHz. The peak height decreased in liquid by ~ 40%, due to added oscillating liquid mass. Adapted from reference(34).

Depending on the dimensions and construction of the composite cantilever, resonance frequencies were observed in the range of 0.7 to 1.2 MHz. The resonance spectra (Fig. 1B) in vacuum (~50 mTorr, 23.6 ^{0}C), in air (23.6 ^{0}C) and in phosphate buffered saline (10mM, PBS; 30 ^{0}C) are shown for the dominant high-order mode at 941.5 kHz in air (Q = 41). The parameter Q is a measure of peak sharpness and is equal to the ratio of resonance frequency to half-height peak width. When mounted in a specially constructed sample flow cell and PBS is flowed in at 0.4 to 1.0 mL/min, the resonant frequency decreased (~50 kHz) due to added mass effect and the peak height decreased by ~20 to 60%. At 1 mL/min, Q-value remained sufficiently high (~ 23), that the resonant frequency can be measured with an accuracy of ± 20-40 Hz. Resonant frequency in vacuum increased by 3,100 Hz compared to air. Since the primary difference between these two measurements is the air mass that surrounded the cantilever, the magnitude of increase gives a measure of the mass-change sensing potential of the sensor.

Mass-change Sensitivity of PEMC Sensors

The mass-change sensitivity can be determined experimentally by adding known mass to the sensor and then measuring the resulting resonance frequency decrease. A plot of the mass added versus the frequency change gave a straight line whose slope is termed mass-change sensitivity, and is expressed in the units of g/Hz. We have used paraffin wax dissolved into hexane (*25*) and sub-microliter amounts of the wax solution of known concentration were dispensed on the sensor surface. After allowing the solvent to evaporate at room temperature in vacuum, the resonance frequency was measured. The sensitivity of PEMC sensors lie in the range of 0.3 to 2 fg/Hz (*25,34*). The implication of such a high sensitivity is the attachment of a single bioterrorism agent such as a *Bacillus anthracis* spore which weighs ~ 1 pg would be detectable in near real time.

Detection of *Bacillus anthracis* in Batch Samples

In batch detection, Ab-immobilized sensor was dipped in one mL of sample containing the bioterrorism agent *Bacillus anthracis-Sterne strain* (BA; formalin-killed; a generous contribution from Professor Rest, DU College of Medicine), and the resulting resonance frequency decrease was measured (*24*). Figure 2 illustrates the sensor response to samples containing $3x10^2$, $3x10^3$, $3x10^4$ and $3x10^6$ BA spores/mL. In all cases, we note that the response showed a rapid decrease during the first few minutes followed by a slower change reaching a constant resonance frequency. For the highest spore sample ($3x10^6$ spores/mL), the rate of decrease was more rapid compared to the lower concentration sample (300 spores/mL). This is an expected response, as the binding rate is proportional to concentration. The total change was 2696 ± 6 Hz (n=2) for 3 x 10^6/mL sample and 92 ± 7 Hz (n=3) for the 300 spores/mL sample. Sample containing 30 spores/mL gave a response of 31 Hz in one out of three trials, and is not shown. Several control experiments with antibody-functionalized sensor were conducted in PBS. The response showed resonance

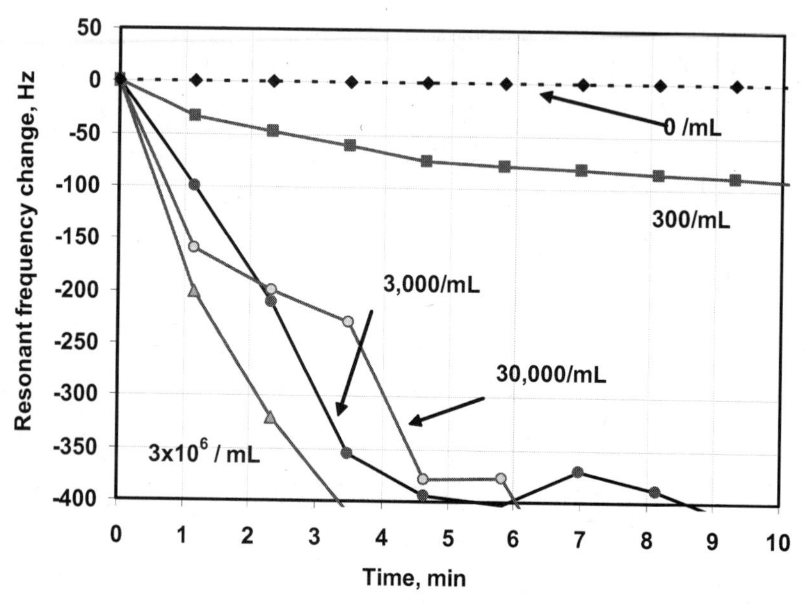

Figure 2. Resonant frequency shift of the second flexural mode of PEMC sensor upon binding of Bacillus anthracis spores at various sample concentrations to antibody functionalized cantilever. The results showed that the binding rate strongly depends on concentration. The control was an antibody-immobilized cantilever immersed in PBS. Adapted from reference(24).

frequency fluctuations of ± 5 Hz which is considerably lower than the signal obtained with 300 spores/mL sample. The noise level in measurement was ~ 20 Hz for the low spore count samples, and thus limit of detection is estimated as ~30 spores per mL. With higher spore concentration samples the change in resonance frequency was significantly higher than noise. The observed resonance frequency fluctuation of ± 5 Hz for the final steady state value indicates that the cantilever resonance characteristic is quite stable under liquid immersion.

In order to examine selectivity properties of PEMC sensor, an antibody-immobilized sensor was exposed to BA spores in presence of copious amount of *Bacillus thuringiensis* (BT). Response to mixed samples containing *Bacillus thuringiensis* (BT) spores were obtained and is given in Fig 3. We compare in Fig 3 the response to pure BA samples with samples that contained BT in various amounts: Sample-A (BA : 3 x 10^6/mL, BT: 0 /mL), Sample-B (BA : 2.4 x 10^6/mL, BT: 3.0 x10^8 /mL), Sample-C (BA : 2.0 x 10^6/mL, BT: 5.0 x10^8 /mL)

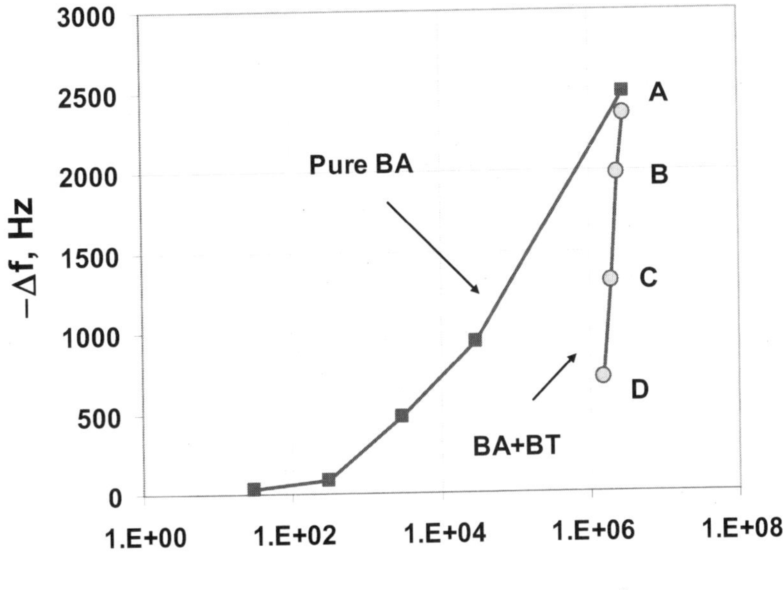

Figure 3. PEMC sensor response to pur BA and mixed BA+BT samples in batch measurement. Zero BA and pure BT responses were zero and are not shown. Composition of samples were: A (BA : 3 x 106/mL, BT: 0 /mL), B (BA : 2.4 x 106/mL, BT: 3.0 x108 /mL), C (BA : 2.0 x 106/mL, BT: 5.0 x108 /mL) and D (BA : 1.5 x 106/mL, BT: 7.5 x108 /mL). Presence of non-antigenic BT reduced sensor response in batch measurement. As shown in a later section flow mitigates this limitation. Data from reference(24).

and Sample-D (BA : 1.5 x 10^6/mL, BT: 7.5 x10^8 /mL). Sample-A gave the highest frequency change (2360 Hz; n=1) in 1 h. Samples containing increasing BT spores exhibited lower the resonance frequency change - 1980, 1310, and 670 Hz for the Samples B, C and D, respectively. Pure BT spore sample (1.5 x 10^9/mL) gave only a 10 Hz change suggesting non-specific binding, if any, was not significant. Sample D contained 1.5 x 10^6 BA spores/mL and therefore, one would expect a higher resonance frequency response in comparison to the pure *Bacillus anthracis* spore sample of 3 x 10^4 BA spores/mL. However, the opposite result was observed as shown in Fig 3. That is, the pure BA sample (3 x 10^4 BA spores/mL) gave a much higher overall resonance frequency change (1030 Hz) than Sample D containing 7.5 x 10^8 /mL BT spores (670 ± 10Hz). These results suggest that the presence of large amount of BT spores hindered attachment of BA spores, in batch mode. We hypothesized that BT in these experiments "crowded" the sensor surface due its very high relative concentration and reduced antigen transport to sensor surface. The crowding effect can be overcome by the introduction of flow and is discussed in the next section.

It is noted that the steady state resonance frequency response of the PEMC sensor showed good correlation with log of BA concentration. That is, the results in Figs 2 and 3 suggest that relationships for estimating BA concentration can be stated as:

$$\log(C_{b0}) = \frac{(-\Delta f)_{ss} + A}{B} \tag{1}$$

where the parameters A and B are constants, and depend on cantilever dimensions, antibody type, and immobilization method. In the above $(-\Delta f)_{ss}$ is the steady state resonance frequency change and C_{b0} is pathogen concentration in sample. The above mathematical relationship has been successfully used to relate sensor response to various types of target analytes: toxin (*32*), pathogen (*23-29*), biomarker (*37*) and self-assembling thiol molecules (*38*).

Effect of Flow on Detection Sensitivity

In batch stagnant samples, no active transport of the target analyte exists. Transport is primarily due to diffusion to the sensor surface. Furthermore, spores being of higher density than aqueous medium tends to settle due to gravity thereby compromising sensor-analyte contact. This deficiency can be partly mitigated by allowing the sample to flow across the sensor suface (*23*). We used an experimental arrangement shown in Fig 4, which consisted of a sensor flow cell that was connected to reagent reservoirs and a flow circuit facilitated by a peristaltic pump. Flow rates in the range of 0.2 to 17 mL/min was used, and corresponds to bulk velocity in the sensor flow cell in the range of 0.02 to 0.85 cm/s. The sensor flow cell is maintained at a constant temperature within ± 0.1 ^{o}C. The sensor response to a samples containing 300 BA spores/mL gave almost twice[23] (162 Hz (n=2)) the response compared to batch measurements (90 Hz (n=2)). The binding kinetics was modeled by Langmuir first order kinetics using the relationship $(\Delta f) = (\Delta f_{\infty})(1 - e^{-k_{obs}\tau})$, where (Δf) is the change in resonance frequency at time, τ and (Δf_{∞}) is the steady state resonance frequency change. The parameter k_{obs} is the observed binding rate constant.

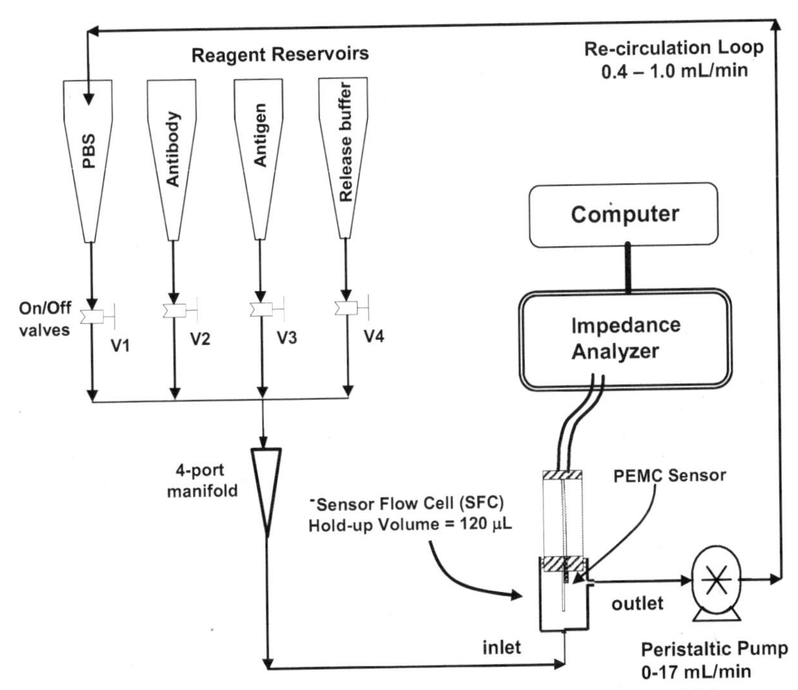

Figure 4. Experimental apparatus. A reagent reservoir manifold containing four cylindrical chambers was connected via a four port manifold into the inlet of the sample flow cell (SFC). A peristaltic pump, connected to the SFC outlet, maintained constant flow rate between 0.4 and 1.0 ml/min. The experimental apparatus allows for a single pass through the SFC as well as for recirculation during antibody immobilization. For the E. coli detection experiments, a fifth reagent reservoir (not shown) was added for hydroxylamine or Protein G. Adapted from reference(34).

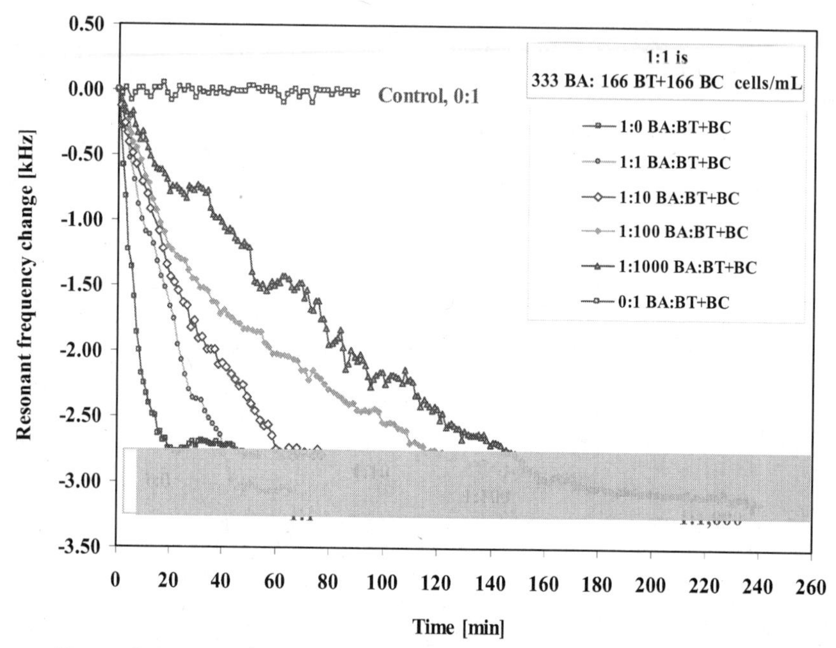

Figure 5: Transient response of PEMC sensor to the binding of 333 BA spores/mL from a solution containing various amounts of other Bacillus species. The control response shown is that of the anti-BA functionalized PAPEMC exposed to a mixture of BT and BC spores, at concentrations of 166 BT spores/mL and 166 BC spores/mL, so as to establish the baseline frequency change of the sensor. Data presented is from reference (25).

The rate constant, k_{obs}, was higher under flow (0.263 min^{-1}) than under stagnant condition (0.195 min^{-1})[23].

In a subsequent study (*27*), PEMC sensor showed limit of detection for *Escherichia coli* (EC) O157:H7 in 1 L samples at 1 cells/mL. The PEMC sensor used in this study exhibited resonance frequencies at 186.5, 883.5, and 1,778.5 kHz in air and 162.5, 800.0, and 1,725.5 kHz in liquid flow conditions. A one-liter sample containing 1,000 EC cells was introduced at various flow rates (1.5, 2.5, 3, and 17 mL/min), and the resulting sensor response was monitored. The 800 kHz mode showed a decrease in resonance response of 2,230 ± 11, 3,069 ± 47, 4,686 ± 97, and 7,188 ± 52 Hz at sample flow rates of 1.5, 2.5, 3, and 17 mL/min, respectively. Kinetic analysis showed that the binding rate constant (k_{obs}) were 0.009, 0.015, and 0.021 min^{-1} at 1.5, 2.5, 3 mL/min, respectively. The significance of these results is that flow can both enhance kinetics of attachment and give higher level of attachment.

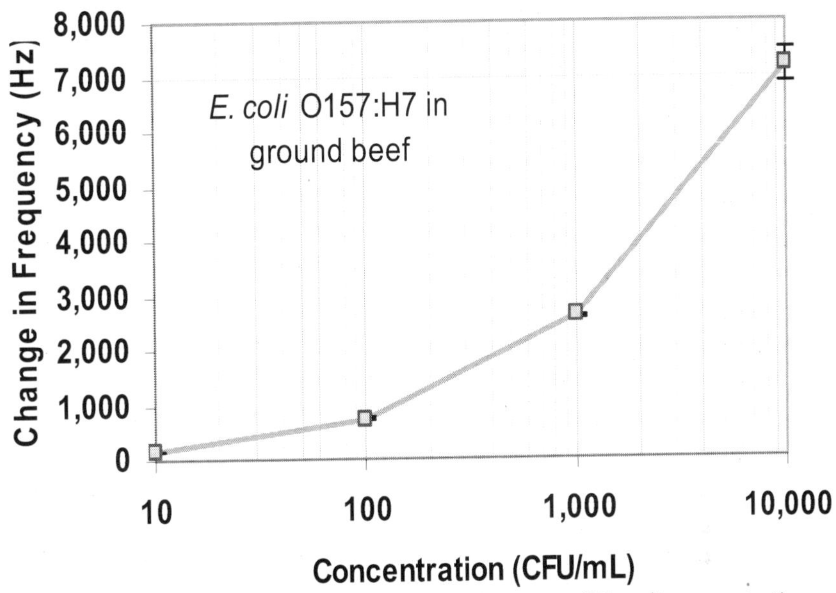

Figure 6. Resonant frequency change as a function of E. coli concentration. Plot of the sensor response data versus the log of E. coli concentration in the beef samples. Linear behavior is observed for sample concentrations ranging from 100 – 10,000 cells/mL. Data are from reference (29).

Selective Detection of *Bacillus anthracis* Spores in Presence of *Bacillus cereus* and *Bacillus thuringiensis*

In practical applications, selectivity of sensor to the target is an absolute requirement. Two types of experiments were carried out in the author's laboratory. First group of experiments consisted of detection of pathogenic bacteria in presence of non-pathogenic ones in clean buffers (*26,27,34*). Second type consists of detection in complex food matrices (*29,30*). In Figure 5 the responses of PEMC sensor to the binding of *Bacillus anthracis* (BA) spores at a concentration of 333 BA spores/mL are presented in presence of two non-pathogenic Bacillus sp- *Bacillus thuringiensis* (BT) and *Bacillus cereus* (BC). In each experiment, the sensor responded with a rapid decrease in resonance frequency before reaching nearly the same final steady state value, but at different time periods. For the pure BA sample (333 spores/mL) the resonance frequency decreased most rapidly and reached steady state in 27 minutes. As the concentration of the non-antigenic *Bacillus* species BT and BC were increased the rate of sensor response decreased. The response took a longer time to reach steady state. Steady states of 2,742 ± 38 (n=3), 3,053 ± 19 (n=2), 2,777 ± 26 (n=2), 2,953 ± 24 (n=2), and 3,105 ± 27 (n=2) Hz were obtained for the 1:0, 1:1, 1:10, 1:100, and 1:1000 BA:BT+BC samples, respectively, in 27, 45, 63, 154, and 219 minutes. The average frequency decrease was 2926 ± 162 (n=11) Hz. The deviation of ± 162 Hz from eleven experiments is small and may be treated as negligible. The results suggest that the presence of non-antigenic components in the sample does not affect the steady state response in flow conditions, but alters the kinetics of detection. The non-antigenic *Bacillus* species (BT and BC) hindered the transport of the BA spores to the sensor surface, but never completely prevented attachment of the antigenic spore. This is in sharp contrast to the results reported in Fig 3 obtained in batch stagnant system. Along with each detection experiment a control one was conducted consisting of exposing anti-BA functionalized cantilever to a sample of only BT and BC at the same concentration. The resonance frequency change fluctuated about zero for the control experiment. The sensor response (14 ± 3 (n=11) Hz) to the control containing BT and BC presented in Figure 5 was typical.

Selective Detection of *E. coli O157:H7* in Ground Beef wash

In another study, we showed detection of 10 cells/mL of *E. coli* O157:H7 in spiked raw ground beef samples which required just 10 minutes (*29*). The PEMC sensors with an area of 2 mm^2 were prepared by immobilizing polyclonal antibody specific to *E. coli* O157:H7 (EC). Ground beef (2.5 g) was spiked with EC at 10 – 10,000 cells/mL in phosphate buffered saline. After a mixing step, and a wait period of ten minutes, one mL of supernatant was used to perform the detection experiments. The total resonance frequency change (See Fig 6) obtained for the inoculated samples was 138 ± 9, 735 ± 23, 2,603 ± 51, and 7,184 ± 606 Hz, corresponding to EC concentrations of 10, 100, 1,000, and 10,000 cells/mL, respectively. Positive detection of EC in the sample solution was observed within the first 10-minutes. The responses of the sensor to the three controls (cells present, but no antibody on sensor, absence of cells with antibody-immoblized sensor, and buffer) were 36 ± 6, 27 ± 2, and 2 ± 7 Hz, respectively. Positive verification of *E. coli* O157:H7 attachment was confirmed by low-pH buffer (PBS/HCl pH 2.2) release, microscopy analysis, and second antibody binding post-EC detection. These results and our earlier study[28] suggest that it is feasible to detect *E. coli* O157:H7 at less than 10 cells/ml in 10 minutes without sample preparation, and with label-free reagents.

Concluding Remarks

We conclude that detecting a small number of cells in buffers and complex matrices and in presence of various other biological agents is feasible using PEMC sensors immobilized with the antibodies against the specific antigenic target. We showed in this paper the detection of *Bacillus anthracis* spores and *E. coli O157:H7*. A typical bioterrorism agent of 1 μm in size has a mass of one picogram. Since the sensitivity of PEMC sensors is ~femtograms, in principle, the attachment of a single pathogen would cause significant signal higher than measurement noise. We showed that this is indeed the case through the two examples – *Bacillus anthracis* in presence of copious number of other Bacillus sp., and the case of *E. coli O157:H7* in meat samples.

Acknowledgement

The authors are grateful for support through an EPA STAR Grant R-833007 and USDA grant 2006-51110-03641 during the writing of this manuscript. The *E. coli* work was supported by the USDA grant and the sensor design was supported by EPA grant, while the early work on BA was supported by United States Department of Transportation under Grant PA-26-0017-00 (Federal Transit Administration, in the interest of information exchange). The United States Government assumes no liability for the contents or use thereof. The United States Government does not endorse products or manufacturers. Trade or manufacturers' names appear herein solely because they are considered essential to the contents of the report.

References

1. Abdel-Hamid, I.; Ivnitski, D.; Atanasov, P.; Wilkins, E. *Analytica Chimica Acta* **1999**, *399*, 99-108.
2. Ivnitski, D.; Abdel-Hamid, I.; Atanasov, P.; Wilkins, E. *Biosensors and Bioelectronics* **1999**, *14*, 599-624.
3. Abdel-Hamid, I.; Ivnitski, D.; Atanasov, P.; Wilkins, E. *Biosensors and Bioelectronics* **1999**, *14*, 309-316.
4. Aguilar, Z. P.; Sirisena, M. *Analytical and Bioanalytical Chemistry* **2007**, *389*, 507-515.
5. Boyer, A. E.; Quinn, C. P.; Woolfitt, A. R.; Pirkle, J. L.; McWilliams, L. G.; Stamey, K. L.; Bagarozzi, D. A.; Hart, J. C.; Barr, J. R. *Analytical Chemistry* **2007**, *79*, 8463-8470.
6. Campbell, G. A.; Delesdernier, D.; Mutharasan, R. *Sensors and Actuators B-Chemical* **2007**, *127*, 376-382.
7. Carter, D. J.; Cary, R. B. *Nucleic Acids Research* **2007**, *35*, 11.
8. Davila, A. P.; Jang, J.; Gupta, A. K.; Walter, T.; Aronson, A.; Bashir, R. *Biosensors & Bioelectronics* **2007**, *22*, 3028-3035.
9. Gierczynski, R.; Zasada, A. A.; Raddadi, N.; Merabishvili, M.; Daffonchio, D.; Rastawicki, W.; Jagielski, M. *Fems Microbiology Letters* **2007**, *272*, 55-59.
10. Hoile, R.; Yuen, M.; James, G.; Gilbert, G. L. *Forensic Science International* **2007**, *171*, 1-4.
11. Kenar, L.; Ortatatli, M.; Karayilanoglu, T.; Yaren, H.; Sen, S. *Military Medicine* **2007**, *172*, 773-776.
12. Naja, G.; Bouvrette, P.; Hrapovich, S.; Liu, Y.; Luong, J. H. T. *Journal of Raman Spectroscopy* **2007**, *38*, 1383-1389.
13. Pal, S.; Alocilja, E. C.; Downes, F. P. *Biosensors & Bioelectronics* **2007**, *22*, 2329-2336.
14. Pohanka, M.; Skladal, P.; Kroea, M. *Defence Science Journal* **2007**, *57*, 185-193.
15. Qiao, Y. M.; Guo, Y. C.; Zhang, X. E.; Zhou, Y. F.; Zhang, Z. P.; Wei, H. P.; Yang, R. F.; Wang, D. B. *Biotechnology Letters* **2007**, *29*, 1939-1946.

16. Saikaly, P. E.; Barlaz, M. A.; de los Reyes, F. L. *Applied and Environmental Microbiology* **2007**, *73*, 6557-6565.
17. Wan, J. H.; Johnson, M. L.; Guntupalli, R.; Petrenko, V. A.; Chin, B. A. *Sensors and Actuators B-Chemical* **2007**, *127*, 559-566.
18. Prescott, L. M.; Harley, J. P.; Klein, D. A. *Microbiology*, 6th edition ed.; McGraw-Hill Education: Boston, 2005.
19. Craighead, H. G. *Journal of Vacuum Science & Technology* **2003**, *A21*, S216-S221
20. Ilic, B.; Craighead, H. G.; Krylov, S.; Senaratne, W.; Ober, C.; Neuzil, P. *Journal of Applied Physics* **2004**, *95*, 3694-3703.
21. Ziegler, C. *Analytical and Bioanalytical Chemistry* **2004**, *379*, 946-959.
22. Lavrik, N. V.; Sepaniak, M. J.; Datskos, P. G. *Review of Scientific Instruments* **2004**, *75*, 2229-2253.
23. Campbell, G. A.; Mutharasan, R. *Biosensors and Bioelectronics* **2006**, *22*, 78-85.
24. Campbell, G. A.; Mutharasan, R. *Biosensors and Bioelectronics* **2006**, *21*, 1684-1692.
25. Campbell, G. A.; Mutharasan, R. *Analytical Chemistry* **2007**, *79*, 1145-1152.
26. Campbell, G. A.; Mutharasan, R. *Biosensors and Bioelectronics* **2005**, *21*, 462-473.
27. Campbell, G. A.; Mutharasan, R. *Environ. Sci. Technol.* **2007**, *41*, 1668 - 1674.
28. Campbell, G. A.; Uknalis, J.; Tu, S.-I.; Mutharasan, R. *Biosensors and Bioelectronics* **2007**, *22*, 1296-1302.
29. Maraldo, D.; Mutharasan, R. *Journal of Food Protection* **2007**, *70*, 1670-1677.
30. Maraldo, D.; Mutharasan, R. *Journal of Food Protection* **2007**, *70*, 2651-2655.
31. Campbell, G. A.; Medina, M. B.; Mutharasan, R. *Sensors and Actuators B-Chemical* **2007**, *126*, 354-360.
32. Maraldo, D.; Mutharasan, R. *Analytical Chemistry* **2007**, *79*, 7636-7643.
33. Rijal, K.; Mutharasan, R. *Analytical Chemistry* **2007**, *79*, 7392-7400.
34. Maraldo, D.; Rijal, K.; Campbell, G.; Mutharasan, R. *Analytical Chemistry* **2007**, *79*, 2762-2770.
35. Campbell, G. A.; Mutharasan, R. *Sensors and Actuators A: Physical* **2005**, *122*, 326-334.
36. Hermanson, G. T. *Bioconjugate Technique*; Elsevier, San Diego, CA, 1996.
37. Maraldo, D.; Garcia, F. U.; Mutharasan, R. *Analytical Chemistry* **2007**, *70*, 763-7690.
38. Rijal, K.; Mutharasan, R. *Langmuir* **2007**, *23*, 6856-6863.

Chapter 4

Conducting Polymer and Polymer/CNT Composite Nanofibers by Electrospinning

Minoo Naebe[1], Tong Lin[1], Lianfang Feng[2], Liming Dai[2,3], Alexis Abramson[4], Vikas Prakash[4], Xungia Wang[1]

[1]Centre for Material and Fibre Innovation, Deakin University, Geelong, VIC 3217, Australia;
[2]State Key Laboratory of Chemical Engineering, Zhejiang University, Hangzhou, Zhejiang 310027, China;
[3]Chemical and Materials Engineering, University of Dayton, Dayton OH 45469, USA;
[4]Mechanical and Aerospace Engineering, Case Western Reserve University, Cleveland, OH 44106, USA

Polymer fibers with diameters in the micrometer range have been extensively used in a large variety of applications ranging from textile fabrics to fiber-reinforced materials. With the rapid development in nanoscience and nanotechnology, there is an ever increasing demand for polymer fibers with diameters down to the nanometer scale. Conducting polymer nanofibers are of particular interest for various applications, ranging from chem-/bio-sensors to electronic devices. Electrospinning has been used to produce nanofibers from various polymers. In particular, electrospun nanofibers based on polymer/carbon nanotube (CNT) composites are also very attractive multifunctional nanomaterials for many potential applications; the confinement-enhanced CNT alignment within such nanofibers have been shown to greatly improve the fiber mechanical and electrical properties. In this article, we summarize recent research progress on electrospun polymers and polymer/CNT composite nanofibers, with an emphasis on those based on conducting polymers and CNTs for potential applications.

Electrospinning and Electrospun Nanofibers

The electrospinning technique has not been well studied until the last decade, even though it was invented in 1934 (*1*). As shown in Figure 1a, the basic set-up for a solution-based electrospinning system consists of a polymer solution container, a high-voltage power supply, the spinneret (needle), and an electrode collector (*2*). The electrospinning process involves the application of a high electric field between a droplet of polymer fluid at the spinneret and a metallic collection screen at a distance of 5-50 cm from the polymer droplet. The electrically charged jet flows from the polymer droplet toward the collection screen when the voltage reaches a critical value (typically, ca.5-20 KV), at which the electrical forces overcome the surface tension of the polymer droplet (*3*). Under the action of high electric field, the polymer jet starts whipping around undergoing jet instability. Solvent evaporation from the jet results in dry/semi-dry fibers, which in most cases, are randomly deposited onto the ground electrode forming a nonwoven nanofiber web composed of a single fiber many kilometers in length (Figure 1b) (*4*).

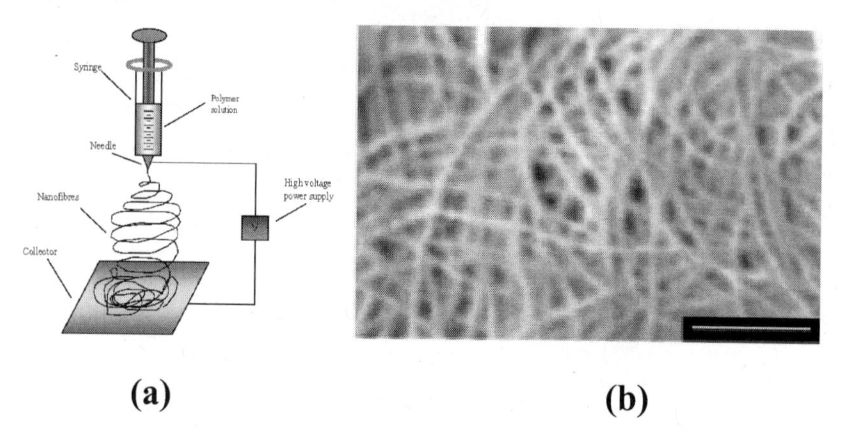

Figure 1. (a) Basic apparatus for electrospinning (2). (b) A typical SEM image of electrospun polystyrene fibers with an average diameter of 43 nm (4). Scale bar: 1000 nm.

Extensive research, including theoretical modeling (*5-12*), has been carried out to understand how the operating parameters (*e.g.* applied voltage, feeding rate, distance between the nozzle and collector), material properties (*e.g.* types of polymer and molecular weight, solvent boiling point/volatility, solution viscosity, surface tension and conductivity) (*13-15*), and polymer-solvent interactions (*15-30*) affect the electrospinning process and fiber morphology (*16*). The electrospun fibers could have diameters ranging from several microns down to tens of nanometers or less. The small fiber diameter and large aspect ratio give electrospun nanofibers extremely high surface-to-volume (weight) ratio, making them very useful for many applications (*31*, *32*).

Conducting Polymer Electrospun Nanofibers

The electrospun fiber mats possess a high surface area per unit mass, low bulk density, and high mechanical flexibility. These features make these materials very attractive for a wide range of potential applications, as mentioned above. Among them, the use of electrospun polymer fibers for electronic applications is of particular interest. The high surface-to-volume ratio associated with electrospun conducting polymer fibers makes them excellent candidates for electrode materials since the rate of electrochemical reaction is proportional to the surface area of an electrode and diffusion rate of the electrolyte.

Reneker and co-workers (*33*) have reported the pyrolysis of electrospun polyacrylonitrile nanofibers into conducting carbon nanofibers. MacDiamid and co-workers (*34*) prepared camphorsulfonic acid doped polyaniline, HCSA-PANI, and polyethylene oxide composite thin fibers (ca.0.95–2 μm in diameter) by the electrospinning technique from a mixed solution. Using the electrospinning technique, Dai and co-workers (*35*) have prepared lithium perchlorate (LiClO4) doped-polyethylene oxide (PEO) electrospun fibers and camphosulfonic acid (HCSA) doped-polyaniline (PANI)/polystyrene (PS) electrospun fibers. The diameters of these as-prepared polymeric fibers are in the range of 400-1000 nm. Figure 2 shows a typical SEM image of the as-spun HCSA-PANI/PS fibers, in which fiber-like structures with diameters in the ranges of 800-1000 nm are clearly evident. Owing to their large surface area and good electrical properties, these fine functional polymer fibers have been demonstrated to be useful for humidity and glucose sensing with a significantly enhanced sensitivity, when compared to their corresponding film-type counterparts (vide infra).

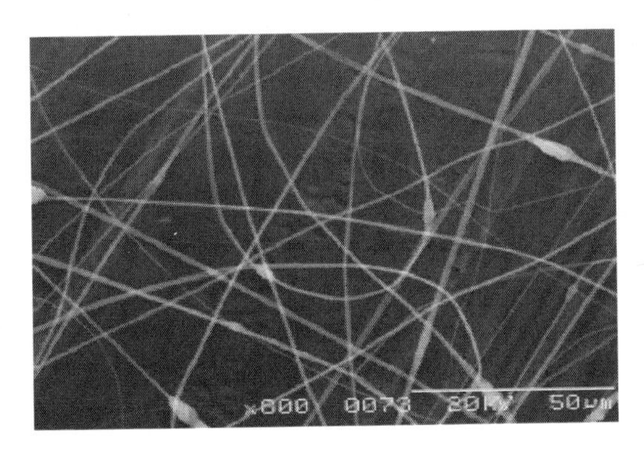

Figure 2. A typical scanning electron micrograph of HCSA doped-PANI/PS fibers (35).

Properties of Conducting Polymer Electrospun Nanofibers

In the past, the four-point probe method (*36*) has been used to measure the conductivity of the non-woven fiber *mat* and crosschecked with measurements of the conductivity of cast films produced from the same solution. Figure 3 shows the effect of the weight percentage content of HCSA-PANI on the room temperature conductivity obtained from the HCSA-PANI/PEO electrospun fibers and the film (*34*)..

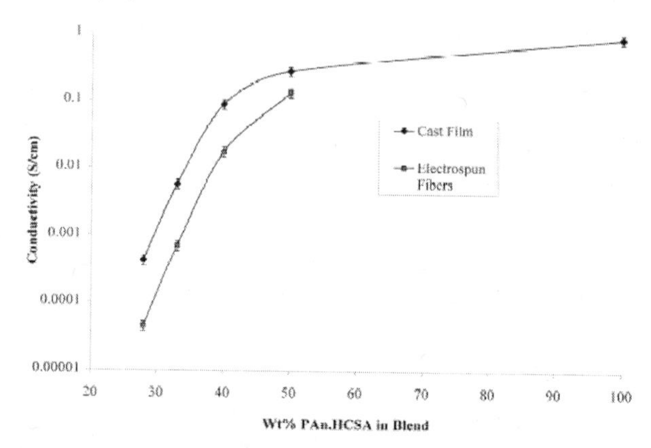

Figure 3. The effect of the weight percentage content of HCSA-PANI on the room temperature conductivity obtained from the HCSA-PANI/PEO electrospun fibers and cast films (34).

As seen in Figure 3, the conductivity of the electrospun fiber mat was lower when compared to that measured for the cast film, though they have very similar UV-visible absorption characteristics. The lower conductivity for the electrospun fibers can be attributed to the porous nature of the non-woven electrospun fiber mat as the four-probe method measures the volume resistivity rather than the conductivity of an individual fiber. Although the measured conductivity for the electrospun mat of conducting fibers is relatively low, their porous structures, together with the high surface-to-volume ratio, enables faster de-doping and re-doping in both liquids and vapors

In an attempt to measure the conductivity of an individual nanofiber, MacDiarmid and co-workers (*4*) collected a *single* electrospun HCSA-PANI/PEO nanofiber on a silicon wafer coated with a thin layer of SiO_2 and deposited two separated gold electrodes (a distance 60.3 μm apart) on the nanofiber, and used the two-probe method. The current-voltage (I - V) curves thus measured for single 50 wt% HCSA-PANI/PEO fibers with diameters of 600 and 419 nm are shown in Figure 4a (*i.e.* Fiber 1 and Fiber 2, respectively); these curves indicate a more or less straight line for the I-V characteristics, with a conductivity of *ca.*0.1 S/cm. The temperature dependence of the conductivity for a single 72 wt% HCSA-PANI/PEO electrospun fiber with a diameter of 1.32 μm (as shown in Figure 4b) also indicates a linear plot with a room-temperature conductivity of 33 S/cm (295K). This value of conductivity is much higher than the corresponding value of 0.1 S/cm for a cast film (Figure 3), suggesting a highly aligned nature of the PANI chains in the electrospun fiber.

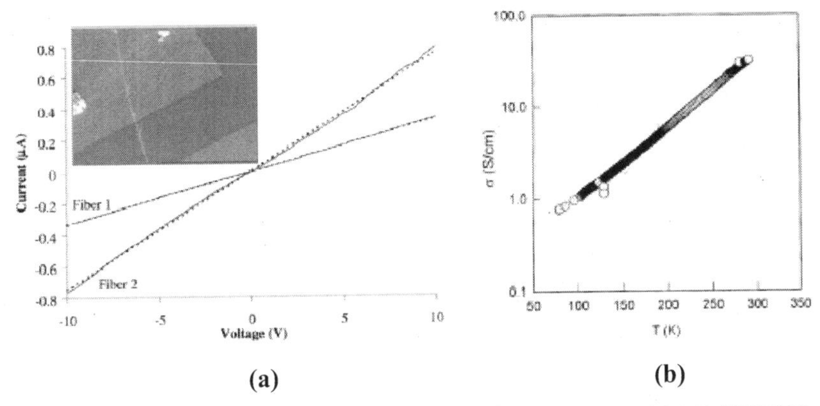

(a) **(b)**

Figure 4. (a) Current-voltage curve for a single 50 wt% HCSA-PANI/PEO nanofiber. (b) Temperature-dependence of the conductivity for a single 72wt% HCSA-PANI/PEO nanofiber (4).

Conducting Polymer Electrospun Nanofibers for Chem-/Bio-Sensing

In response to the pressing needs for cheaper, quicker, simpler, and more reliable detections, there has been tremendous progress in the development of chemical and biological sensors with high sensitivities. In this regard, the electrospinning process has been used to produce polymeric nanofibers for sensing applications. For instance, it has recently been demonstrated that optical sensors based on certain electrospun fluorescent polymer nanofibers showed sensitivity up to three orders of magnitude higher than that obtained from thin film sensors for detection of nitro compounds and ferric and mercury ions (*37-39*). In particular, Wang *et al.*, (*40*) reported a sensitive optical sensor by assembling fluorescent probes onto electrospun cellulose acetate nanofibers, which showed fluorescence quenching upon exposure to an extremely low concentration (ppb) of methyl viologen cytochrome in aqueous solutions.

Apart from the optical transduction, the conductivity changes of certain electrospun conducting polymer nanowires have also been exploited for sensing chemicals, as exemplified by electrospun polyaniline (PANI) nanowires that showed a rapid and reversible resistance change upon exposure to NH_3 gas at concentration as low as 0.5 ppm (*41*). The unusually high sensitivities observed for these electrospun nanofibers can be attributed to their high surface-to-volume ratios. In this regard, Dai and co-workers (*35*) have demonstrated the potential use of the $LiClO_4$-doped PEO ionically-conducting electrospun nanofibers and electronically-conducting camphosulfonic acid (HCSA) doped-PANI/PS nanofibers for humidity and glucose sensing, respectively.

$PEO/LiClO_4$ nanofiber humidity sensors

Figure 5 represents changes of the resistance in the logarithm scale as a function of % humidity for $PEO/LiClO_4$ nanofiber (line a) and film (line b) sensor, respectively. A linearly inverse proportional relationship between the resistance and the % humidity was observed in both cases. However, the rate of resistance reduction with the % humidity is much higher for the $PEO/LiClO_4$ nanofiber sensor than for the corresponding $PEO/LiClO_4$ film sensor, as indicated by the different slope values of the two lines. The greater absolute value of the slope for the nanofiber sensor (0.06 vs. 0.01) indicates a higher sensitivity, presumably due to the larger surface area.

As also seen in Figure 5, the initial resistance of the electrospun nanofiber mat was higher than that of the cast film, although they have been made from the same starting solution. The lower conductivity values for the electrospun nanofiber mats than those of the cast films can be attributed to the porous nature of the non-woven nanofiber mats, as the present method measures the volume resistivity rather than the conductivity of an individual fiber.

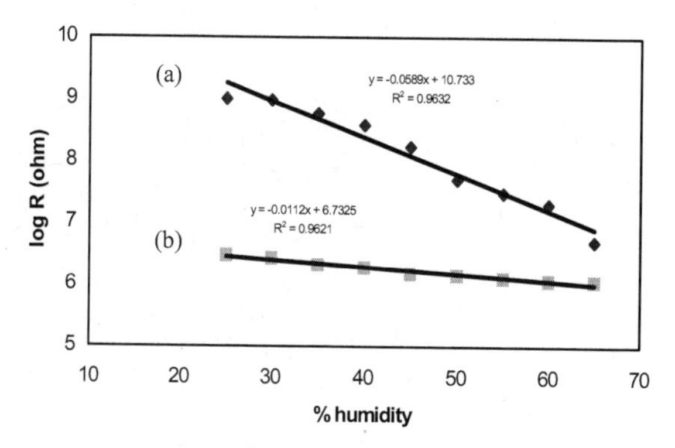

Figure 5. Humidity dependence of resistivity for $PEO/LiClO_4$ sensors: (a) the nanofiber sensor and (b) the film sensor (35).

Although the measured conductivity for the electrospun nanofiber mats is relatively low, their porous structures together with the high surface-to-volume ratio and good electrical properties, have been demonstrated to be of significant benefit for the development of advanced humidity sensors with a high sensitivity.

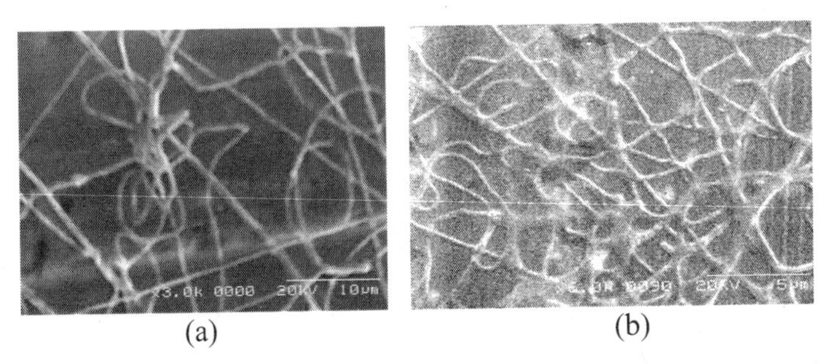

(a) (b)

Figure 6. SEM images of PEO/LiClO$_4$ nanofibers; (a) before and (b) after the humidity measurement (35).

After the humidity sensing measurements, some deformation of the nanofiber mats was observed, as shown by the SEM images in Figure 6. The observed morphological changes in Figure 6a and Figure 6b resulted from the swelling effect of water-soluble PEO/LiClO$_4$ nanofibers during the humidity measurements. Consequently, the *as-prepared* PEO/LiClO$_4$ nanofiber sensors might be utilized only as a disposable humidity sensor. However, further modification of the material/device design may circumvent this problem.

HCSA-PANI/PS nanofiber glucose sensors

For glucose sensing, the HCSA-PANI/PS nanofiber containing glucose oxidase (GOX, 15,500 units/g) and the corresponding film sensors were used to monitor concentration changes of H_2O_2, released from the glucose oxidation reaction, by measuring redox current at the oxidative potential of H_2O_2 (*i.e.* the amperometric method). Prior to any sensing measurement, both the HCSA-PANI/PS nanofiber and film sensors were dried in a vacuum oven at 60°C for 2 hr. Thereafter, GOX was immobilized onto the HCSA-PANI/PS nanofiber and film electrodes by electrodeposition from an acetate buffer (pH = 5.2) solution containing 2.5 mg/ml GOX at a potential of 0.4 V and a scan rate of 100 mV/s for 1 hr. Prior to the glucose sensing measurements, however, the HCSA-PANI/PS nanofiber and film sensors were tested with the pristine H_2O_2.

The redox current at the oxidative potential of H_2O_2 from the GOX-immobilized HCSA-PANI/PS nanofibers for glucose sensing was found to increase with increasing glucose concentrations (Figure 7a). The amperometric response from the GOX-immobilized HCSA-PANI/PS nanofiber sensor is much higher than that of the HCSA-PANI/PS film sensor, as shown in Figure 7b. Since nanofibers contain a large specific surface area, the amount of GOX immobilized on the nanofiber sensor was much higher than that for the film sensor, leading to a higher sensitivity for the current response. Thus, the HCSA-PANI/PS electrospun nanofibers could be promising for fabricating novel glucose sensors with a higher sensitivity. Unlike the PEO/LiClO$_4$ nanofiber sensor, there is no obvious swelling observed for the water insoluble HCSA-PANI/PS nanofiber sensors, suggesting reusability of the HCSA-PANI/PS based nanofiber sensors.

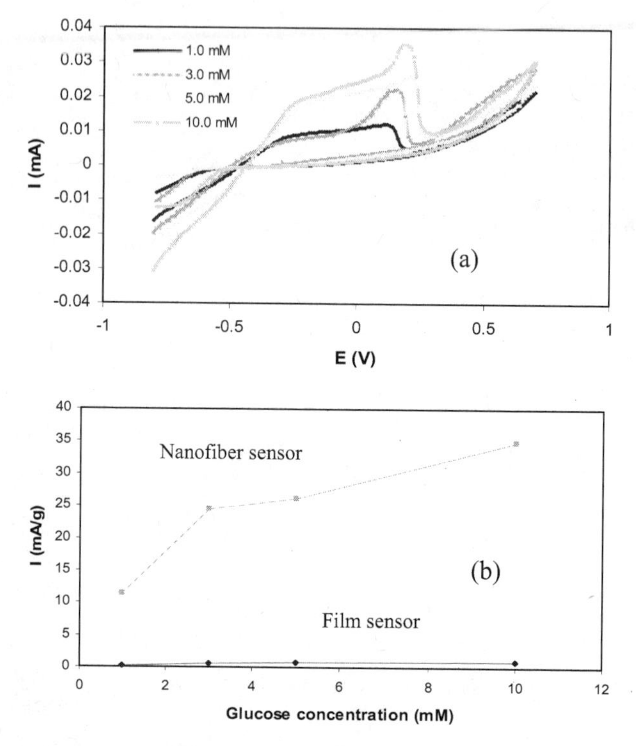

Figure 7. (a) Cyclic voltammetric (CV) spectra for the GOX-immobilized HCSA-PANI/PS nanofiber sensors at different glucose concentrations, and (b) The current response of the GOX-immobilized HCSA-PANI/PS nanofiber and film sensors to various glucose concentrations. Note the current response has been scaled by the weight of the polymeric material deposited on the electrodes (35).

Polymer/CNT Electrospun Nanofibers

The use of the electrospinning technique to incorporate carbon nanotubes (CNTs) into polymer nanofibers is also of great scientific and practical importance. The confinement effects associated with the constituent electrospun thin fibers are expected to induce alignment of carbon nanotubes within the polymer matrix, leading to high-strength, high-modulus, and even high electrical conductivity.

Carbon Nanotubes

As a new form of carbon having a seamless hollow cylindrical graphite structure, carbon nanotubes (CNTs) have attracted a great deal of interest since Iijima's report in 1991 (*42*). According to the number of graphite layers, CNTs can be divided into single-walled carbon nanotubes (SWNTs) and multiwalled carbon nanotubes (MWNTs). SWNTs consist of only one layer of graphite sheet with both ends being capped by fullerene-like molecular hemispheres of a diameter in the range from 0.4 to 3nm. On the other hand, MWNTs contain more than one co-axial graphite-cylinders with a diameter in the range of 1.4 to 100 nm (*43*). Depending on their chiral angles (*44*), SWNTs can be further grouped into armchair, zigzag and intermediate structures, as shown in Figure 8. The armchair SWNTs exhibit very high electrical conductivity, while the zigzag SWNTs are semi-conducting. Several techniques have been developed to synthesise SWNTs and MWNTs, including arc-discharge (*42*), laser-ablation (*45*) and chemical vapor deposition (*46*).

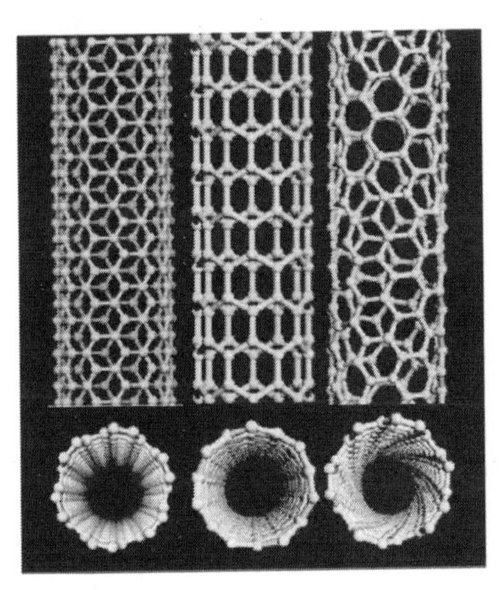

Figure 8. Schematic illustration of armchair, zigzag and chiral SWNT structures (from left to right) (44).

The extraordinary molecular structure with C-C sp^2 bonding among all the carbon atoms around the cylinder-graphite core sets carbon nanotubes apart from other forms of carbon and many other materials. CNTs combine outstanding mechanical, electronic, and thermal properties with a Young's modulus larger than 1 TPa (*47, 48*) and a tensile strength as high as 63 GPa (*49*). The tensile strength of a SWNT could be more than 5 times higher than that of a steel fiber with the same diameter, yet with only one-sixth of its density (*50*, *51*). Carbon nanotubes can also be conductive both electrically and thermally. These excellent properties of CNTs have led to extensive development work on the use of CNTs in structural and functional composite materials (*51-54*). However, the large-scale fabrication of CNTs and polymer/CNT composites of a consistent quality at a low cost has been a formidable challenge to the realization of their commercial applications.

Some critical issues governing the polymer/CNT composites preparation include: uniform dispersion of CNTs and control of the nanotube orientation within the polymer matrix by regulating the polymer-CNT interaction. CNTs tend to aggregate into bundles due to their strong inter-tube interactions and high surface/volume ratio (*55-61*). The uneven dispersion could adversely affect the mechanical properties of the resultant polymer/CNT composites. Several techniques have been developed to improve the CNT dispersion. Examples include the use of ultrasonication (57), solution-evaporation (56), and/or surfactants (*62*, *63*) to de-aggregate CNT bundles and force them to disperse throughout the polymer matrix uniformly. In-situ polymerization (*64*, *65*) in the presence of CNTs has also been used to improve the dispersion while other dispersion methods, such as melt-spinning (*66*), mechanical/ magnetic stretching (*67-69*), and chemically modification (*70*), were also reported.

Techniques commonly used for fabricating polymer/CNT composite fibers include the casting/spinning from solution (*71*, *72*), melt processing (*73*, *74*), and electrospinning (*75*). Among them, electrospinning is the most efficient and cost effective process for large scale production of polymer/CNT nanofibers (*76*, *77*). Indeed, there has been growing interest in applying the electrospinning technique to produce nanofibers with CNTs well-dipersed/orinted within various polymer nanofibers (*76*).

Electrospun Polymer/CNT Composite Nanofibers

It has been established that electrospinning a polymer solution containing well-dispersed carbon nanotubes leads to nanocomposite fibers with the embedded carbon nanotubes oriented parallel to the nanofiber axis due to the large shear forces in a fast fiber-drawing process. Table 1 lists most of the polymer/CNT composite nanofibers produced by electrospining, along with their fiber diameters and tensile properties.

Table 1. Electrospun Polymer/CNT Composite Nanofibres

Polymers	CNT	CNT (%)	Diameter (nm)	Tensile strength (Modulus, MPa)	Ref.
Polyacrylonitrile	SWNT/MWNT	1	180	-	(*78*)
	MWNT	1-20	50~300	285~312 (6.4~14.5GPa)	(*79*)
	MWNT	1.5, 7	20~140	-	(*80*)
	MWNT	1	200~2000	-	(*81*)
	SWNT	1-10	50~400	20~30	(*82*)
	SWNT	1~4	50~200	(140GPa)	(*75*)
Polyacrylonitrile	MWNT	2-35	100~300	37~80(2~4.4GPa)	(*83*)
Polyvinyl alcohol	MWNT	8	-	-	(*84*)
	MWNT	4.5	295~429	4.2~12.9	(*85*)
Polyethylene oxide	SWNT	3	-	(0.7~1.7GPa)	(*86*)
Polymethyl methacrylate	MWNT	0.5-2	200~6000	-	(*87*)
	MWNT	1-5	100~800	-	(*88*)
Polyurethane	SWNT	1	50~100	10~15	(*89*)
Polycaprolactone	MWNT	7-15	100~550	-	(*90*)
Polylactic acid	SWNT	1-5	1000	-	(*75*)
Regenerated silk fibroin	SWNT	0.5-5	147	2.84~7.4(180~705MPa)	(*91*)
Polybutylene terephthalate	MWNT	5	250~3500	(1.79GPa)	(*92*)
Polycarbonate	MWNT	4	350	-	(*94*)
Nylon 6,6	MWNT	2-20	150~200	-	(*94*)
Polystyrene	MWNT	0.8,1.6	300, 4500	-	(*95*)

Structure of Electrospun Polymer/CNT Composite Nanofibers

Dror et al. (*96*) proposed a theoretical model on rod-like particles in the electrospinning jet, and predicted that the possibility for carbon nanotubes to be aligned along the streamlines of an electrospun jet. It was shown that nanotube alignment was determined strongly by the quality of the nanotube dispersion (*86, 97*). When MWNTs were individually dispersed in electrospun polymer nanofibers, they predominantly aligned themselves along the fiber axis. Nevertheless, twisted, bent and non-uniform nanotubes have also be found in electrospun polymer fibers (Figure 9a).

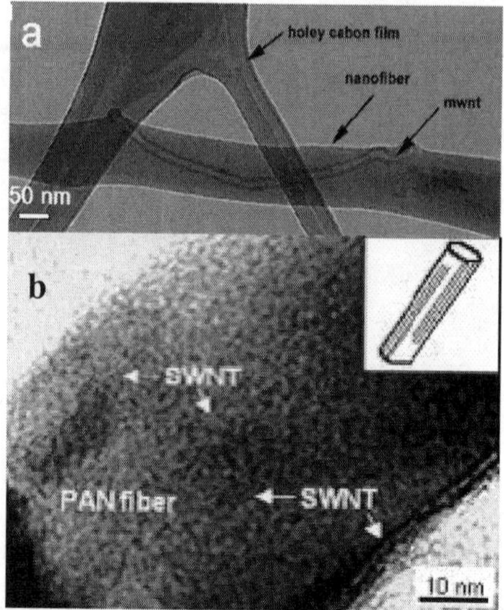

Figure 9. (a) TEM image of a protruded MWNT from the fibre (106). (b) High resolution TEM image of a SWNT/PAN fiber (75).

In an independent study, Ko *et al.* (*75*) found that better CNT alignment could be formed in polyacrylonitrile (PAN) nanofiber than in polylactic acid (PLA) nanofiber, indicating the importance of polymer-CNT interactions to the nanotube alignment (Figure 9b). Besides, Ge *et al.* (*79*) prepared electrospun PAN/MWNT nanofiber sheets by collecting the composite nanofibers onto a winder with a surface velocity higher than the velocity of electrospun nanofibers. They found that highly oriented CNTs formed within the polymer nanofibers by the fast electrospinning and slow relaxation of CNTs (*79*). More recently, Pan *et al.* (*95*) reported the preparation of hollow multilayered polystyrene nanofibers containing MWNTs by selective removal of the component part(s) in the template through a combination of the layer-by-layer technique with electrospinning.

Properties of Electrospun Polymer/CNT Composite Nanofibers

Mechanical properties

It has been demonstrated that CNTs can enhance crystallinity of certain host polymers (*72*, *98*), leading to improved mechanical properties. The CNT-enhanced nucleation crystallinity has been reported to occur in various polymer systems, including poly(vinyl alcohol) (PVA), polycarbonate (PC), polypropylene (PP), and poly(m-phenylenevinylene-co-2,5-dioctyloxy-p-phenylenevinylene) (PmPV) (*98-111*). In addition to these synthetic polymers, certain proteins, such as streptavidin, were also found to form crystalline structures around CNTs (*100*, *102*, *104*). Of particular interest is that polyvinyl alcohol (PVA), which showed significant improvement in mechanical properties upon reienforcement with CNTs (*72*, *112-117*). The CNT-induced nucleation crystallization in PVA composite films was confirmed by microscopic investigations of fractured PVA/CNT composite films (*117*), and the presence of these ordered polymer dominates was demonstrated to be responsible for the reinforcement effects (*85*, *114*).

The need for mechanical reinforcement has been the driving force for most of the reported work on polymer/CNT composites. In an attempt to investigate the mechanical properties of electrospun PAN/SWNT nanofibers, Ko *et al.* (*75*) have used an atomic force microscope (AFM) to measure the elastic modulus of the electrospun composite nanofibers. The obtained fiber modulus was 140 GPa, a value which is much higher than that of conventional PAN fibers (60 GPa) (*75*). In a somewhat related but independent study, Mathew *et al.* (*92*) also used AFM to measure the mechanical properties of electrospun polybutylene/MWNT terephthalate nanofibers. Elastic deformation of MWNTs in electrospun PEO/MWNT and PVA/MWNT nanofibers was studied by Zhou and co-workers (84), and was found to increase with an increase in the modulus of the polymer matrix. In the same study, a simplified model was also proposed to estimate the elastic modulus ratio of MWNT and polymers. To confirm the validity of their model, these authors compared the model predictions with experimental data obtained from AFM measurements.

The mechanical properties of SWNT reinforced polyurethane (PU) electrospun nanofibers were studied by Sen *et al.* (*89*). Stress-strain analysis showed that the tensile strength of the PU/SWNT nanofiber membrane was enhanced by 46% compared to pure PU nanofiber membrane. This value of tensile strength was further increased by 104% for ester functionalized PU/SWNT membranes because of the improved SWNT dispersion and the enhanced PU-SWNT interfacial interaction.

Hou and Reneker (*83*) investigated mechanical properties of thick PAN/MWNT electrospun nanofiber sheets. They found that tensile modulus increased by 144% with the increase in the concentration of MWNT from 0 to 20 wt%. However, the tensile strength increased with the increase in MWNT concentration up to 5w t% and reduced with further increase in the MWNT content. Meanwhile, the strain at break value reduced with an increase in the MWNT concentration. This was attributed to the poor dispersion of MWNTs and a reduced PAN/MWNT interfacial contact at the higher MWNT contents. To investigate the effect of polymer morphology on mechanical properties of polymer/CNT composite nanofibers, Naebe *et al* (*85*) performed a series of post-electrospinning treatments to PVA/MWNT composite nanofibers to increase the PVA crystallinity and/or crosslinking degree. The tensile strength of the PVA/MWNT composite nanofibers was improved significantly by regulating the polymer morphology through the post–electrospinning treatment. The CNT-induced nucleation crystallization of PVA was also observed in the PVA/MWNT composite nanofibers.

Ye *et al.* (*118*) used TEM to study the reinforcement and rupture behavior of both PAN/SWNT and PAN/MWNT electrospun nanofibers. These authors found a two-stage rupture behavior under tension, including crazing of the polymer matrix and pulling-out of the carbon nanotubes. In this case, CNTs reinforced the polymer fibers by hindering crazing extension, leading to a reduced stress concentration and increased energy dissipation. The underlying mechanism for the observed reinforcement was believed to be attributable to the good CNT dispersion in the polymer matrix and strong polymer-CNT interfacial interaction.

Kim *et al.* (*93*) have recently studied *in-situ* mechanical deformation of single electrospun composite nanofiber. The bombardment of electron beam onto the fibers in a TEM chamber led to local thermal expansion, and hence initiating the tensile deformation. The strain rate can be controlled by adjusting the electron beam flux on the fibers (*56*, *118*). As the strain increased, the fiber elongated and even underwent necking often at the positions where the nanotube ends were located due to the slippage of the MWNTs along the applied tensile stress. More recently, Prakash and co-workers (*119*) developed a novel characterization device for nanomechanical testing of individual electrospun polymer nanofibers. The tool consists of a nanomanipulator, a transducer and associated probes and is operated inside a scanning electron microscope (Figure 10a). The three plate capacitive transducer independently measures force and displacement with a micronewton and nanometer resolution, respectively. Tensile testing of an electrospun polyaniline fiber (diameter ~1 μm) demonstrated the capabilities of the system. Engineering stress versus strain curves exhibited two distinct regions (Figure 10b); the Young's modulus of the latter region was approximately 5.9GPa. Failure at the probe-specimen weld occurred at ~67MPa, suggesting a higher yield stress for polyaniline microfibers when compared with bulk.

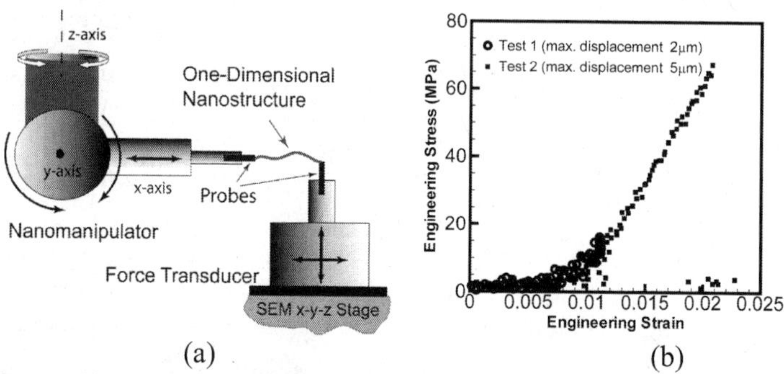

Figure 10. (a) Schematic of the nanomechanical characterization device showing the magnified view *of the fiber specimen between two probes and the available degrees of freedom. (b) Engineering stress and strain for test 1 (circles) and test 2 (squares); the latter experiment involved loading the specimen to a maximum displacement of 5 μm; failure at the probe-specimen weld occurred prior to reaching maximum displacement (119).*

Electrical properties

Although the potential applications of electrospun polymer/CNT nanofibers have yet to be fully exploited, conducting electrospun polymer/CNT nanofibers have been demonstrated to be attractive for a large variety of potential applications, including in optoelectronic and sensing devices (*43*). Owing to the excellent electrical properties intrinsically characteristic of CNTs, the addition of CNTs into appropriate polymer matrices could impart electrical properties to certain electrospun composite nanofibers. In this context, Sundaray and co-workers (*87*) have measured DC electrical conductivity of individual PMMA/MWNT electrospun nanofibers. It was found that the percolation threshold for these PMMA/MWNT composite electrospun nanofibers was well below 0.05%w/w of the CNT loading and conductivity of the electrospun composite nanofibers increased with the MWNT content (Figure 11). Ge *et al.* (*79*) have also studied electrical properties of the PAN/MWNT composite electrospun nanofibers, and found that the presence of MWNTs in the nanofiber increased its conductivity up to ~1.0 S/cm.

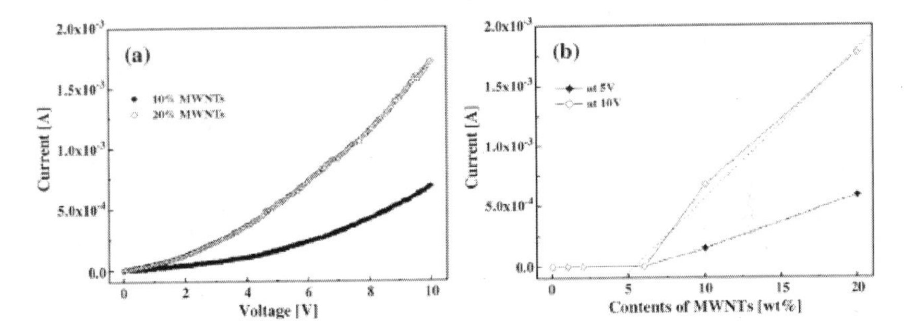

Figure 11. (a) I-V characteristics for the nylon electrospun nanofibers loaded with 10 and 20 wt.%. (b) Plot of the current as a function of the CNT wt.% at 5 and 10 V (94).

Jeong and co-workers (*94*) have electrospun nylon/MWNT nanofibers onto an ITO coated glass and measured the *I-V* characteristics of the CNT-containing nanofiber membrane supported by the conducting glass. The measured *I-V* characteristic showed a non-ohmic behavior over the MWNT contents from 0 to 20wt%, and the non-ohmic behavior was enhanced with increased MWNT contents. Similar electrical behavior was also reported for PVDF/SWNT (*120*) and PEO/MWNT (*121*) composite nanofibers.

Summary

We have presented a brief overview of the recent progress in research and development of conducting polymer and polymer/CNT composite electrospun nanofibers. These materials possess interesting electronic and mechanical properties of use for many potential applications. Conducting polymer nanofibers have large surface area and good electrical properties, making them promising materials for chem-/bio-sensing applications with a high sensitivity. The use of the electrospinning technique to incorporate CNTs into polymer nanofibers has been shown to induce possible alignment of the nanotubes within the polymer matrix, leading to a high-strength, high-modulus, and even electrical conductivity. Electrical and mechanical properties of both the polymer and polymer/CNT composite electrospun nanofiber mats and the constituent individual nanofibers have been measured also. In order to realize their commercial applications, considerable further work is still required. This includes a thorough understanding of the structure–property relationship for various electrospun polymer nanofibers, the effective incorporation of carbon nanotubes into polymer fibers with a high loading content, and large scale production of carbon nanotubes and their composites of consistent quality and at a low cost.

Acknowledgements

We thank the National Science Foundation for support (NSF-CTS 0438389).

References

1. Formhals A. US Patent No. 1,975,504, 1934
2. Dai L. *Intelligent Macromolecules for Smart Devices: From Materials Synthesis to Device Applications*, Springer-Verlag, Berlin, 2004
3. Taylor GI. *Proc R Soc London,Ser A* **1969**, 313, 453
4. MacDiarmid AG, Jones WE Jr, Norris ID, Gao J, Johnson AT Jr, Pinto NJ, Hone J, Han B, Ko FK, Okuzaki H, Llaguno M, *Synthetic Metal* **2001**, 119, 27
5. Reneker DH, Yarin AL, Fong H, Koombhongse S. *Journal of Applied Physics* **2000**, 87, 4531
6. Hohman MM, Shin M, Rutledge G, Brenne MP. *Physics of Fluids* **2001**, 13, 2201
7. Saville DA. *Annual Review of Fluid Mechanics* **1997**, 29, 27
8. Feng JJ. *Physics of Fluids* **2002**, 14, 3912
9. Hohman MM, Shin M, Rutledge G, Brenner MP. *Physics of Fluids* **2001**, 13, 2221
10. Ganän-Calvo AM. *Journal of Fluid Mechanics* **1997**, 335,165
11. Hartman RPA, Brunner DJ, Camelot DMA, Marijnissen JCM, Scarlett. B. *Journal of Aerosol Science* **1999**, 30, 823
12. Yarin AL, Koombhongse S, Reneker DH. *Journal of Applied Physics* **2001**, 89, 3018
13. Doshi J, Reneker DH. *Journal of Electrostatics* **1995**, 35, 151
14. Chronakis IS. *Journal of Materials Processing Technology* **2005**, 167, 283
15. Reneker D, Chun I. *Nanotechnology* **1996**, 7, 216
16. Bognitzki M, Hou H, Ishaque M, Frese T, Hellwig M, Schwarte C. *Advanced Materials* **2000**, 12, 637
17. Bognitzki M, Czado W, Frese T, Schaper A, Hellwig M, Steinhart M. *Advanced Materials* **2001**, 13, 70
18. Buchko CJ, Chen LC, Shen Y, Martin DC. *Polymer* **1990**, 40, 7397
19. Chen Z, Foster MD, Zhou WS, Fong H, Reneker DH, Resendes R. *Macromolecules* **2001**, 34, 6156
20. Theron A, Zussman E, Yarin AL. *Nanotechnology* **2001**, 12, 384

21. Megelski S, Stephens JS, Rabolt JF , Bruce CD. *Macromolecules* **2002**, 35, 8456
22. Deitzel J, Kosik W, McKnight SH , Ten NCB, Desimone JM, Crette S. *Polymer* **2002**, 43, 1025
23. Ding B, Kim H-Y, Lee S-C, Shao C-L, Lee D-R, Park S-J, Kwag G-B, Choi K-J. *Journal of Polymer Science: Part B: Polymer Physics* **2002**, 40, 1261
24. Kim J-S, Reneker DH. *Polymer Composites* **1999**, 20, 124
25. Koombhongse S, Liu WX, Reneker DH. *Journal of Polymer Science: Part B: Polymer Physics* **2001**, 39, 2598
26. MacDiarmid AG, Jones WE Jr, Norris ID, Gao J, Johnson AT, Pinto NJ, Hone J, Han B, Ko FK, Okuzaki H, Llaguno M. *Synthetic Metals* **2001** 119, 27
27. Matthews J, Wnek GE, Simpson DG, Bowlin GL. *Biomacromolecules* **2002**, 3, 232
28. Zong X, Kim K, Fang D, Ran S, Hsiao BS, Chu B. *Polymer* **2002**, 43, 4403
29. Fong H, Liu W-D, Wang C-S, Vaia R. *Polymer* **2002**, 43, 775
30. Fong H, Chun I, Reneker DH. *Polymer* **1999**, 40, 4585
31. Huang Z, Zhang YZ, Kotaki M, Ramakrishna S. *Composites Science and Technology* **2003**, 63, 2223
32. Greiner A,Wendorff JH. *Angew. Chem. Int.* **2007**, 46, 5670
33. Chun I, Reneker DH, Fong H, Fang X, Deitzel J, Tan NB, Kearns KJ. *Adv. Mater.* **1996,** 31, 36
34. Norris ID, Shaker MM, Ko FK, MacDiarmid AG, *Synth. Met.* **2000,** 114, 109 .
35. Aussawasathien D, Dong JH, Dai L. *Synth. Met.* **2005**,154, 37
36. Van der Pauw LJ, *Philips Research Reports* **1958**, 13, 1
37. Lee SH, Ku BC, Wang X, Samuelson LA, Kumar J. *Mat. Res. Soc. Symp. Pro.* **2002,** 708, 403
38. Wang XY, Lee SH, Drew C, Senecal KJ, Kumar JK, Samuelson LA. *Mat. Res. Soc. Symp. Pro.* **2002,** 708, 397
39. Wang XY, Drew C, Lee SH, Senecal KJ, Kumar J, Samuelson LA. *Nano Lett.* **2002**, 2, 1273
40. Wang X, Kim YG, Drew C, Ku BC, Kumar J, Samuelson LA. *Nano Lett.* **2004**, 4, 331.
41. Liu H, Kameoka J, Gzaplewski DA, Craighead HG, *Nano Lett.* **2004**, 4, 671.
42. Iijima S. *Nature* **1991**,,354, 56
43. Dai L. (Ed), *"Carbon Nanotechnology: Recent Developments in Chemistry, Physics, Materials Science and Device Applications*", Elsevier, Amsterdam (2006)
44. Bachilo SM, Strano MS, Kittrell C, Hauge RH, Smalley RE, Weisman RB. *Science* **2002**, 298, 2361
45. T. Guo PN, Nikolaev, P, Rinzler AG, Tomanek D, Colbert DT, Smalley RE. *J.Phys.Chem.* **1995**, 99, 10694
46. Ren ZF, Huang ZP, J. W. Xu, J. H. Wang, P. Bush, M.P. Siegal, Provencio PN. *Science* **1998**, 282,1105
47. Treacy MMJ, Ebbesen TW, Gibson JM. *Nature* **1996**, 381, 678
48. Wong EW, Sheehan PE, Lieber CM. *Science* **1997**, 277, 1971

49. Yu M, Lourie O, Dyer M, Kelly T, Ruoff R. *Science* **2000**, 287, 637
50. Berber S, Know YK, Tomanek D. Physics Review Letters **2000**, 84, 4613
51. Andrews R, D.Jacques, Rao AM, Rantell T, Derbyshire F, Chen Y, Chen J, Haddon RC. Applied Physics Letters **1999**, 75, 1329
52. Coleman JN, Khan U, Blau WJ, Gunko YK. Carbon **2006**, 44, 1624
53. Coleman JN, Khan U,Gunko YK. Advanced Materials **2006**, 18, 689
54. Ajayan PM. Chemistry Review **1999**, 99, 1787
55. Tucknott R,Yaliraki SN. Chemical Physics **2002**, 281, 455
56. Qian D, Dickeya EC, Andrews R,Rantell T. Applied Physics Letters **2000**, 76, 2868
57. Biercuk MJ, Liaguno MC, Radosavljevic M, Hyun JK, Fischer JE, Johnson AT. Applied Physics Letters **2002**, 80, 2767
58. Liu L, Barber AH, Nuriel S, Wagner HD. Advanced Functional Materials **2005**, 15, 975
59. Kumar S, Dang TD, Arnold FE, Bhattacharyya AR, Min BG, Zhang X, Vaia RA, Park C, Adams WW, Haughe RH, Smalley RE, Ramesh S, Willis PA. Macromolecules **2002**, 35, 9039
60. Ruan SL, Gao P, Yang XG,Yu TX. Polymer **2003**, 44, 5643
61. Paiva MC, Zhou B, Fernando KAS, Lin Y, Kennedy JM, Sun YP. carbon **2004**, 42, 2849
62. Vigolo B, Nicaud AP, Coulon C, Sauder C, Pailler R, Journet C, Bernier P, Poulin P. Science **2000**, 290, 1331
63. Shaffer MSP,Windle AH. advanced materials **1999**, 11, 937
64. Gao J, Yu A, Itkis ME, Bekyarova E, Zhao B, Niyogi S, Haddon RC. Journal of American Chemical Society **2004**, 126, 16698
65. Zhao C, Hu G, Justice R, Schaefer DW, Zhang S, Yang M, Han CC. Polymer **2005**, 46, 5125
66. Kumar S, Doshi H, M. Srinivasarao, Park JO, Schiraldi DA. Polymer **2002**, 43
67. Jin L, Bower C, Zhou O. Applied Physics Letters **1998**, 73, 1197
68. Bower C, Rosen R, Jin L, Han J, Zho O. Appled Physics Letters **1999**, 74, 3317
69. Smith BW, Benes Z, Luzzi DE, Fischer JE, Walters DA, Casavant MJ, Schmidt J, Smalley RE. Applied Physics Letters **2000**, 77, 663
70. Schlittler RR, Seo JW, Gimzewski JK, Durkan C, Saifullah MSM,Welland ME. Science **2001**, 292, 1136
71. Kilbride BE, Coleman JN, Fournet P, Cadek M, Drury A, Blau WJ. Journal of Applied Physics **2002**, 92, 4024
72. Cadek M, Coleman JN, Barron V, Hedicke K, Blau WJ. Applied Physics Letters **2002**, 81, 5123
73. Meincke O, Kaempfer D, Weickmann H, Friedrich C, Vathauer M, Warth H. Polymer **2004**, 45, 739
74. Sandler JKW, Pegel S, Cadek M, Gonjy F, Vanes M, Lohmar J, Blau WJ, Schulte K, Windle AH, Shaffer MSP. Polymer **2004**, 45, 2001
75. Ko F, Gogotsi Y, Ali A, Naguib N, Ye H, Yang G, Li C,Wilis P. Advanced Materials **2003**, 15, 1161
76. Li D, Xia Y. Advanced Materials **2004**, 16, 1151
77. Dzenis Y. Science **2004**, 304, 1917

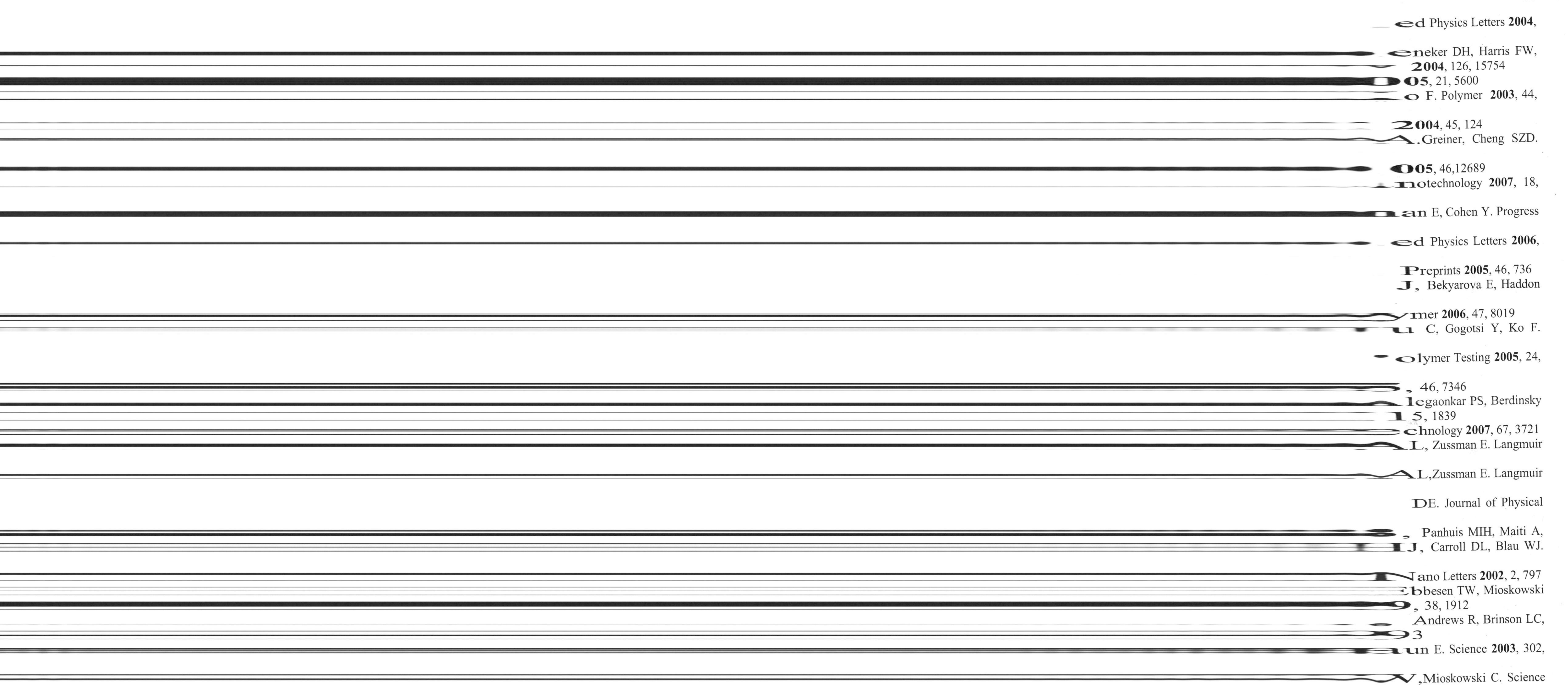

ed Physics Letters **2004**,
eneker DH, Harris FW,
2004, 126, 15754
005, 21, 5600
o F. Polymer **2003**, 44,
2004, 45, 124
A. Greiner, Cheng SZD.
005, 46,12689
notechnology **2007**, 18,
an E, Cohen Y. Progress
ed Physics Letters **2006**,
Preprints **2005**, 46, 736
J, Bekyarova E, Haddon
mer **2006**, 47, 8019
u C, Gogotsi Y, Ko F.
olymer Testing **2005**, 24,
, 46, 7346
legaonkar PS, Berdinsky
15, 1839
chnology **2007**, 67, 3721
AL, Zussman E. Langmuir
AL,Zussman E. Langmuir
DE. Journal of Physical
, Panhuis MIH, Maiti A,
J, Carroll DL, Blau WJ.
Nano Letters **2002**, 2, 797
bbesen TW, Mioskowski
9, 38, 1912
Andrews R, Brinson LC,
93
un E. Science **2003**, 302,
V, Mioskowski C. Science

105.Ryan KP, Cadek M, Nicolosi V
H, Fonseca A, Nagy JB, Mase
Science and Technology **2007**,
106.Bhattacharyya AR, Sreekumar
RH, Smalley RE. Polymer **2003**,
107.Valentini L, Biagiotti J, Kenny
Technology **2003**, 63, 1149
108.Valentini L, Biagiotti J, Kenny J
Science **2003**, 87, 708
109.Valentini L, Biagiotti J, Kenny
Polymer Science **2003**, 89, 2657
110.Assouline E, Lustiger A, Barb
Wagner HD. Journal of Polyme
41, 520
111.Sandler J, Broza G, Nolte M, S
of Macromolecular Science, Phy
112.Coleman JN, Blau WJ, Dalton
M, Vieiro G, Baughman RH. Ap
113.Probst O, Moore EM, Resasco D
114.Coleman JN, Cadek M, Ryan K
MS. Polymer **2006**, 47, 8556
115.Cadek M, Coleman JN, Ryan K
JB, Szostak K, Beguin F, Blau W
116.Dalton AB, Collins S, Munoz
Coleman JN, Kim BG, Baughma
117.Coleman JN, Cadek M, Blake R
A, Nagy JB, Gunko YK, Blau W
14
118.Ye H, Lam H, Titchenal N, Gogo
85, 1775
119.Singh U, Prakash V, Abramson
Lett. **2006**, 89, 073103
120.Seoul C, Kim YT,Baek CK. Jou
Physics **2003**, 41, 1572
121.Lim JY, C.K.Lee, S.J.Kim, I.Y.
Science, Part A: Pure and Applie

78. Ye H, Lam H, Titchenal N, Gogotsi Y, Ko F. Applied Physics Letters **2004**, 85, 1775
79. Ge JJ, Hou H, Li Q, Graham MJ, Greiner A, Reneker DH, Harris FW, Cheng SZD. Journal of American Chemical Society **2004**, 126, 15754
80. Kedem S, Schmidt J, Paz Y, Cohen Y. Langmuir **2005**, 21, 5600
81. Titchenal N, Naguib N, Ye H, Gogotsi Y, Liu J, Ko F. Polymer **2003**, 44, 115
82. Lam H, H.Ye, Y.Gogotsi, F.Ko. Polymer Preprints **2004**, 45, 124
83. Hou H, Ge JJ, Zeng J, Li Q, Reneker DH, A.Greiner, Cheng SZD. Chemistry of Materials **2005**, 17, 967
84. Zhou W, Wu Y, Wei F, Luo G, Qian W. Polymer **2005**, 46,12689
85. Naebe M, Lin T, Tian W, Dai L,Wang X. Nanotechnology **2007**, 18, 225605
86. Dror Y, Salalha W, Hintzen WP, Yarin AL, Zussman E, Cohen Y. Progress in Colloid and Polymer Science **2005**, 130, 64
87. Sundaray B, Subramanian V, Natarajan TS. Appled Physics Letters **2006**, 88, 143114
88. Kim HS, Sung JH , Choi HJ, Chin I, Jin H. Polmer Preprints **2005**, 46, 736
89. Sen R, Zhao B, Perea D, Itkis ME, Hu H, Love J, Bekyarova E, Haddon RC. Nano Letters **2004**, 4, 459
90. Saeed K, Park SY, Lee HJ, Baek JB, Huh WS. Polymer **2006**, 47, 8019
91. Ayutsede J, Gandhi M, Sukigara S, Ye H, Hsu C, Gogotsi Y, Ko F. Biomacromolecules **2006**, 7, 208
92. Mathew G, Hong JP, Rhee JM, Lee HS, Nah C. Polymer Testing **2005**, 24, 712
93. Kim GM, Michlera GH, Pötschke P. Polymer **2005**, 46, 7346
94. Jeong JS, Jeon SY, Lee TY, Park JH, Shin JH, Alegaonkar PS, Berdinsky AS, Yoo JB. Diamond & Related Materials **2006**, 15, 1839
95. Pan C, Ge LQ, Gu ZZ. Composites Science and Technology **2007**, 67, 3721
96. Dror Y, Salalha W, Khalfin RL, Cohen Y, Yarin AL, Zussman E. Langmuir **2003**, 19, 7012
97. Salalha W, Dror Y, Khalfin RL, Cohen Y, Yarin AL,Zussman E. Langmuir **2004**, 20, 9852
98. Grady BP, Pompeo F, Shambaugh RL, Resasco DE. Journal of Physical Chemistry **2002**, 106, 5852
99. McCarthy B, Coleman JN, Czerw R, Dalton AB, Panhuis MIH, Maiti A, Drury A, Bernier P, Nagy JB, Lahr B, Byrne HJ, Carroll DL, Blau WJ. Journal of Physical Chemistry B **2002**, 106, 2210
100. Barraza HJ, Pompeo F, O'Rear EA, Resasco DE. Nano Letters **2002**, 2, 797
101. Balavoine F, Schultz P, Richard C, Mallouh V, Ebbesen TW, Mioskowski C. Angewandte Chemie-International Edition **1999**, 38, 1912
102. Ding W, Eitan A, Fisher FT, Chen X, Dikin DA, Andrews R, Brinson LC, Schadler LS, Ruoff, RS. Nano Letters **2003**, 3, 1593
103. Keren K, Berman RS, Buchstab E, Sivan U, Braun E. Science **2003**, 302, 1380
104. Richard C, Balavoine F, Schultz P, Ebbesen TW,Mioskowski C. Science **2003**, 300, 775

105. Ryan KP, Cadek M, Nicolosi V, Blond D, Ruether M, Armstrong G, Swan H, Fonseca A, Nagy JB, Maser WK, Blau WJ, Coleman JN. Composites Science and Technology **2007**, 67, 1640
106. Bhattacharyya AR, Sreekumar TV, Liu T, Kumar S, Ericson LM, Hauge RH, Smalley RE. Polymer **2003**, 44, 2373
107. Valentini L, Biagiotti J, Kenny JM, Santucci S. Composites Science and Technology **2003**, 63, 1149
108. Valentini L, Biagiotti J, Kenny JM, Santucci S. Journal of Applied Polymer Science **2003**, 87, 708
109. Valentini L, Biagiotti J, Kenny JM,Manchado MAL. Journal of Applied Polymer Science **2003**, 89, 2657
110. Assouline E, Lustiger A, Barber AH, Cooper CA, Klein E, Wachtel E, Wagner HD. Journal of Polymer Science: Part B: Polymer Physics **2003**, 41, 520
111. Sandler J, Broza G, Nolte M, Schulte K, Lam Y-M, Shaffer MSP. Journal of Macromolecular Science, Physics **2003**, B42, 479
112. Coleman JN, Blau WJ, Dalton AB, Collins S, Kim BG, Razal J, Selvidge M, Vieiro G, Baughman RH. Applied Physics Letters **2003**, 82, 1682
113. Probst O, Moore EM, Resasco DE,Grady BP. Polymer **2004**, 45, 4437
114. Coleman JN, Cadek M, Ryan KP, Fonseca A, Nagy JB, Blau WJ, Ferreira MS. Polymer **2006**, 47, 8556
115. Cadek M, Coleman JN, Ryan KP, Nicolosi V, Bister G, Fonseca A, Nagy JB, Szostak K, Beguin F, Blau W. Nano letters **2004**, 4, 353
116. Dalton AB, Collins S, Munoz E, Razal JM, Ebron VH, Ferraris JP, Coleman JN, Kim BG, Baughman RH. Nature **2003**, 423, 703
117. Coleman JN, Cadek M, Blake R, Nicolosi V, Ryan KP, Belton C, Fonseca A, Nagy JB, Gunko YK, Blau WJ. Advanced Functional Materials **2004**, 14
118. Ye H, Lam H, Titchenal N, Gogotsi Y, Ko F. Applied Physics Letters **2004**, 85, 1775
119. Singh U, Prakash V, Abramson AR, Chen W, Qu L. Dai L. Appl. Phys. Lett. **2006**, 89, 073103
120. Seoul C, Kim YT,Baek CK. Journal of Polymer Science: Part B:Polymer Physics **2003**, 41, 1572
121. Lim JY, C.K.Lee, S.J.Kim, I.Y.Kim,S.I.Kim. Journal of Macromolecular Science, Part A: Pure and Applied Chemistry **2006**, 43, 785

Chapter 5

Device Structures Composed of Single-Walled Carbon Nanotubes

Pornnipa Vichchulada, Deepa Vairavapandian and Marcus D. Lay

Department of Chemistry, and Nanoscale Science and Engineering Center (NanoSEC), University of Georgia, Athens, GA 30602

Due to the great sensitivity of single-walled carbon nanotubes (SWNTs) to molecular adsorption events, they have been widely studied for use in various gas sensing applications. However, useful applications of SWNTs are limited by the lack of reproducibility in devices is based on a single SWNT. The use of two dimensional networks of SWNTs provides a new material that maintains the great sensitivity of SWNTs while producing a material with more reproducible characteristics. Such networks can be grown by high temperature techniques, or room-temperature liquid deposition techniques. Liquid deposition techniques allow the deposition of pre-treated, highly aligned SWNT networks on a wide variety of substrates.

Introduction

Nanotechnology investigations of the development of new sensing materials is of great current interest. Single walled carbon nanotubes (SWNTs) are of particular interest due to the fact that these hollow, nanoscale, molecular conductors are composed entirely of surface atoms (*1, 2*). This makes them particularly sensitive to the adsorption of gas phase molecules. A major impediment to the use of SWNTs in practical sensor device structures is the great variability between individual SWNTs; depending on chirality, or rolling vector, an SWNT may be a semiconductor or ametallic conductor (*3-6*).

Further, the semiconducting SWNTs have bandgaps that vary inversely with the diameter. Investigations that involved the separation of the two forms of SWNTs have met with limited success for very small masses of materials (*7-11*). Therefore, separation of SWNTs based on metallic or semiconductive behavior will still not result in greater control over device response, as there is great variability among semiconductive SWNTs due to very small changes in chirality (*12-14*).

One way to take advantage of the enhanced properties of SWNTs while surmounting the difficulties involved with this material is to use two dimensional networks (*15-19*). In such a network, device properties can be controlled by the overall density of the network rather than the physical properties of any individual SWNT. At densities where semiconducting pathways dominate, high mobility thin-film switching devices can be fabricated.

Such devices also have higher current drive and are easier to fabricate than devices based on a single SWNT. A well established method of fabricating micro and nano-scaled devices is the process known as photolithography. This manuscript will present a brief background and description of how this fabrication method is provided entailing how nano-scaled device fabrication can be achieved using already in place and well understood techniques.

A Brief History of Carbon Nanotube Material

Soot, or amorphous carbon, is the material that remains upon decomposition of hydrocarbons. Though it has long been known that soot is produced during the decomposition of hydrocarbons over Fe catalysts, it wasn't until 1953 that Davis, Slawson and Rigby also observed the presence of carbon "fibers"(*20*). These fibers existed within a sample of amorphous carbon that formed from the decomposition of CO at Fe surfaces. These fibers were described as worm-like structures that form during the decomposition of CO at about 450°C in the presence of Fe. The iron could be small iron particles present in, or even in macroscopic iron samples. It was clear that the Fe played the role of catalyst in the formation of these carbon structures. Today, the process of growing carbon nanotubes from the decomposition of a hydrocarbon precursor at an iron catalysts is commonly known as thermal chemical vapor deposition (CVD).

A subsequent report in 1978 by Wiles and Abrahamson described the formation of bundles of "carbon fibers" on graphite electrodes used in arch discharge experiments (*21*). Transmission electron microscopy (TEM) was used to ascertain that these carbon fibers had diameters as small as 10 nm and clumped together to form bundles of between 200 to 1000 fibers. In 1991, Iijima observed that these carbonaceous fibers were hollow concentric tubes of carbon (*22*). This report began a new era of research into what became known as multiwalled carbon nanotubes (MWNTs). Separate reports from the Iijima and Bethune research groups in 1993 independently confirmed the existence of SWNTs, which were observed to be composed of tubes composed of a single layer of carbon atoms. The years following these discoveries have seen an exponential increase in publications related to both types of nanotubes.

Properties of SWNTs

There is great interest in SWNT for a wide variety of applications due to their enhanced physical properties. These molecular wires on the strongest known material (*8,23-31*). They have been observed to be hundreds of times stronger than the highest grade carbon steel currently available (*28,30,32-34*). They can be stretched over five times their original length with nearly 100% memory and undetectable damage.

SWNTs also exhibit enhanced electronic properties, which include near ballistic transport (*4,35-40*). Therefore, SWNTs are investigated for a wide variety of potential applications, including conductive and high-strength composites, paint additives, nanoscale test tubes (*41,42*), hydrogen storage media (*43,44*), batteries (*45*), field emission materials (*43,46-52*), tips for scanning probe microscopy (*53-59*), transistors (*60*), diodes (*61*), and sensors (*1, 2,18,40,62-72*).

Sensors composed of SWNTs show a fast response and have a significantly higher sensitivity than that of existing solid-state sensors. Due to the fact that SWNTs are composed entirely of surface atoms, they have the highest known surface area of any material at about 1600 m^2/g (*73,74*). This high surface area presents a huge sensing area for the adsorption of gas phase analytes. Therefore SWNTs show great potential as the building of nano- and macro- scaled electronic sensors of the near future.

Graphite is an allotrope of carbon in which each carbon atom is bound to three other carbon atoms and has a fourth electron which is able to move about between carbon atoms. This structure accounts for the excellent electrical conductivity of graphite. Another important property of graphite is that it is a layered material composed of multilple layers of single sheets, called graphene, which are bound together by weak van der Wall attractions.

An SWNT is formed by taking a single graphene sheet and rolling it into a seamless cylinder (*75-78*). This cylinder is then composed of a single hexagonal array of carbon atoms that are sp^2 hybridized, as found in graphite. As each carbon atom is bound to three other carbon atoms through a covalent bond and has one other electron that is in an extended π-bonding network, SWNTs exhibit the low electrical resistance observed in graphite. Additionally, quantum confinement effects, due to the rolling vector and/or size of the SWNT, causes roughly $1/3^{rd}$ of SWNTs to behave as metallic conductors while $2/3^{rd}$ behave as semiconductors.

An SWNT's rolling vector, or chirality, has the most profound effect on its electronic properties. Figure 1 shows the three possible types of carbon nanotubes as defined by chirality. If a carbon nanotube is formed by connecting one carbon atom in the graphene lattice to another carbon atom that is located directly along one of the unit vectors of the surface, the result is what is known as an armchair carbon nanotube. If an SWNT is formed in such a way that one atom connects and atom which is 30° from the zigzag direction, the result is an armchair carbon nanotube. Any connection possibility between those two directions on the graphene lattice results in a chiral carbon nanotube. The chirality of a carbon nanotube is determined by measuring the angle of the rows of carbon atoms with respect to the tube axis. This measurement can be obtained directly with scanning tunneling microscopy (STM).

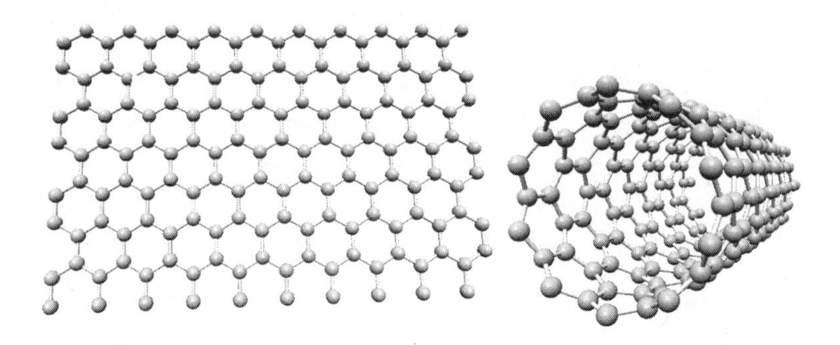

Figure 1. (a) A single graphene sheet; a van der Waals material with covalent bonds occurring only within each layer. Each C atom is bound to 3 other C atoms and has one delocalized electron; (b) a single-wall carbon nanotube (SWNT) results from the rolling of one graphene sheet into a seamless cylinder.

Chemical Sensing with SWNTs

Under ambient conditions, SWNTs are p-type semiconductors. This means that the charge carriers are holes, or the absence of electrons. This class of semiconductors shows an increase in conductivity when electron density is withdrawn and a decrease in conductivity when electron density is donated. This behavior has been observed experimentally for individual SWNTs connected to a set of electrodes. This allows for chemical detection with a simple chemiresistor. This is a device composed of two electrodes which are connected by some material, in this case a single-walled carbon nanotube, which changes resistance upon exposure to a particular analyte.

Dai and coworkers were the first to show that individual SWNTs worked as highly sensitive chemical sensors (*1,2*). They reported the use of SWNT chemiresistors to detect NO_2 and NH_3. Resistance changes of 2-3 orders of magnitude upon exposure. The exposure of an SWNT to NO_2 gas resulted in a decrease in the resistivity of the SWNT, while exposure to an NH_3 gas resulted in any increase in resistivity. This behavior is due to the fact that NO_2 is an electron withdrawing molecule. Therefore, upon adsorption into the SWNT surface, it withdraws electron density and increases the conductivity of the SWNT by increasing its density of charge carriers. On the other hand, ammonia behaves as a very strong electron donating group. Therefore, the adsorption of ammonia results in a very strong decrease in conductivity as the charge carriers within the SWNT are depleted. SWNTs were found to be highly sensitive in these studies as detection ranges in the ppm were observed (*15,52,62,79,80*).

Collins also reported that SWNT response was extremely sensitive to environmental conditions (*1*). They found the electrical resistance and thermal electric power response were greatly affected by whether the experiment was performed in the presence of oxygen or under vacuum. Numerous subsequent studies have demonstrated the promise of SWNT sensors for a wide variety of gaseous analytes.

Necessity of 2-D Networks of SWNTs

Though individual SWNTs have shown great promise as gas sensors, devices based on a single SWNT are impractical for 2 important reasons. First, there is currently no method for forming large numbers of highly aligned SWNTs on a variety of substrates. Such methods are necessary in order to form device structures on a large scale in a way that can be manufactured. Second, there is wide variability in electrical properties of individual SWNTs. Changes in electrical conductivity including metallic behavior, semiconductive behavior and bandgap energy will lead to wildly different responses from varying devices to the same analyte. The result is that while SWNTs are interesting from a research perspective their applications are limited.

One way to surmount these two difficulties with SWNT material is the use of a 2-D network of SWNTs. Such networks present a new material with qualities that result from the average properties of many individual SWNTs and are not dominated by any one particular form. Therefore, device properties are largely a result of the density of the SWNT network. Low-density networks behaving as thin-film semiconductors and high-density networks exhibit metallic behavior. This is more likely to be a manufacturable approach to SWNT device structures, as it greatly increases the reproducibility in device performance. Networks such as these are expected to behave reproducibly for the types of large-area devices that can be formed from SWNT networks; as the size of the network increases, the variability in device performance should increase due to the averaging of properties of more SWNTs.

The Application of Percolation Theory to SWNT Networks

Percolation theory can be applied to carbon nanotube networks.(*81-83*) As mentioned earlier, one third of carbon nanotubes or metallic and two thirds or semiconductive. If one has a random arrangement of SWNTs on a non-conducting surface at very low density, there are no electrical pathways. However, as the number of SWNTs increases, one will first reach the percolation threshold for the semiconductive carbon nanotubes. At this point, there is the possibility of having fully semiconductive pathways throughout the thin-film. However metallic pathways are not yet favored as metallic carbon nanotubes compose only 1/3rd of the sample (Figure 2).

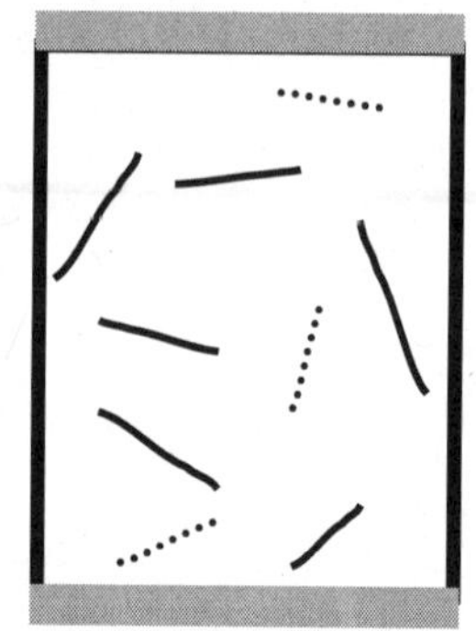
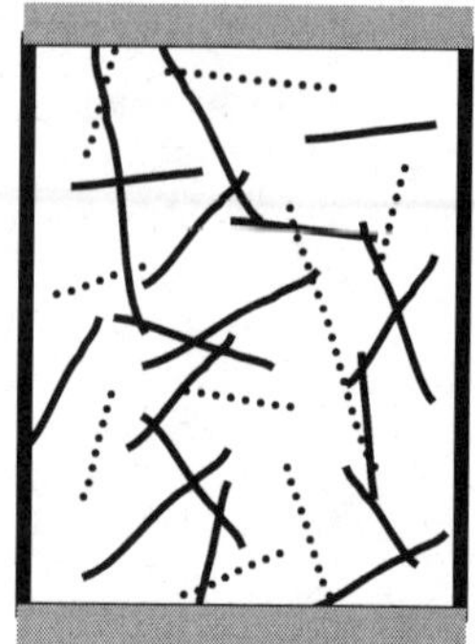

Figure 2. Solid lines are semiconductive SWNTs and dashed lines are metallica SWNTs. At low SWNT densities, conducting pathways do not exist (left). Semicondutive pathways are reached prior to formation of metallic pathways (right).

As the number of SWNTs continues to increase, the percolation threshold for metallic SWNTs is reached. At the density of SWNTs continues to increase, the number of metallic pathways increases commensurately. The result is thin-film which conducts electricity with metallic behavior (further decreasing resistance) and less semiconductive behavior (sensitivity to molecular adsorption events).

As a result, a two dimensional network composed of a random distribution of metallic and semiconducting SWNTs will behave as a thin-film semiconductor in the limit between between the percolation thresholds for the semiconductive and metallic SWNTs. For densities of SWNTs in this range, semiconductive pathways dominate the film and metallic pathways are not favored until the percolation threshold for metallic SWNTs.

Further, partially metallic pathways have little effect on electronic properties on electronic properties as semiconductive SWNTs in such a network will show response to changes in conductivity upon exposure to an analyte thus acting at the bottleneck to conductivity throughout the partially metallic network. Also, metallic/metallic and the semiconductive/semiconductive SWNT junctions have enhanced electrical connections. While, metallic/semiconductive SWNT junctions have a high potential energy barrier to charge transfer, known as a Schottky contact (*84*). This barrier results in greatly decreased conductivity, thus aiding in the removal of partially metallic pathways from the network. The result is that metallic SWNTs only become an issue in device performance above their percolation threshold.

SWNT Network Sensors

Chemiresistive SWNT network based sensing devices were first reported by Dai and coworkers (*2*). The devices were composed of random networks of SWNTs that bridged two metal electrodes on a dielectric surface. They demonstrated that SWNT networks functioned as highly sensitive protein detectors in aqueous solution. Other groups have subsequently reported the formation of highly sensitive SWNT network based sensors for a variety of gaseous and fluid analytes (*52,62,79,80*). There are two main approaches to the formation of SWNT networks: high temperature growth techniques like CVD, as well as various liquid deposition techniques.

High Temperature SWNT Network Growth

The CVD growth technique was the first used to grow carbon nanotube networks for sensing applications. SWNT networks are typically grown on silicon wafer substrates due to this substrates tolerance to high temperatures. The CVD has proved to be an excellent way to grow randomly oriented networks of carbon nanotubes. Various conditions such as catalysts, temperature, and growth of time may be varied in order to achieve desired network properties.

Though the growth conditions may vary slightly by research group, they are commonly variations on a method first reported by Lieber and coworkers (*85*). The silicon wafer fragment is dipped into a catalyst solution. That catalyst is typically an iron salt which has been dissolved in isopropanol. After the fragment is dipped in the solution it is rinsed with solvent. This process leaves behind a very tiny amount of iron ions on the surface. Next, the substrate is placed in a quartz tube furnace in which an inert atmosphere is maintained by the flow of high purity argon gas. The quartz tube and substrate are then heated to 600 to 800°C. During this heating, the surface bound metal catalyst ions are reduced by introducing hydrogen gas into the system. This results in the formation of iron nanoparticles which will act as a catalyst and during the growth of a carbon nanotubes. Lastly, a hydrocarbon source is introduced and SWNT growth results from the decomposition of this hydrocarbon source at the iron catalyst nanoclusters.

Room Temperature SWNT Network Formation

One of the major advantages of liquid deposition techniques for the formation of SWNT networks is that they occur at room temperature this allows inclusion of heat sensitive substrates like polymers plastics and glass (*17*). A second major advantage of liquid deposition techniques is the they deconstruct network formation into two separate steps, therefore SWNT growth and deposition occurred in two separate steps. This facilitates the deposition of SWNTs onto a wide variety of substrates. Additionally, this also allows the introduction of purification and/or modification steps between SWNT growth and deposition.

Purification steps are important because amorphous carbon is typically formed along with SWNTs regardless of the growth process. This amorphous carbon will change the electrical properties of the network by adding additional junctions between SWNTs and may also act to increase device noise. Therefore methods of removing carbonaceous impurities are a large part of SWNT research. Work in this group has involved the use of centrifugation to remove impurities prior to the SWNT network formation (Figure 3).

The ability to insert a step in which the SWNTs are chemically modified is another major advantage of liquid deposition techniques. Such modification is frequently necessary in order to impart selectivity for highly sensitive sensor response of SWNTs.

Regardless of whether purification or modification steps are employed liquid deposition of allows the formation of SWNT networks on a wide variety of substrates at room temperature without the use of complicated equipment. Further, these networks can be formed on transparent substrates like glass or plastic quickly (*17,86-89*). This property results in numerous novel sensor applications and such devices are expected to have a huge impact on the electronics industry in general (*18,90-95*).

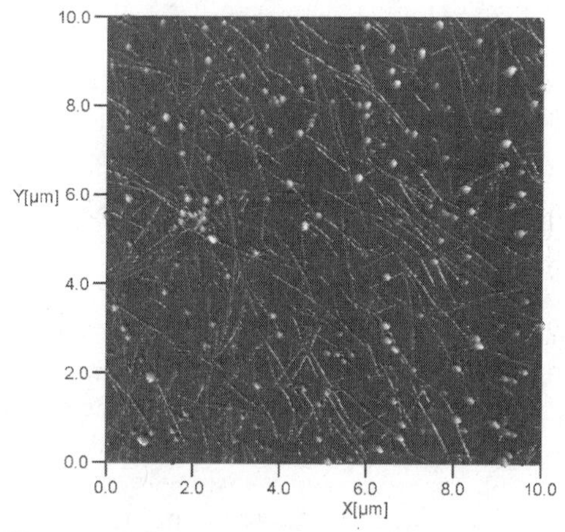

Figure 3. SWNT network formed by room-temperature deposition method developed in this group.

Liquid Deposition of Highly Aligned Networks of SWNTs

Liquid deposition of SWNT networks presents another major step toward the inclusion of SWNTs into manufacturable device structures. This is due to the fact that liquid deposition, combined with a laminar flow drying technique allows the formation of highly aligned SWNTs electrical networks (*15, 16*). In a manner similar to molecular combing of DNA, high purity air is applied to the air/liquid interface in a manner which aligns the SWNTs in aqueous suspension and then deposits them on a desired substrate.

In theory, this method allows formation of highly aligned SWNTs on any arbitrary substrate. Therefore in addition to creating practical sensor device structures, this method may be used to create electronic device structures composed of highly aligned SWNTs on a wide variety of substrates.

A network of highly aligned SWNTs is a macroscopic material formed from nanoscopic building blocks; the size of the network is not limited only limited by the size of the substrate. Therefore, this is an ideal fabrication method for sensing applications where large-area substrates are needed.

It has been recently reported that networks composed of highly aligned SWNTs have anisotropic electrical properties (*16*). Thin-film transistors composed of such networks display on/off ratios (the relative magnitude of the on-state current to the off-state current) that were dictated largely by the orientation of the network with respect to electron flow through the network. The on/off ratio can be considered a measure of the sensitivity of a chemieresistor. This is because the greater the on/off ratio, the more sensitive it will be to molecular adsorption from gas phase analytes; a high onto off ratio indicate the semiconductive SWNTs dominate the performance of the device and only semiconductive SWNTs show sensor response. This research group is currently investigating the application of percolation theory to this phenomenon.

Sensing Applications of SWNT Networks

SWNT networks formed from aqueous suspensions of carbon nanotubes have proved to be useful for a variety of sensor applications. Figure 4 demonstrates how these networks can be used for sensing a variety of volatile organic vapors. Saturated vapors of each of the analytes were added to a flowing stream of purified air in order to determine the response of the carbon nanotube network.

A simple chemiresister setup was used; this was composed of two metal electrodes, 1 cm apart, separated by the liquid deposited carbon nanotube network. The dimension of the network was 1 cm^2. The density of SWNTS was approximately 2μm^{-2}. A voltage of 100 mV was applied between the two metal electrodes and current versus time data was recorded for each device. Exposure to saturated vapors of acetone, methanol and 2-propanol resulted in a decrease in current for all devices. The data presented in figure 4 has been plotted as $\Delta R/R$ in order to facilitate the comparison of various devices with differing initial resistances.

The acetone molecule is a keytone composed of a functional group which contains an oxygen atom having two loan pairs of electrons. This enables acetone to behave as a very strong electron donor upon adsorption on SWNTs within the network. This electron density doantion causes the greatest decrease in conductivity upon exposure, resulting in the largest $\Delta R/R$. Conversley, in alcohols, a hydrogen atom is bound directly to the oxygen atom, resulting in the presence of only one lone-pair of electrons on this very electro-negative element. This accounts for the lower device response to methanol and isopropanol. All of the devices were observed to fully recover although in some cases the recovery was slow. Current efforts in this group involve the determination of the effect of a gate potential on device performance and refreshing. Additionally, we are investigating the effect of the silane, used to adhere the SWNTs to the substrate, on the overall device performance.

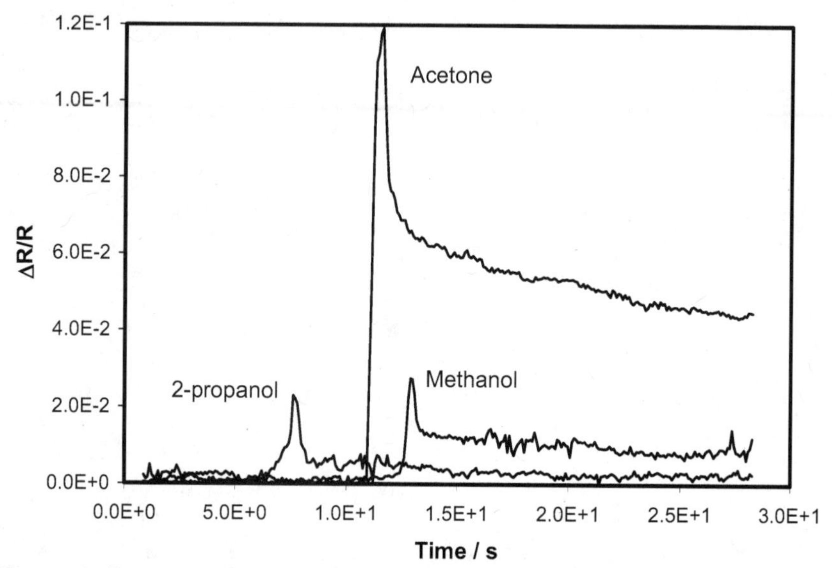

Figure 4. Responsive room temperature or deposited carbon nanotube networks to vapors of various volatile analytes.

Studies in this group currently involve determining the effect of density and level of alignment of SWNTs on overall device performance. For semiconductive networks of SWNTs, the devices are expected to have a greater response at the number of SWNTs increases. This will occur until the percolation threshold for metallic SWNTs is reached and then response will level off as metallic pathways will show little change in conductivity upon exposure.

The fact that the carbon nanotubes can be aligned between the electrodes yields a further degree of control in device formation; initial results have indicated that the alignment of the SWNTs with respect to the metal electrodes affects overall response. This allows an additional degree of control, or tunability, in device behavior.

Conclusion and Future Directions

The use of liquid deposited networks of aligned SWNTs in sensing applications presents a new frontier in the creation of lightweight and flexible sensor device structures. This allows the formation of devices composed of either two electrodes or of arrays of interdigitated electrodes.

Additionally, recent work has shown that networks of aligned SWNTs have anisotropic electrical behavior; the degree of sensitivity to affect conductivity of these materials varies according to the direction in which an electrical current is passed through a network of SWNTs. Efforts in this group are focused currently on determining the ideal degree of alignment and type of alignment for sensor applications.

References

1. P. G. Collins, K. Bradley, M. Ishigami and A. Zettl, *Science*, **2000**, 287, 1801.
2. J. Kong, N. R. Franklin, C. W. Zhou, M. G. Chapline, S. Peng, K. J. Cho and H. J. Dai, *Science*, **2000**, 287, 622.
3. M. Liebau, A. P. Graham, G. S. Duesberg, E. Unger, R. Seidel and F. Kreupl, *Fullerenes, Nanotubes, Carbon Nanostruct.*, **2005**, 13, 255.
4. M. S. Dresselhaus and H. Dai, *MRS Bull.*, **2004,** 29, 237.
5. H. J. Dai, *Surf. Sci.*, **2002**, 500, 218.
6. W. Kim, H. C. Choi, M. Shim, Y. M. Li, D. W. Wang and H. J. Dai, *Nano Lett.*, **2002**, 2, 703.
7. K. Balasubramanian, R. Sordan, M. Burghard and K. Kern, *Nano Lett.*, **2004**, 4, 827.
8. H. J. Huang, R. Maruyama, K. Noda, H. Kajiura and K. Kadono, *J. Phys. Chem. B*, **2006**, 110, 7316.
9. M. Zheng, A. Jagota, M. S. Strano, A. P. Santos, P. Barone, S. G. Chou, B. A. Diner, M. S. Dresselhaus, R. S. McLean, G. B. Onoa, G. G. Samsonidze, E. D. Semke, M. Usrey and D. J. Walls, *Science*, **2003**, 302, 1545.
10. M. S. Strano, C. A. Dyke, M. L. Usrey, P. W. Barone, M. J. Allen, H. W. Shan, C. Kittrell, R. H. Hauge, J. M. Tour and R. E. Smalley, *Science*, **2003**, 301, 1519.
11. R. Krupke, F. Hennrich, H. von Lohneysen and M. M. Kappes, *Science*, **2003**, 301, 344.
12. L. C. Venema, J. W. Janssen, M. R. Buitelaar, J. W. G. Wildoer, S. G. Lemay, L. P. Kouwenhoven and C. Dekker, *Phys. Rev. B*, **2000**, 62, 5238.
13. L. C. Venema, V. Meunier, P. Lambin and C. Dekker, *Phys. Rev. B*, **2000**, 61, 2991.
14. J. W. G. Wildoer, L. C. Venema, A. G. Rinzler, R. E. Smalley and C. Dekker, *Nature*, **1998**, 391, 59.
15. P. Vichchulada, Q. Zhang and M. D. Lay, *Analyst*, **2007**, 132, 709.
16. M. D. Lay, J. P. Novak and E. S. Snow, *Nano Letters*, **2004**, 4, 603.
17. N. Saran, K. Parikh, D. S. Suh, E. Munoz, H. Kolla and S. K. Manohar, *Journal of the American Chemical Society*, **2004**, 126, 4462.
18. K. Parikh, K. Cattanach, R. Rao, D. S. Suh, A. M. Wu and S. K. Manohar, *Sensors and Actuators B-Chemical*, **2006**, 113, 55.
19. R. J. Chen, S. Bangsaruntip, K. A. Drouvalakis, N. W. S. Kam, M. Shim, Y. M. Li, W. Kim, P. J. Utz and H. J. Dai, *Proceedings of the National Academy of Sciences of the United States of America*, **2003**, 100, 4984.
20. W. R. Davis, R. J. Slawson and G. R. Rigby, *Nature*, **1953**, 171, 756.
21. P. G. Wiles and J. Abrahamson, *Carbon*, **1978**, 16, 341.
22. S. Iijima and T. Ichihashi, *Nature*, **1993**, 363, 603.
23. E. P. S. Tan and C. T. Lim, *Compos. Sci. Technol.*, **2006**, 66, 1102.
24. S. Bhattacharyya, J. P. Salvetat and M. L. Saboungi, *Appl. Phys. Lett.*, **2006**, 88 .
25. P. J. F. Harris, *Int. Mater. Rev.*, **2004**, 49, 31.
26. P. Laborde-Lahoz, W. Maser, T. Martinez, A. Benito, T. Seeger, P. Cano, R. G. de Villoria and A. Miravete, *Mech. Adv. Mater. Struct.*, **2005**, 12, 13.

27. B. Lukic, J. W. Seo, R. R. Bacsa, S. Delpeux, F. Beguin, G. Bister, A. Fonseca, J. B. Nagy, A. Kis, S. Jeney, A. J. Kulik and L. Forro, *Nano Letters*, **2005**, 5, 2074.
28. R. S. Ruoff and D. C. Lorents, *Carbon*, **1995**, 33, 925.
29. M. M. J. Treacy, T. W. Ebbesen and J. M. Gibson, *Nature*, **1996**, 381, 678.
30. M. F. Yu, O. Lourie, M. J. Dyer, K. Moloni, T. F. Kelly and R. S. Ruoff, *Science*, **2000**, 287, 637.
31. Y. Zhang, G. L. Yu and J. M. Dong, *Phys. Rev. B*, **2006**, 73 .
32. M. Daenen, R. D. de Fouw, B. Hamers, P. G. A. Janssen, K. Schouteden and M. A. J. Veld, The Wondrous World of Carbon Nanotubes: a review of current carbon nanotube technologies, in, p. 1, Eindhoven University of Technology, Eindhoven, The Netherlands (2003).
33. M. F. Yu, B. S. Files, S. Arepalli and R. S. Ruoff, *Physical Review Letters*, **2000**, 84, 5552.
34. A. Krishnan, E. Dujardin, T. W. Ebbesen, P. N. Yianilos and M. M. J. Treacy, *Physical Review B*, **1998**, 58, 14013.
35. B. Obradovic, R. Kotlyar, F. Heinz, P. Matagne, T. Rakshit, M. D. Giles, M. A. Stettler and D. E. Nikonov, *Applied Physics Letters*, **2006**, 88 .
36. Y. Ouyang, Y. Yoon, J. K. Fodor and J. Guo, *Applied Physics Letters*, **2006**, 89 .
37. J. Guo, E. C. Kan, U. Ganguly and Y. G. Zhang, *Journal of Applied Physics*, **2006**, 99 .
38. H. C. d'Honincthun, S. Galdin-Retailleau, J. See and P. Dollfus, *Applied Physics Letters*, **2005**, 87 .
39. A. Javey, J. Guo, Q. Wang, M. Lundstrom and H. J. Dai, *Nature*, **2003**, 424, 654.
40. S. Polizu, O. Savadogo, P. Poulin and L. Yahia, *J. Nanosci. Nanotechnol.*, **2006**, 6, 1883.
41. H. Orikasa, N. Inokuma, S. Okubo, O. Kitakami and T. Kyotani, *Chemistry of Materials*, **2006**, 18, 1036.
42. R. Gasparac, P. Kohli, M. O. Mota, L. Trofin and C. R. Martin, *Nano Letters*, **2004**, 4, 513.
43. M. Terrones, *Annual Review of Materials Research*, **2003**, 33, 419.
44. A. C. Dillon, K. M. Jones, T. A. Bekkedahl, C. H. Kiang, D. S. Bethune and M. J. Heben, *Nature*, **1997**, 386, 377.
45. B. Gao, C. Bower, J. D. Lorentzen, L. Fleming, A. Kleinhammes, X. P. Tang, L. E. McNeil, Y. Wu and O. Zhou, *Chemical Physics Letters*, **2000**, 327, 69.
46. R. P. Raffaelle, B. J. Landi, J. D. Harris, S. G. Bailey and A. F. Hepp, *Materials Science and Engineering B-Solid State Materials for Advanced Technology*, **2005**, 116, 233.
47. S. J. Tans, A. R. M. Verschueren and C. Dekker, *Nature*, **1998**, 393, 49.
48. T. Feng, J. H. Zhang, X. Wang, X. H. Liu, S. C. Zou, Q. Li and J. F. Xu, *Surf. Rev. Lett.*, **2005**, 12, 733.
49. C. M. Hsu, C. H. Lin, H. L. Chang and C. T. Kuo, *Thin Solid Films*, **2002**, 420, 225.
50. M. H. Moon, D. J. Kang, J. H. Jung and J. M. Kim, *J. Sep. Sci.*, **2004**, 27, 710.

51. O. A. Nerushev, M. Sveningsson, L. K. L. Falk and F. Rohmund, *J. Mater. Chem.*, **2001**, 11, 1122.
52. J. J. Zhao, *Curr. Nanosci.*, **2005**, 1, 169.
53. E. S. Snow, P. M. Campbell and J. P. Novak, *J. Vac. Sci. Technol. B*, **2002**, 20, 822.
54. E. S. Snow, P. M. Campbell and J. P. Novak, *Appl. Phys. Lett.*, **2002**, 80, 2002.
55. S. S. Wong, E. Joselevich, A. T. Woolley, C. L. Cheung and C. M. Lieber, *Nature*, **1998**, 394, 52.
56. S. Carnally, K. Barrow, M. R. Alexander, C. J. Hayes, S. Stolnik, S. J. B. Tendler, P. M. Williams and C. J. Roberts, *Langmuir*, **2007**, 23, 3906.
57. W. P. Huang, H. H. Cheng, S. R. Jian, D. S. Chuu, J. Y. Hsieh, C. M. Lin and M. S. Chiang, *Nanotechnology*, **2006**, 17, 3838.
58. A. J. Austin, C. V. Nguyen and Q. Ngo, *Journal of Applied Physics*, **2006**, 99.
59. J. Martinez, T. D. Yuzvinsky, A. M. Fennimore, A. Zettl, R. Garcia and C. Bustamante, *Nanotechnology*, **2005**, 16, 2493.
60. C. W. Zhou, J. Kong and H. J. Dai, *Applied Physics Letters*, **2000**, 76, 1597.
61. J. U. Lee, *Applied Physics Letters*, **2005**, 87 .
62. J. Andzelm, N. Govind and A. Maiti, *Chem. Phys. Lett.*, **2006**, 421, 58.
63. A. B. Artyukhin, M. Stadermann, R. W. Friddle, P. Stroeve, O. Bakajin and A. Noy, *Nano Lett.*, **2006**, 6, 2080.
64. K. P. Gong, Y. M. Yan, M. N. Zhang, L. Su, S. X. Xiong and L. Q. Mao, *Anal. Sci.*, **2005**, 21, 1383.
65. J. Koehne, J. Li, A. M. Cassell, H. Chen, Q. Ye, H. T. Ng, J. Han and M. Meyyappan, *J. Mater. Chem.*, **2004**, 14, 676.
66. Y. Lin, S. Taylor, H. P. Li, K. A. S. Fernando, L. W. Qu, W. Wang, L. R. Gu, B. Zhou and Y. P. Sun, *J. Mater. Chem.*, **2004**, 14, 527.
67. Y. J. Lu, J. Li, J. Han, H. T. Ng, C. Binder, C. Partridge and M. Meyyappan, *Chem. Phys. Lett.*, **2004**, 391, 344.
68. G. L. Luque, N. F. Ferreyra and G. A. Rivas, *Microchim. Acta*, **2006**, 152, 277 .
69. Q. F. Pengfei, O. Vermesh, M. Grecu, A. Javey, O. Wang, H. J. Dai, S. Peng and K. J. Cho, *Nano Lett.*, **2003**, 3, 347 .
70. T. Someya, J. Small, P. Kim, C. Nuckolls and J. T. Yardley, *Nano Lett.*, **2003**, 3, 877.
71. S. Q. Wang, E. S. Humphreys, S. Y. Chung, D. F. Delduco, S. R. Lustig, H. Wang, K. N. Parker, N. W. Rizzo, S. Subramoney, Y. M. Chiang and A. Jagota, *Nat. Mater.*, **2003**, 2, 196.
72. Q. Zhao, Z. H. Gan and Q. K. Zhuang, *Electroanalysis*, **2002**, 14, 1609.
73. M. Cinke, J. Li, B. Chen, A. Cassell, L. Delzeit, J. Han and M. Meyyappan, *Chemical Physics Letters*, **2002**, 365, 69.
74. B. Yu and M. Meyyappan, *Solid-State Electron.*, **2006**, 50, 536.
75. X. J. Zhou, J. Y. Park, S. M. Huang, J. Liu and P. L. McEuen, *Physical Review Letters*, **2005**, 95 .
76. O. Zhou, H. Shimoda, B. Gao, S. Oh, L. Fleming and G. Yue, *Acc Chem Res*, **2002**, 35, 1045.
77. Y. C. Tseng, K. Phoa, D. Carlton and J. Bokor, *Nano Letters*, **2006**, 6, 1364.

78. E. S. Snow, J. P. Novak, M. D. Lay, E. H. Houser, F. K. Perkins and P. M. Campbell, *Journal of Vacuum Science & Technology B*, **2004**, 22, 1990.
79. E. Bekyarova, M. Davis, T. Burch, M. E. Itkis, B. Zhao, S. Sunshine and R. C. Haddon, *Journal of Physical Chemistry B*, **2004**, 108, 19717.
80. X. Feng, S. Irle, H. Witek, K. Morokuma, R. Vidic and E. Borguet, *Journal of the American Chemical Society*, **2005**, 127, 10533.
81. M. Sato and M. Sano, *Langmuir*, **2005**, 21, 11490.
82. D. J. Frank and C. J. Lobb, *Phys. Rev. B*, **1988**, 37, 302.
83. M. A. Dubson and J. C. Garland, *Phys. Rev. B*, **1985**, 32, 7621.
84. M. S. Fuhrer, J. Nygard, L. Shih, M. Forero, Y. G. Yoon, M. S. C. Mazzoni, H. J. Choi, J. Ihm, S. G. Louie, A. Zettl and P. L. McEuen, *Science*, **2000**, 288, 494.
85. J. H. Hafner, C. L. Cheung, T. H. Oosterkamp and C. M. Lieber, *Journal of Physical Chemistry B*, **2001**, 105, 743.
86. G. Gruner, *Journal of Materials Chemistry*, **2006**, 16, 3533.
87. Y. X. Zhou, L. B. Hu and G. Gruner, *Applied Physics Letters*, **2006**, 88.
88. P. Vichchulada, Q. Zhang and M. D. Lay, *Analyst*, In Press (2008).
89. J. P. Novak, M. D. Lay, F. K. Perkins and E. S. Snow, *Solid-State Electronics*, **2004**, 48, 1753.
90. H. Lee, S. H. Hong, K. Y. Yang and K. W. Choi, *Appl. Phys. Lett.*, **2006**, 88.
91. Q. Cao, S. H. Hur, Z. T. Zhu, Y. Sun, C. J. Wang, M. A. Meitl, M. Shim and J. A. Rogers, *Adv. Mater.*, **2006**, 18, 304.
92. A. Maliakal, H. Katz, P. M. Cotts, S. Subramoney and P. Mirau, *J. Am. Chem. Soc.*, **2005**, 127, 14655.
93. M. Zhang, S. L. Fang, A. A. Zakhidov, S. B. Lee, A. E. Aliev, C. D. Williams, K. R. Atkinson and R. H. Baughman, *Science*, **2005**, 309, 1215.
94. E. Artukovic, M. Kaempgen, D. S. Hecht, S. Roth and G. GrUner, *Nano Lett.*, **2005**, 5, 757.
95. K. Nomura, H. Ohta, A. Takagi, T. Kamiya, M. Hirano and H. Hosono, *Nature*, **2004**, 432, 488.

Chapter 6

Hand Held Biowarfare Assays

Rapid Biowarfare Detection Using the Combined Attributes of Microfluidic *in vitro* Selections and Immunochromatographic Assays

Letha J. Sooter[1], Dimitra N. Stratis-Cullum[1*], Yanting Zhang[2], Jeffrey J. Rice[2], John T. Ballew[3], Hyongsok T. Soh[3], Patrick S. Daugherty[3], Paul Pellegrino[1], Nancy Stagliano[2]

[1]US Army Research Laboratory, AMSRD-ARL-SE-EO, 2800 Powder Mill Road, Adelphi, MD 20783
[2] CytomX, LLC., 460 Ward Drive Suite E-1, Santa Barbara, CA 93111
[3] University of California, Santa Barbara, Santa Barbara, CA 93106

Immunochromatography is a rapid, reliable, and cost effective method of detecting biowarfare agents. Gold nanoparticles provide a visual indication of target presence. A fundamental limitation for this type of assay is the availability of molecular recognition elements (MRE's) which bind the target. Typical *in vitro* selection methods to produce the MRE's require weeks or months of work. To alleviate this problem, microfluidic sorting chips have been developed to rapidly screen for binders.

Many people are familiar with immunochromatography (*1-3*). Over-the-counter pregnancy tests are a prime example. These assays have a number of advantages. Two of their most important characteristics is that they are inexpensive to produce and disposable. They are also small, portable, and have a long shelf life. These assays are easy to use because they can be read visually. One purple line indicates the assay is functioning properly and two purple lines indicate the target molecule is present. An example of this can be seen in Figures 1 and 2. The Army is interested in producing these immunochromatrography tests rapidly when new chemical or biowarfare threats are identified. In order to do this, *in vitro* selections must be performed in a timely manner and they must produce strong, selective binders for the target of interest. Typical selections take from weeks to months to be completed. However, a microfluidic sorting chip has been developed that rapidly screens libraries to identify binders.

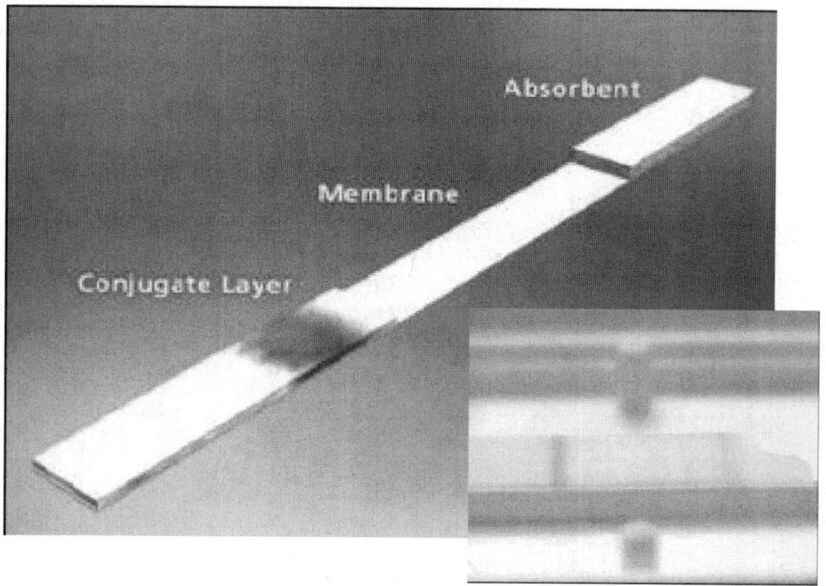

Figure 1. Typical hand held assay test (4).
(see page 1 of color inserts)

Introduction

Immunochromatography

A hydrophobic membrane is one of the primary components of an immunochromatographic hand held assay (*4*). As shown in Figure 2, two lines of white latex beads, the control line and the sample line, are immobilized through hydrophobic interactions on one end of the membrane strip. The control line is composed of latex beads conjugated to one member of a two molecule binding pair, such as an immunoglobulin (IgG) and ProteinA. The latex beads in the sample line are similar in that they are conjugated to a binding molecule, but this molecule binds one of multiple epitopes on a target. A target may be a small chemical, a large protein, or something as complex as cells or spores. The binding molecules are typically antibodies, but they may also be peptides or nucleic acids.

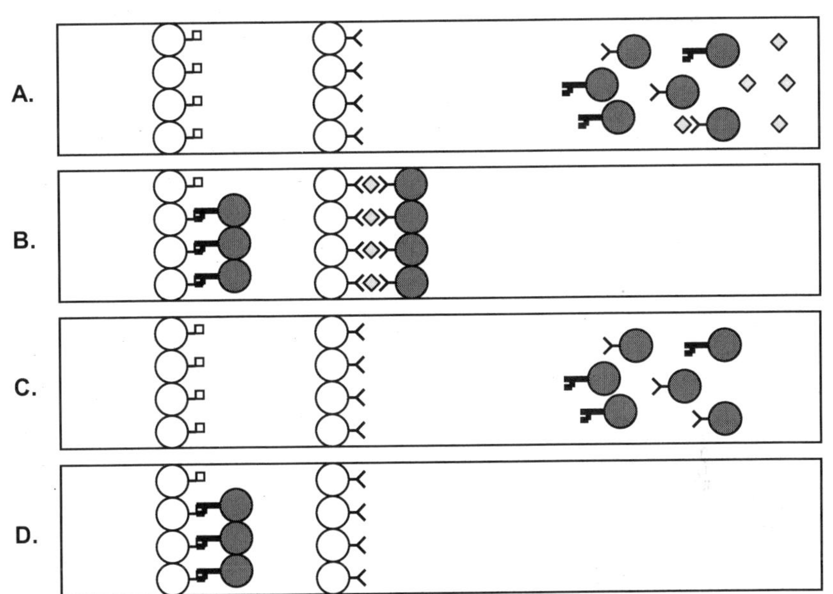

Figure 2. Schematic of positive and negative hand held assay tests. A. Target is present in the applied sample. B. The presence of target results in the formation of two purple lines on the test. C. No target is present in the applied sample. D. The lack of target results in the formation of one purple line on the test (4). (see page 1 of color inserts)

On the opposite end of the membrane is a mixture of gold nanoparticles. Gold nanoparticles are a purple color and provide a visible signal when immobilized on the line of latex beads. The nanoparticles are composed of two populations. One complements the two member binding pair of the control line. The other binds a second epitope on a target. Therefore, regardless of whether a target species is present, the control line will turn purple upon application of a liquid to the membrane. When a target is present, it flows through the gold nanoparticles conjugated to specific binders and becomes attached. The nanoparticle/target pair then travels through the membrane via capillary action. The sample line will turn purple when the conjugated latex bead binds a second epitope on the target. In the absence of target in the sample, the gold particles will flow past the latex beads and no signal will be visible.

In Vitro Selections

In vitro selections are a powerful means of producing ligands with high affinity and specificity for a target of choice. The molecules which bind the target of interest in an immunochromatographic hand held assay are typically isolated through *in vitro* selection. These selections may utilize, for example, DNA, RNA, peptides, or antibodies, each with their own attributes. *E. coli* bacterial libraries displaying peptides have a particularly short time period required for library generation/regeneration, see Figure 3. As an example of an *E. coli* surface display peptide selection, the process begins with a library. Each bacteria displays a different peptide to create a population, or library, of approximately 10^{10} unique cells (*2*). This library is incubated with a labeled target for a period of time to allow binding between the peptide and the target. Following this incubation, the assay is typically loaded into a Fluorescence Activated Cell Sorter (FACS). This process separates the two populations which are present: *E. coli* and *E. coli* bound to target. The *E. coli* bound to the target is collected and placed in growth media allowing for amplification of the selected population for additional rounds of selection.

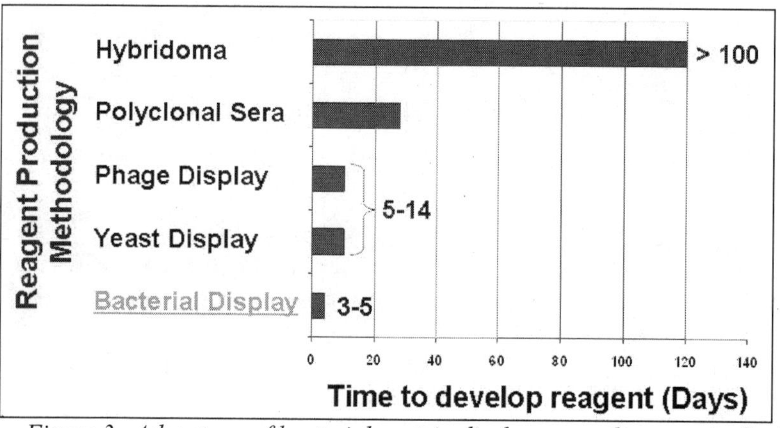

Figure 3. Advantage of bacterial protein display over other protein display methods (4).

The instrumentation required for FACS is large and quite expensive. It may also take a significant time to sort samples. To alleviate these problems, a microscope slide sized, microfluidic sorter using a technique known as continuous trapping magnetic activated cell sorting (CTMACS) was developed as a new means of affinity reagent discovery, see Figure 4.

Figure 4. CTMACS instrument.

The combination of bacterial display libraries with CTMACS allows the rapid production of affinity reagents, see Figure 5. These, in turn, are used in the production of immunochromatographic hand held assays. The result is high throughput production and deployment of simple to use assays for the most relevant biological and chemical warfare agents.

Experimental Methods

Immunochromatography

Immunochromatographic hand held assays are conceptually straightforward, but require significant optimization (*4*). Whatman FUSION5 membranes were chosen because of their "one membrane, five functions" technology. These membranes eliminate the need for multiple layers of different membranes to construct a laminar flow, immunochromatographic, hand held assay. They are highly absorbent and require little or no blocking. A BioDot platform equipped with a Biojet can be used to dispense the lines of latex beads. The result is clear, sharp signal lines. A mixture of the conjugated gold beads are deposited in a spot at the other end of the membrane. Then, the membrane is allowed to dry.

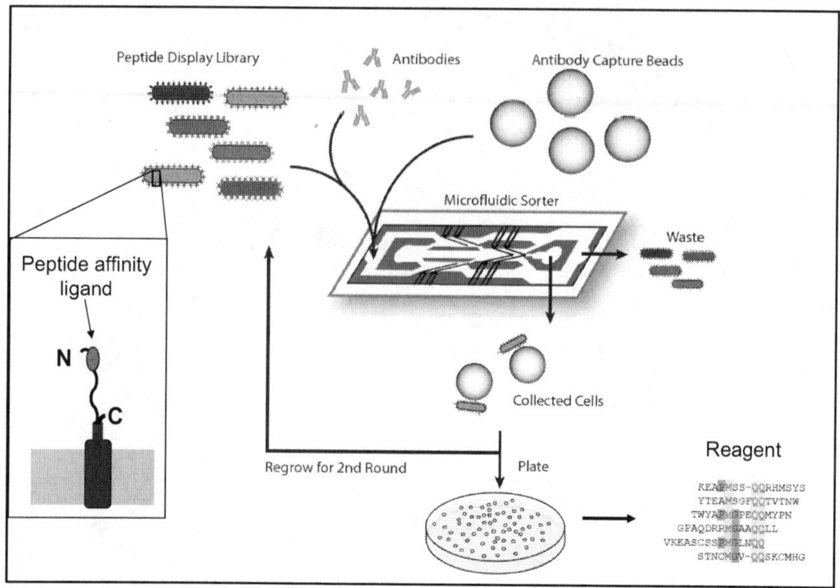

Figure 5. Bacterial display selection using a microfluidic sorting device (1).

A target sample solution and a negative control are pipetted onto separate hand held assays near the gold nanoparticles. The sample with target should produce two purple lines and the negative control should produce one purple line. Any non-bound nanoparticles will continue to flow to the end of the membrane strip.

In Vitro Selections with CTMACS

Continuous trapping magnetic activated cell sorting devices are approximately the size of a microscope slide. In order to perform selections using these devices, the target of interest must be conjugated to a magnetic particle. The bacterial display library is then incubated with the conjugated targets to allow binding. Then the bacterial library and target solution is applied to the CTMACS device, see Figure 6. Buffer channels move the assay across a series of magnets, where magnetic particles with bacteria bound targets are captured at the edges between

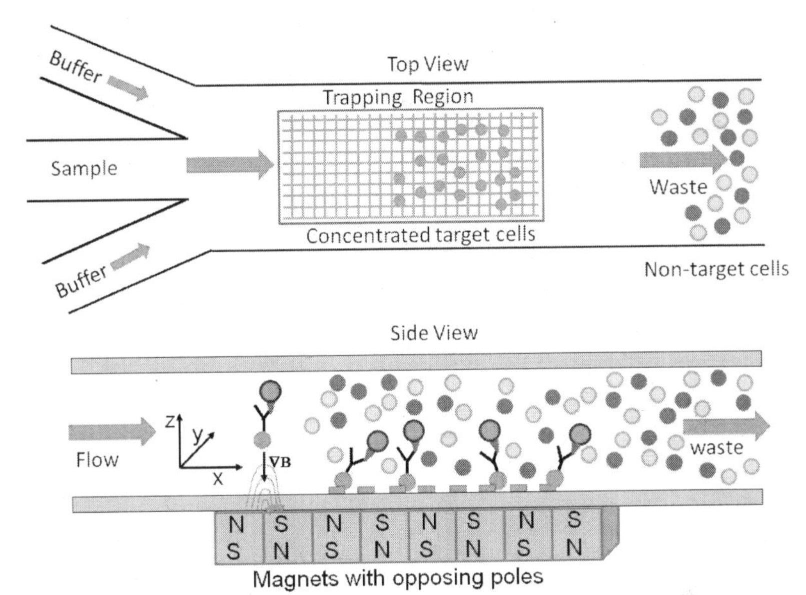

Figure 6. Diagram of a selection using a CTMACS device. *(see page 2 of color inserts)*

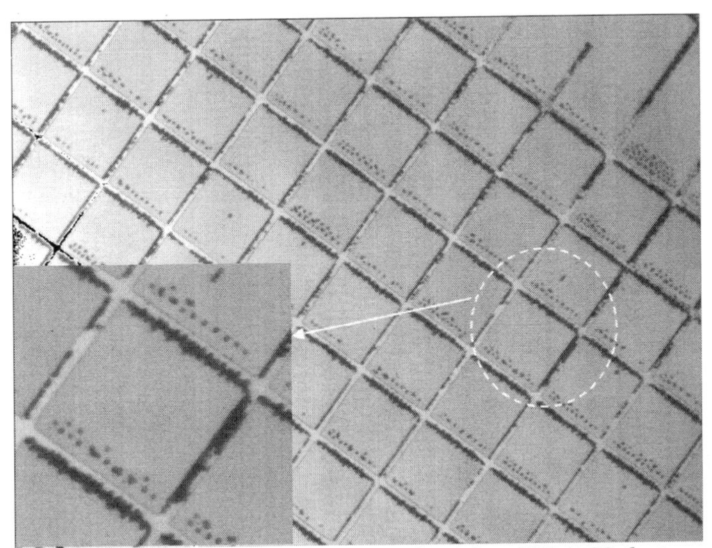

Figure 7. Magnetic particles trapped inside a CTMACS device.

opposing magnetic poles, see Figure 7. Unbound bacterial cells will flow through the chip and into waste. The buffer composition and flow rate ensure that target binding occurs with high affinity and specificity. Following washing of the bound magnetic particles, they are released and collected. The captured bacteria are allowed to grow, from which the peptide binding sequences are obtained.

Control mixtures were run through the CTMACS device to measure the enrichment of magnetic particles. In Figure 8, a solution which was 0.004% magnetic particles and 99.996% polystyrene beads was run through the CTMACS device. The recovered mixture was 97.26% magnetic particles.

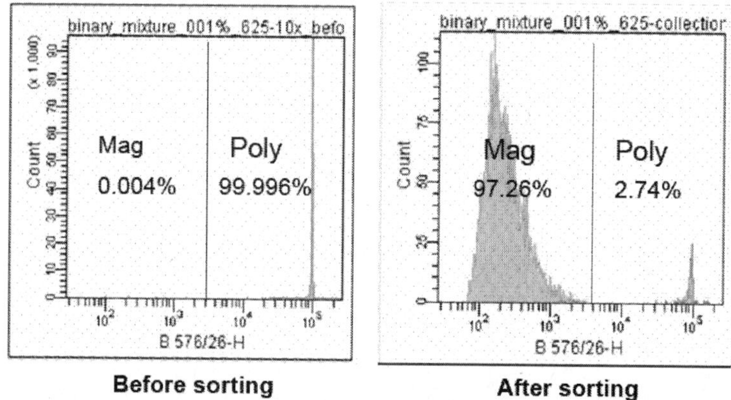

Figure 8. Enrichment of magnetic particles using a CTMACS device.

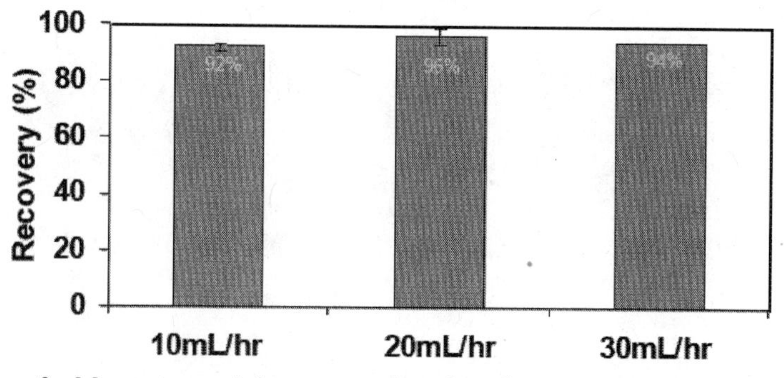

Figure 9. Magnetic particle recovery based on flow rate in CTMACS device.

Next, the recovery of magnetic particles was measured based on flow rate, see Figure 9. The recovery remained relatively constant despite an increase from 10mL/hr to 30mL/hr.

Results

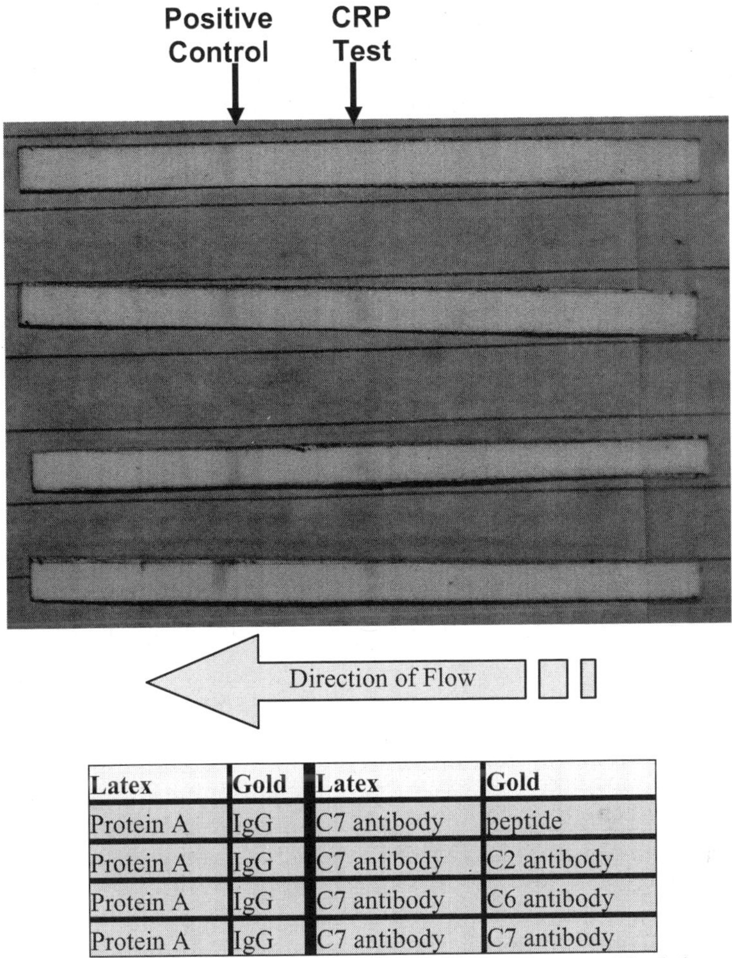

Latex	Gold	Latex	Gold
Protein A	IgG	C7 antibody	peptide
Protein A	IgG	C7 antibody	C2 antibody
Protein A	IgG	C7 antibody	C6 antibody
Protein A	IgG	C7 antibody	C7 antibody

Figure 10. Hand held assay test for C-Reactive Protein. The leftmost line is the positive control and the rightmost line is the sample line (4).

Figure 10 shows the results of hand held assays constructed to detect C-reactive protein (CRP). CRP was chosen as an initial target due to the degree to which it has been studied and the availability of multiple antibodies against it. Figure 11 demonstrates an alternative method of immobilizing a binding molecule. Instead of conjugating the binder to a latex bead, it is expressed on the surface of bacterial cells. The cells remain immobilized in the Whatman FUSION5 membrane where they are deposited.

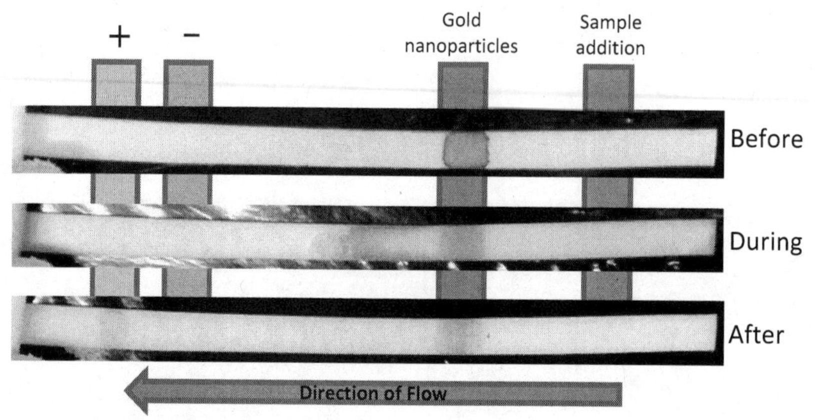

Figure 11. Cell based hand held assay for the detection of T7 antibodies. *(see page 2 of color inserts)*

In this particular assay, cells expressing the T7 peptide on their surface are immobilized at the positive detection mark (green) and cells expressing a negative control peptide are immobilized at the negative mark (red). Streptavidin coated gold nanoparticles are placed and dried at the blue mark. An antibody solution containing the biotinylated T7-antibody is added at the gray mark and PBS wash buffer is added at the same location to allow for transport and migration of the nanoparticles and protein solution. The purple signal line indicates the presence of the target molecule.

Conclusions

Each type of hand held assay has its advantages. When latex beads are used, the binding molecule must be isolated from the bacterial cell and conjugated to the bead. This is time consuming. However, the resultant stability of the assay makes it ideal for long term storage or exposure to harsher environmental conditions. The cell based hand held assays may not be as stable to harsh environments, but they can be developed much more rapidly. They are ideal for medical based diagnostics where they would be constructed and used in a controlled environment within a short time period. When coupled with the CTMACS technology, a powerful method of chemical and biological warfare detection is the result.

References

1. Bessette, P. H.; Hu, X.; Soh, H. T.; Daugherty, P. S. *Anal Chem.* **2007,** 79, 2174.
2. Rice, J. J.; Schohn, A.; Bessette, P. H.; Boulware, K. T.; Daugherty, P. S. *Protein Sci.* **2006,** 15, 825.
3. Chiao, D. J.; Shyu, R. H.; Hu, C. S.; Chiang, H. Y.; Tang, S. S. *J Chromatogr B Analyt Technol Biomed Life Sci.* **2004,** 809, 37.
4. Sooter, L.J.; Stratis-Cullum, D.N.; Zhang, Y,; Daugherty, P.S.; Soh, H.T.; Pellegrino, P.; Stagliano, N. Smart Biomedical and Physiological Sensor Technology V, edited by Brian M. Cullum, D. Marshall Porterfield. Proceedings of SPIE Vol. 6759, 67590A. **2007**.

Chapter 7

Portable Analytical Systems for On-Site Diagnosis of Exposure to Pesticides and Nerve Agents

Yuehe Lin[*], Jun Wang, Guodong Liu, Charles Timchalk

Pacific Northwest National Laboratory, Richland, WA 99352.

In this chapter, we summarize recent work in our laboratory on the development of sensitive portable analytical systems for use in on-site detection of exposure to organophosphate (OP) pesticides and chemical nerve agents. These systems are based on various nanomaterials functioning as transducers; recognition agents or labels and various elelectrochemical/immunoassay techniques. The studied nanomaterials included functionalized carbon nanotubes (CNT), zirconia nanoparticles (NPs) and quantum dots (QDs). Three biomarkers e.g. the free OPs, metabolites of OPs and protein-OP adducts in biological matrices have been employed for biomonitoring of OP exposure with our developed system. It has been found that the nanomaterial-based portable analytical systems have high sensitivity for the detection of the biomarkers, which suggest that these technologies offer great promise for the rapid and on-site detection and evaluation of OP exposure.

Introduction

Organophosphate (OP) pesticides have been extensively used as insecticides in modern agriculture, but their potential to produce neurotoxicity in animals and humans have raised concerns about their potential adverse impact on both the environment and human health (*1,2*). Some OP compounds have also been developed as extremely potent neurotoxic chemical warfare agents (CWAs) (*3,4*). Because of OP's toxicity, its use in the environment, and the danger presented by its accidental or intentional release in populated areas, the development of a rapid, sensitive, and incxpensive method to detect exposure to OP pesticides and nerve agents is crucial.

The acute toxicity resulting from OP contact is well documented and understood (*5*). Because they are powerful inhibitors of carboxylic ester hydrolases, including cholinesterase (ChE), OP exposure can result in acute cholinergic syndrome, which is characterized by a variety of symptoms including rhinorrhea, salivation, lachrymation, tachycardia, headache, convulsions, and death (*6*). There are three major biological pathways following human and animal exposure to OPs (Figure 1 presents typical results for human exposure): (a). when OP compounds are absorbed into the human body, for a limited period of time, OPs may exist as the parent compound (unbound form) in biological matrices, such as blood,and saliva; (b). the majority of the OP compounds will be metabolized by both enzymatic or non-enzymatic routes. For example, the OP pesticide chlorpyrifos is directly metabolized by the liver forming the major metabolite trichloropyridinol (TCP); (c) OP compounds can also covalently interact with a broad range proteins and form an OP adduct protein complex. In particular, OPs can stoichiometrically covalently bind to a broad family of cholinesterase (ChE) enzymes, such as acetylcholinesterase (AChE), butyrylcholinesterase (BChE) and carboxylesterases (CaE). Of particular concern in the binding and associated inhibition of AChE which is particularly critical for central and peripheral nervous-system functions, where significant enzyme inhibition results in acute toxicity and potentially death (*7,8*).

Biomonitoring provides an efficient approach for detecting and evaluating the exposure to Ops (*1,2*). Three types of markers, which result from the three pathways shown in Figure 1, have been widely used for the detection of exposure to OPs and nerve agents.

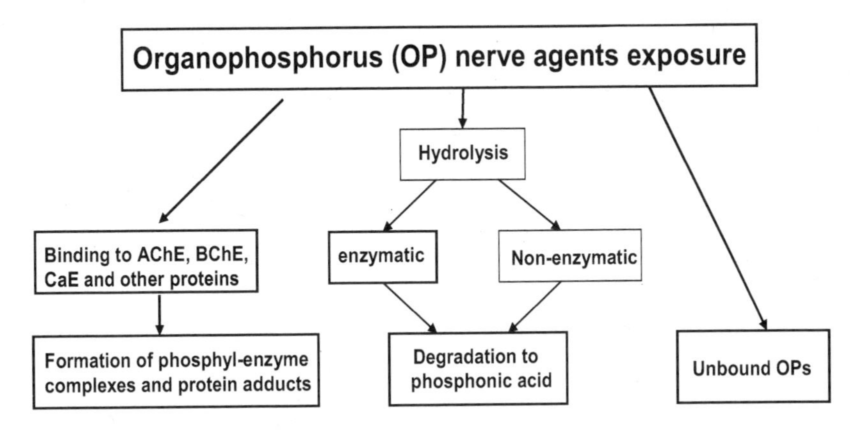

Figure 1. The biological pathway of organophosphate pesticides and nerve agents in humans.

Measuring free OP and associated metabolites in biological matrixes has been widely developed by coupling liquid chromatography with mass spectrometry (LC-MS), gas chromatography (GC)-MS, or high pressure liquid chromatography (HPLC)-MS (*9-12*). However, these methods are quite sophisticated, very expensive, time consuming, laboratory oriented, and need a pre-treatment of the sample and highly qualified technicians. Alternative enzyme based competitive immunoassay of metabolites such as TCP (major metabolite of chlorpyrifos) have been developed; these antibody based systems represent a less cumbersome approach than the conventional laboratory based analytical methods.

The measurement of blood (RBC and plasma) ChE activity has historically been employed as a biomarker for biomonitoring of OP exposure. The principle is based on the well known inhibition of enzyme activity by OPs, where ChE activities are conventionally measured with the Ellman assay (*13,14*). Some alternative approaches, such as radiometric assay (*15*) and mass spectroscopy (*16*), have been reported for sensitively detecting enzyme activities. However, these assays are tedious and/or time-consuming, and they require expensive and sophisticated instruments—which makes this approach unsuitable for rapid and on-site applications.

Electrochemical techniques are very attractive for developing a simple and inexpensive approach to rapid and onsite biomonitoring of exposure to OPs and nerve agents because of their simplicity, low cost, high sensitivity, and ease of miniaturization (*17-23*). The sensitivity of electrochemical chemo/biosensors can be further enhanced by using various nanotechnology-based amplifications (*24,25*). Carbon nanotubes (CNTs) and QDs have been widely used for electrochemical chemo/biosensors (*26-30*). For example, CNTs have been used as modified electrodes to detect small molecules, such as H_2O_2, because CNTs can improve the direct electron transfer reaction. QDs have also been used as electroactive labels (for electrochemical bioassays of DNA and proteins) by measuring metallic components after the acid dissolution step (*31,32*). Recently, nanomaterial based electrochemical biosensors and bioassays have shown great promise for detection of small molecules and trace biomolecules by using versatile amplification approaches (*22,33-37*).

In this chapter, we report recent advances in our laboratory in the development of portable analytical systems for rapid, sensitive biomonitoring exposure to OPs and chemical nerve agents (*19,25,38-45*). The CNTs and layer-by-layer technique (LBL) have been used to assemble AChE/CNT-based biosensors for the detection of free OPs in biological matrixes (*40,43*). Zirconia nanoparticles have been used as selective recognition agents of OPs for electrochemical detection (*39*). We also developed an electrochemical sensor using the magnetic beads and enzyme labels for competitive assay of TCP, a metabolite of organophosphate pesticides, chlorpyrifos (*41,42*). Furthermore, two approches have been developed for the electrochemical detection of ChE-OP exposure biomarkers: (i) Nanoparticle label/electrochemical immunoassay of ChE-OP based on zirconia NPs as recognition agents and quantum dots as labels; and (ii) electrochemical detection of ChE enzyme activity using carbon nanotube sensor (*44,45*). These methods are sensitive and resulting analytical systems are inexpensive, portable, and quite sutable for onsite applications and emergency use.

Carbon Nanotubes-Based Amperometric Biosensors for Detection of Free OPs

Layer-by-layer (LBL) technique was used for assembling AChE on a CNT transducer (Figure 2). The principle of the biosensor is based on the inhibition of enzyme activity by OPs. The inhibition can be electrochemically measured. Positively charged poly(diallyldimethylammonium chloride) (PDDA) and negatively charged AChE enzyme were assembed on the CNT surface by LBL technique. Here, CNTs play dual significant roles in the transduction and enzyme-immobilization events. As carriers, CNTs provide a suitable microenvironment to retain the AChE activity. As a transducer, CNTs amplify the electrochemical signal of the product of the enzymatic reaction.

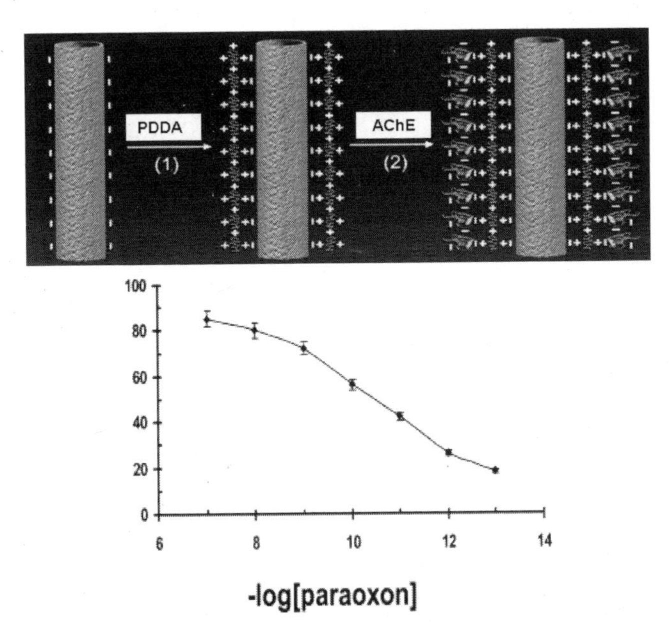

Figure 2. (Top) Schematic of assembling enzyme on CNT by LBL technique. (Bottom) Inhibition curve of the biosensor to different concentrations of paraoxon. (Reproduced from reference 40, Copyright 2006 American Chemical Society)

Under the optimum conditions established, the performance of this biosensor was studied (Figure 2). It was found that the relative inhibition of AChE activity increased with the concentration of paraoxon, ranging from 10^{-13} to 10^{-7} M and is linearly with -log[paraoxon] at the concentration range 1×10^{-12} - 10^{-8}M with a detection limit of 0.4 pM (calculated for 20% inhibition). This detection limit is three orders of magnitude lower than the covalent binding or adsorbing AChE on the CNT-modified screen-printed electrode (SPE) under batch conditions. The reproducibility of the biosensor for paraoxon detection was examined, and a relative standard deviation (RSD) of less than 5.6 % (n= 6) was obtained. The biosensor can be reused as long as residual activity is at least 50% of the original value.

Detection of OP Metabolite Based on Competitive Immunoassay

We developed a magnetic beads-based detection system for competitive assay of metabolites from OPs (Figure 3) (*41*). This method employed a sequentially injecting washing buffer, a mixture of TCP (chlorpyrifos metabolite) containing sample,

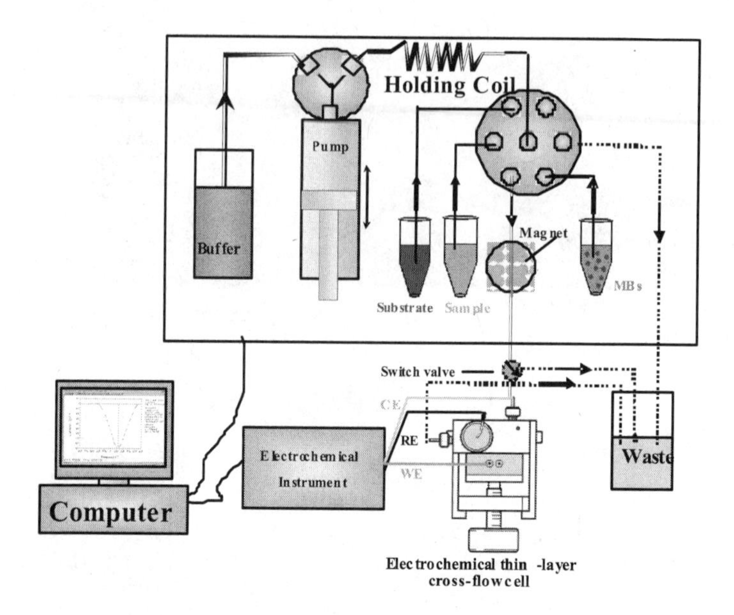

Figure. 3 Sequential injection electrochemical immunoassay system for determination of TCP.and TCP-HRP conjugates for competitive immunoreaction and the substrate solution containing hydrogen peroxide is sent for electrochemical detection. Antibody-modified magnetic beads in the tube were used as a platform for competitive immunoreaction.

The performance of sequential injection electrochemical immunoassay of TCP was studied. The linear measuring range was 0.01 to 0.02 μg/L, and the relative standard deviation was below 3.9% (n = 6). The detection limit of the sensor for TCP at 6 ng/L (ppt) is roughly 50 times lower than that of TCP RaPID Assay kit (at 0.25 μg/L, colorimetric detection).

Nanoparticle Labels/Immunosensors for the Detection of OP-ChE Biomarkers

The prototype of the portable analytical system, shown in Figure 4, consists of a disposable electrode and a portable electrochemical analyzer. This system can also be integrated with a flow injection system or a senquential flow injection system for rapid and on-site detection of OP exposure.

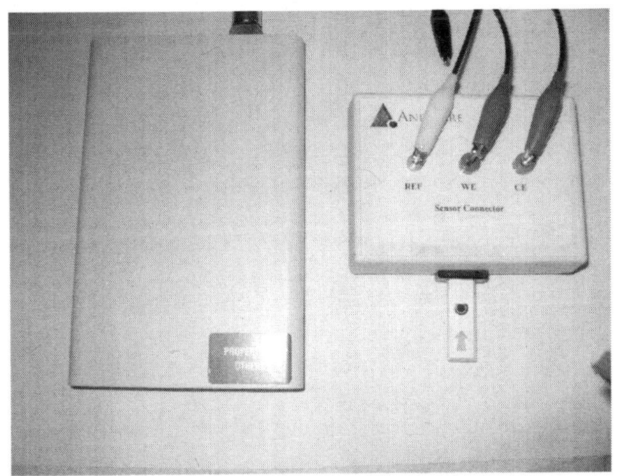

Figure 4. Prototype of the Portable Analytical System

We have developed nanoparticle label-based electrochemical immunoassay of OP exposure biomarkers, OP-ChE (*44*). The principle and procedure of this method is shown in Figure 5. Here, zircornia (ZrO_2) NPs, electrochemically deposited on the electrode surface, which functioned as the capture antibodies in the sandwich immunoassay. The zirconia- NP- modified electrode is then exposed to the sample solution and the ChE-OP in the sample is captured by the ZrO_2 nanoparticles. The QD-tagged, anti-AChE conjugate is introduced to form the sandwich-like complex on the sensor surface. The captured QDs are then dissolved by a drop of acid to release cadmium ions. This is followed by square wave voltammetric (SWV) [abbreviation already introduced but unexplained] detection of the released cadmium ions at an *in situ* plated mercury/bismuth film electrode.

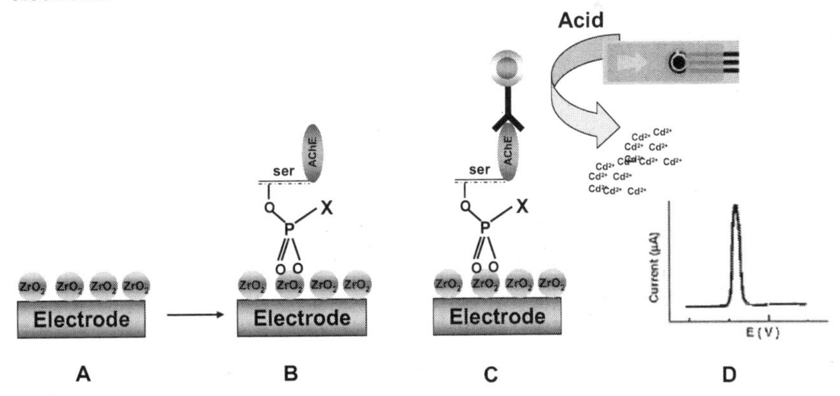

Figure 5 The Principle of electrochemical immunosensing of paraoxon-AChE (A) ZrO_2 nanoparticle modified SPE; (B) selective capturing paraoxon-AChE adducts; (C) immunoreaction between bound paraoxon-AChE adducts and QD-labeled anti-AChE antibody; (D) dissolution of nanoparticle with acid following an electrochemical stripping analysis.

Detection of AChE-OP Biomarkers

In this method, we used ZrO_2 NP as a recognition agent to capture the OP adduct on AChE. To complete the immunoassay sandwich, QD-labeled primary anti-AChE antibodies were used. Here, AChE-OP was prepared by incubating AChE and paraoxon, with a subsequent purification step to remove free paraoxon.

Under optimal conditions, typical electrochemical responses of the sensor with increasing concentrations of AChE-OP (10 pM to 10 nM, from bottom to top) are shown in Figure. 6. Peak current intensities heightened with the increase of AChE-OP concentrations. A negligible signal was observed in the control experiment (in the absence of AChE-OP); such behavior is ascribed to the well blocking step utilizing 1% bovine serum albumin (BSA). The resulting calibration plot of current versus OP-adduct [AChE-OP] is linear over the range of 10 pM to 10 nM, and is suitable for quantitative work. The detection limit is estimated to 8 pM (based on S/N=3) which is comparable to the detection limits using a mass spectrometric analysis of phosphylated cholinesterase adducts (*46*). A series of measurements of 100 pM AChE-OP yielded reproducible electrochemical responses with an RSD of 4.3% (data not shown).

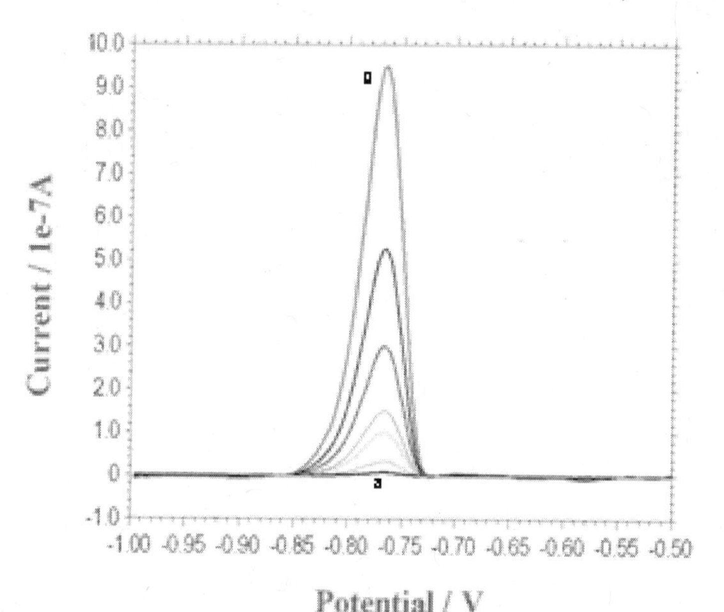

Figure 6. Typical electrochemical responses of the immunosensor with the increasing the AChE-OP concentration (10 pM to 10 nM from a to g).

Detection of BChE-OP Biomarkers

Because of its high concentration in human plasma, BChE is also used as a biomarker for detection of exposure to OP pesticides and nerve agents. Hence, the sensor was likewise evaluated for detection of BChE-OP biomarkers. The detection method is the same as the protocol described previously in section 4.1; however, the chemical nerve agent (structurally similar to Sarin) diisopropyl fluorophosphate (DFP) was chosen as a model OP in this studies.

The BChE-OP adduct was created by incubating BChE and DFP, followed by a purification step to remove any free DFP. Typical electrochemical responses of the sensor with increasing concentrations of BChE-OP (0.1 ng mL^{-1} to 50 ng mL^{-1} from bottom to top) are shown in Figure 7(A). A negligible signal was observed in the control experiment (curve a, in the absence of BChE-OP). It can be seen from this Figureure that the peak current intensities increased concurrently with the increase in concentrations of BChE adducts.

Calibration plot of peak current versus [BChE-OP], which is linear over the range of 0.1 ng mL^{-1} to 50 ng mL^{-1} , is displayed in Figure 7(B). The detection limit for this method is about 0.1 ng mL^{-1} (~ 1.5 pM). A series of measurements of 10 ng mL^{-1} BChE-OP yielded reproducible electrochemical responses with an RSD of 8.0% (data not shown).

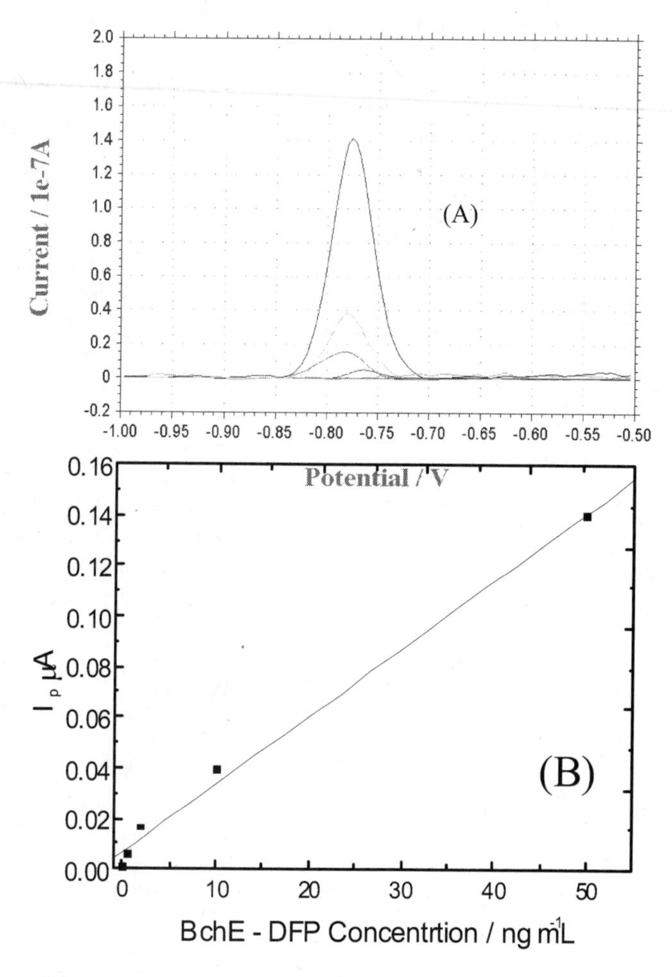

Figure. 7. (A) Typical SWV responses of the sensor with increasing BChE-OP concentration in solution. From bottom to top, the concentrations of BChE-OP solution (a-f) are 0, 0.1, 0.5, 2, 10, 50 ng mL^{-1}, respectively; (B) the plot of peak current vs. concentration of BChE-OP.

Carbon Nanotubes Based Sensors for Biomonitoring ChE Activity in Saliva

We developed a CNT-based electrochemical sensor in conjunction with a flow-injection system to measure ChE activity in saliva. The assay of enzyme activity in saliva is based on excellent electrocatalytic properties of CNTs, which make enzymatic products detectable at a low potential with an extremely high sensitivity. The amount of enzymatic product generated from ChEs depends on enzyme activity; therefore, by detecting the electroactive enzymatic product, this CNT-based electrochemical sensor can measure the ChE activity in saliva.

In the testing procedure, saliva solutions (obtained from naïve rats) are mixed with substrate (acetylthiocholine) for several minutes, and are then injected into the flow cell, where the electrochemical signals are generated and recorded. The magnitude of signals, resulting from the oxidation of enzymatic product thiocholine, reflects enzyme activity in saliva.

The testing system consisted of a carrier, a syringe pump, a sample injection valve, and a laboratory-built flow-through electrochemical cell (Figure 8). The cell was constructed by sandwiching an SPE between a plastic base and a plastic cover outfitted with inlet and outlet holes. A Teflon gasket with a flow channel was mounted between the SPE and the cover to form a flow cell, in which the SPE was exposed to the cell for the electrochemical detection.

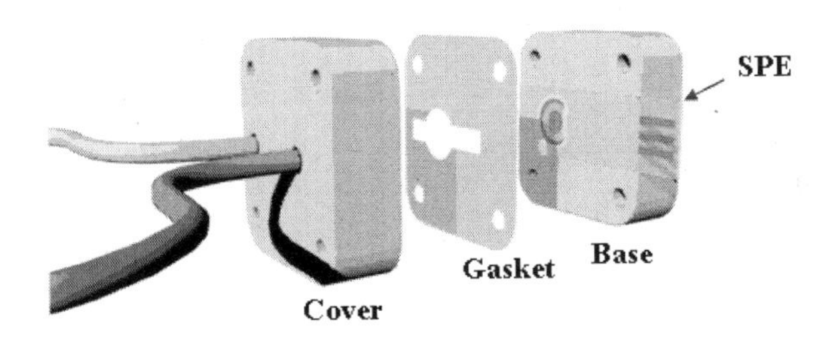

Figure. 8. Electrochemical flow cell.

The CNT-based electrochemical sensor was used for the measurement of ChE enzyme activity in saliva samples. Paraoxon and the 10–fold diluted naïve rat saliva sample were chosen for this study. The inhibition of ChE enzyme activity (with different concentrations of paraoxon) over an increasing incubation time is shown in Figure 9. It can be seen from this Figureure that the enzyme lost almost 97% activity after 0.5 h incubation with 7.0 nM paraoxon, and 80% after 0.5 h incubation with 0.7 nM paraoxon. Thereafter, the enzyme activity corresponding to 0.7 nM paraoxon continues to decrease through 2.0 h post-treatment, while the activity that corresponds to 7.0 nM paraoxon was nearly completely inhibited within 0.5 h This demonstrates that the amperometric sensor with a flow injection system can monitor salivary ChE enzyme activities and detect the exposure to OP pesticides and nerve agents.

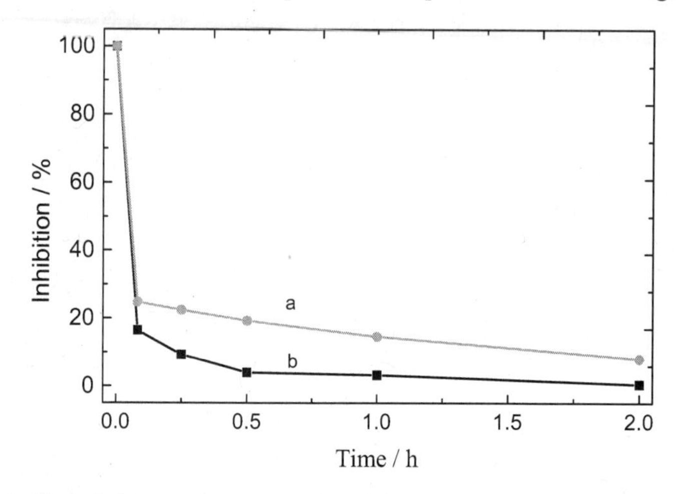

Figure 9. The inhibition of salivary ChE activity with the incubation of different concentrations of paraoxon for different periods; curve a is for 0.7 nM paraoxon and the curve b for 7.0 nM paraoxon. (Reproduced from reference 45, Copyright 2008 American Chemical Society)

Summary

In this chapter, portable analytical systems for detecting various markers of OP exposure have been developed. The methods for the detection of OP exposure take advantages of various exposure biomarkers. The sensing schemes relied upon nanomaterials and various electrochemical and immunoassay techniques. The results demonstrate that these portable analytical systems are promising for on-site diagnosis of exposure to OP insecticides and chemical nerve agents.

Acknowledgements

This work is supported by Grant U01 NS058161-01 from the National Institute of Neurological Disorders and Stroke, and the National Institutes of Health, and Grant R01 OH008173-01 from the Centers for Disease Control and Prevention/National Institute for Occupational Safety and Health. The research described in this paper was performed at the Environmental Molecular Sciences Laboratory, a national scientific user facility sponsored by DOE's Office of Biological and Environmental Research and located at the Pacific Northwest National Laboratory, which is operated by Battelle for DOE under Contract DE-AC05-76RL01830.

References

(1) Bouchard, M.; Carrier, G.; Brunet, R. C.; Dumas, P.; Noisel, N. *Ann. Occupat. Hygiene* **2006**, *50*, 505-515.

(2) Barr, D. B.; Thomas, K.; Curwin, B.; Landsittel, D.; Raymer, J.; Lu, C. S.; Donnelly, K. C.; Acquavella, J. *Environ. Health Perspect.* **2006**, *114*, 936-942.

(3) Giordano, B. C.; Collins, G. E. *Curr. Org. Chem.* **2007**, *11*, 255-256.

(4) Watson, A.; Opresko, D.; Young, R.; Hauschild, V. *J. Toxicol. Environ. Health Part B* **2006**, *9*, 173-263.

(5) George, K. M.; Schule, T.; Sandoval, L. E.; Jennings, L. L.; Taylor, P.; Thompson, C. M. *J. Bio. Chem.* **2003**, *278*, 45512-45518.

(6) Reigart, J. R.; Roberts, J. R. *Recognition and Management of Pesticide Poisonings*; United States Environmental
Protection Agency: Washington, D. C., 1999; Vol. Publication No. 735-R-98-003.

(7) Quinn, D. M. *Chem. Rev.* **1987**, *87*, 955-979.

(8) Jarv, J. *Bioorg. Chem.* **1984**, *12*, 259-278.

(9) Gundel, J.; Angerer, J. *J. Chromatogr. B* **2000**, *738*, 47-55.

(10) Lacorte, S.; Barcelo, D. *Anal. Chim. Acta* **1994**, *296*, 223-234.

(11) Lacorte, S.; Barcelo, D. *Envrion. Sci. Technol.* **1994**, *28*, 1159-1163.

(12) Hernandez, F.; Sancho, J. V.; Pozo, O. J. *Anal. Bioanal. Chem.* **2005**, *382*, 934-946.

(13) Kousba, A. A.; Poet, T. S.; Timchalk, C. *Toxicol.* **2003**, *188*, 219-232.

(14) Gorun, V.; Proinov, I.; Baltescu, V.; Balaban, G.; Barzu, O. *Anal. Biochem.* **1978**, *86*, 324-326.

(15) Henn, B. C.; McMaster, S.; Padilla, S. *J. Toxicol. Environ. Health Part A* **2006**, *69*, 1805-1818.

(16) Shen, Z. X.; Go, E. P.; Gamez, A.; Apon, J. V.; Fokin, V.; Greig, M.; Ventura, M.; Crowell, J. E.; Blixt, O.; Paulson, J. C.; Stevens, R. C.; Finn, M. G.; Siuzdak, G. *ChemBioChem* **2004**, *5*, 921-927.

(17) Wang, J.; and Lin, Y. *Anal. Chim. Acta* **1993**, *271*,53-58

(18) Matsuura, H.; Sato, Y.; Sawaguchi, T.; Mizutani, F. *Sens. Actuators B* **2003**, *91*, 148-151.

(19) Liu, G. D.; Riechers, S. L.; Mellen, M. C.; Lin, Y. H. *Electrochem. Commun.* **2005**, *7*, 1163-1169.

(20) Mitchell, K. M. *Anal. Chem.* **2004** *76*, 1098-1106.

(21) Joshi, K. A.; Prouza, M.; Kum, M.; Wang, J.; Tang, J.; Haddon, R.; Chen, W.; A., M. *Anal. Chem.* **2006**, *78*, 331-336.

(22) Chen, S. H.; Yuan, R.; Chai, Y. Q.; Zhang, L. Y.; Wang, N.; Li, X. L. *Biosen. Bioelectron.* **2007**, *22*, 1268-1274.

(23) Ciucu, A. A.; Negulescu, C.; Baldwin, R. P. *Biosen. Bioelectron.* **2003**, *18*, 303-310.

(24) Wang, J.; Musameh, M.; Lin, Y. *J. Am. Chem. Soc.* **2003**, *125,* 2408-2409.

(25) Lin, Y. H.; Yantasee, W.; Wang, J. *Frontiers in Biosci.* **2005**, *10*, 492-505.

(26) Tu, Y.; Lin, Y.; Ren, Z. *Nano Letters,* **2003**, *3*,107-109.

(27) Deo, R.P.; Wang, J.; Block, I.; Mulchandani, A.; Joshi, K.A.; Trojanowicz, M.; Scholz, F.; Chen, W.; **Lin, Y.;** *Anal. Chim. Acta* **2005**, *530*, 185-189.

(28) Chu, T. C.; Shieh, F.; Lavery, L. A.; Levy, M.; Richards-Kortum, R.; Korgel, B. A.; Ellington, A. D. *Biosen. bioelectron.* **2006** *21*, 1859-1866.
(29) Lin, Y.; Lu, F.; Tu, Y.; Ren. Z. *Nano Letters* **2004,** *4*, 191-195.
(30) Liu, G.; Lin, Y. *Electrochem. Commun.* **2006**, *8,* 251-256.
(31) Liu, G.; Lin, Y.; Wang, J.; Wu, H.; Wai, C.M.; Lin, Y. *Anal. Chem.* **2007**, *79,* 7644-7653..
(32) Wu, H.; Liu, G.; Wang, J.; Lin, Y. *Electrochem. Commun.* **2007**, *9*,1573-1577.
(33) Liu, G.; Wang, J.; Wu, H.; Lin, Y. Y.; Lin, Y. *Electroanalysis*, **2007**, *19*, 777-785.
(34) Wang, J.; Liu, G. D.; Engelhard, M. H.; Lin, Y. H. *Anal. Chem.* **2006**, *78*, 6974-6979
(35) Wang, J.; Liu, G. D.; Lin, Y. H. *Samll* **2006**, *2*, 1134-1138.
(36) Dequairem, M.; Degrand, C.; Limoges, B. *Anal. Chem.* **2000**, *72*, 5521-5528.
(37) Liu, G.; Lin, Y. *Talanta* **2007**,*74,* 308-317.
(38) Liu, G. D.; Lin, Y. H. *Electrochem. Commun.* **2005**, *7*, 339-343.
(39) Liu, G. D.; Lin, Y. H. *Anal. Chem.* **2005**, *77*, 5894-5901.
(40) Liu, G. D.; Lin, Y. H. *Anal. Chem.* **2006**, *78*, 835-843.
(41) Liu, G. D.; Riechers, S. L.; Timchalk, C.; Lin, Y. H. *Electrochem. Commun.* **2005**, *7*, 1463-1470.
(42) Liu, G. D.; Timchalk, C.; Lin, Y. H. *Eelectroanal.* **2006**, *18*, 1605-1613.
(43) Lin, Y. H.; Lu, F.; Wang, J. *Electroanal.* **2004**, *16*, 145-149
(44) Liu, G. D.; Wang, J.; Timchalk, C.; Lin, Y. H. *Anal. Chem.* **2007**, *In revision.*
(45) Wang, J.; Liu, G. D.; Timchalk, C.; Lin, Y. H. *Envion. Sci. Technol.* **2007**, *In Press.*
(46) Fidder, A.; Hulst, A. G.; Noort, D.; de Ruiter, R.; J.;, v. d. S. M.; Benschop, H. P.; Langenberg, J. P. *Chem. Res. Toxicol.* **2002**, *15*, 582-590.

Chapter 8

Novel Nanoarray SERS Substrates Used for High Sensitivity Virus Biosensing and Classification

J. D. Driskell,[1] S. Shanmukh,[1] Y. Liu,[2] S. Chaney,[2] S. Hennigan,[3] L. Jones,[4] D. Krause,[3] R. A. Tripp,[4] Y.-P. Zhao,[2] and R. A. Dluhy[1]

[1]Department of Chemistry, University of Georgia, Athens, GA 30602
[2]Department of Physics and Astronomy, and Nanoscale Science and Engineering Center, University of Georgia, Athens, GA 30602
[3]Department of Microbiology, University of Georgia, Athens, GA 30602
[4]Department of Infectious Diseases, College of Veterinary Medicine, University of Georgia, Athens, GA 30602

Development of diagnostic methods for rapid and sensitive identification of viruses is essential for the advancement of therapeutic and preventive intervention strategies. Current diagnostic methods, e.g. virus isolation, PCR, antigen detection and serology, are time-consuming, cumbersome, or lack the required sensitivity. We have investigated the use of aligned Ag nanorod arrays, prepared by oblique angle deposition (OAD), as surface-enhanced Raman scattering (SERS) substrates for the identification and quantitation of viral pathogens. The OAD method of substrate preparation facilitates the selection of nanorod size, shape, density, alignment, orientation, and composition. The current chapter (??) will address the fundamental nanostructural design of metallic nanorod arrays and their influence on SERS enhancement, as well as the development of a spectroscopic biosensor assay for virus detection based on these unique nanostructured SERS probes. We will also present results of multivariate statistical analysis on the SERS spectra of different viral strains that indicate its feasibility to differentiate and classify viruses based on their intrinsic SERS spectra.

Introduction

A rapid and sensitive diagnostic method is a critical component for defining the emergence of a viral infection, determining the period that preventive measures should be applied, and for evaluating drug efficacy. Current diagnostic methods often have limited sensitivity, are cumbersome, have poor predictive value, or are time-consuming. An ideal diagnostic test would not require target labeling, provide molecular specificity, and eliminate the need for an amplification step. Mounting evidence suggests that surface enhanced Raman scattering (SERS) can provide the necessary level of specificity and sensitivity for biosensing. Evidence from our laboratories shows that the SERS virus spectrum provides a unique "molecular fingerprint" of individual viruses, is extremely sensitive (near single virus particle detection), is rapid (detection in 60 seconds), and does not require modification to the virus for detection. Thus, SERS provides a framework for enhancing viral diagnostics and for evaluating anti-viral disease strategies.

SERS has been employed to solve a multitude of chemical and biochemical problems in the biochemistry and life sciences (*1,2*). This is due in part to its ability to provide a high degree of structural information. The ultrasensitive detection limits provided by SERS, which approach single molecule sensitivity (*3-5*), has initiated research to develop SERS-based sensors for the detection of both bacteria (*6-9*) and viruses (*10-12*). SERS based detection of infectious agents using reporter molecule sandwich assemblies have been shown to be extremely sensitive (*6,7*), but are ultimately limited by nonspecific binding of labels in the absence of analyte, which can not be discriminated from specific target binding. With the development of more strongly enhancing and more reproducible SERS substrates, direct spectroscopic characterization of infectious agents is possible, which prevents false positives due to nonspecific binding (*6-10,12,14-16*).

In spite of its demonstrated potential, a major challenge in the development of SERS-based diagnostics has been the lack of a method for reproducibly producing a substrate with large enhancements (*8*). In predicting the optimal SERS enhancement, the importance of the morphology of the metal particles on the substrate's surface can not be underestimated. Many substrate preparation techniques exist that attempt to form metal surfaces consisting of an ideal SERS surface morphology. Such methods include oxidation-reduction cycles (ORC) (*9*), preparation of metal colloid hydrosols (*10*), laser ablation (*11*), chemical etching (*12*), and roughened and photodeposited films (*13*). In recent years, vapor-deposited Ag metal films have gained favor as SERS substrates due to their increased stability and reproducibility over metal sols and electrochemical roughened surfaces. However, they often suffer from reduced enhancement factors (*14-17*).

It has long been theorized that the electromagnetic field is maximal for points with high curvature resulting in greater SERS enhancement for nanostructures with higher aspect ratios (*18*). Several groups have attempted to support this prediction experimentally (*19-22*), but perhaps the most convencing is the work of Murphy and coworkers on isolated silver and gold nanoparticles of various aspect ratios which provided conclusive evidence that high aspect ratio nanostructures offer greater SERS enhancement factors (*23*). In addition to the aspect ratio, it has also been shown that nanoparticles present in either an aggregated form or as arrays produce greater SERS intensities due to very strong electromagnetic fields present in the interstitial regions as a result of the coupling between adjacent nanoparticles (*24,25*). Fundamental experimental work by VanDuyne, using nanosphere lithography, and by Sepaniak, using electron beam lithography, demonstrate the advantage of interparticle coupling in arrays with respect to SERS enhancements. These fundamental studies on nanostructured arrays inspired the development of a highly uniform, reproducible nanostructured arrays as SERS substrates for sensing applications (*26*).

Consequently, fabrication methods that result in highly ordered arrays of high aspect ratio nanostructures with control over surface morphology are highly desirable. Typically such methods to produce high aspect ratio nanostructures suffer from disadvantages such as elaborate preparation protocols, expensive instrumentation, and the need for highly skilled operators. Oblique angle deposition (OAD) is a physical vapor deposition technique that overcomes some of the difficulties and disadvantages of the previously discussed methods. OAD is a physical vapor deposition technique in which the angle between the incoming vapor from the source and the surface normal of the substrate is set between 0° to 90°, and the substrate can rotate in both polar and azimuthal directions (*27*). In this paper, we summarize the fabrication, characterization, and application of Ag nanorod SERS substrates prepared by the OAD technique.

Experimental

Substrate Preparation

Silver nanorod arrays served as SERS substrates and were fabricated by the oblique angle deposition (OAD) method using a custom-designed electron-beam/sputtering evaporation system that has been described previously (*28*). Glass microscope slides (Gold Seal® Cat No. 3010) were cut into 1 × 1 cm chips and functioned as the base onto which the nanorod array was grown. The glass slide substrates were cleaned using Piranha solution and loaded into the evaporator. A thin layer (20 nm) of Ti (Alfa Aesar, Ward Hill, MA, 99.999%) was first deposited onto the glass to serve as an adhesion layer between the glass and Ag. Next, a 500 nm thin film of Ag (Alfa Aesar, Ward Hill, MA, 99.999%) was deposited on the substrate. During these first two thin layers depositons the incident angle of the vapor was normal to the substrates. The substrates were then rotated to 86° relative to the incident angle for the growth of the silver nanorods. Throughout the deposition, the overall thickness of the metal deposited was monitored by a quartz crystal microbalance positioned at normal incidence to the vapor source.

Substrate Characterization

The Ag nanorods were characterized according to morphology, crystal structure, optical properties, and SERS activity. The size, shape, and spacing of the nanorods were determined with a LEO 982 field emission scanning electron microscope (SEM) (Thornwood, NY).

The SERS activity of the Ag nanorod array substrates were evaluated using a common molecular probe, trans-1,2-bis(4-pyridyl)ethene (BPE, Aldrich, St. Louis, MO, 99.9+%). Serial dilutions of BPE in HPLC grade methanol (Fisher, Pittsburgh, PA) were prepared and 1 μL aliquots of the BPE solutions were applied to the SERS substrates. SERS spectra were acquired using a fiber-optic coupled, confocal Raman microscope interfaced to a CCD-equipped spectrograph (Kaiser Optical Systems Incorporated, Ann Arbor, MI). A fiber-optic interfaced 785 nm near-IR diode laser (Invictus, Kaiser Optical) was used to provide the incoming radiation. Spectra were collected with ~12 mW laser power using 20× (N.A. = 0.40) objective. Post processing of the collected spectra was performed using GRAMS32/AI spectral software package (Galactic Industries, Nashua, NH).

Infectious Agents

Four strains of respiratory syncytial virus (RSV), RSV strain A2, RSV strain A/Long, and RSV strain B1, along with a recombinant wild type strain A2 (6340) with a single G gene deletion (ΔG) were studied. Each RSV was propagated using Vero cells maintained in Dulbecco's Modified Eagles Medium (DMEM; GIBCO BRL laboratories, Grand Island, NY) supplemented with 2% heat-inactivated (56 °C) FBS (Hyclone Laboratories, Salt Lake City, UT). RSV was harvested in serum-free DMEM three days post-infection followed by freeze-thaw cycles (-70 °C/4 °C). The contents were then collected and centrifuged at 4000g for 15 min at 4 °C. The virus titers were determined to be 5×10^6 to 1×10^7 PFU/mL by an immunostaining plaque assay (*29*). The RSV sample was deactivated with a 1 hr exposure to UV radiation.

A CCR5-tropic strain of human immunodeficiency virus (HIV), BaL, was also used to evaluate the SERS-based sensor. This HIV strain was isolated approximately 20 years ago. The HIV was propagated in human 293T cells transfected with pCMV5-based CCR5 expression plasmid. The HIV was propagated and suspended in RPMI with 10% FBS, 1% L-glutamine, 1% pen/strep, and IL-2. The sample was treated with 4% formalin prior to handling for spectroscopic measurements.

Influenza virus strain A/HKx31 was propagated in embryonated chicken eggs. The influenza virus titer ranged from 1×10^7 to 1×10^8 EID50 as determined by hemagglutination assay with chicken red blood cells. Influenza viruses were inactivated by 4% paraformaldehyde treatment at room temperature for 3 hours.

M. pneumoniae strains were stored at -80°C and cultured in 25 ml SP4 medium (*30*) at 37 °C for 4 days. When lawn-like growth was observed by microscopic examination cultures were scraped from the surface of the flasks and transferred to an Oak Ridge tube (*31*). Samples were centrifuge at 6°C for 30 minutes at 11,500 rpm. The supernatant was discarded and the pellet was suspended in 1 ml SP4 medium, vortexed, syringe-passaged 10 times with a 25 gauge needle to disaggregate cells, and stored in a -80°C freezer. To quantify the M. pneumoniae, dilutions were spread on PPLO agar plates and incubated at 37°C for 7 days. In order to visualize mycoplasma colonies, each plate was overlaid with 0.8 ml blood agar 20% sheep blood in 1% Noble agar in saline. The *M. pneumoniae* was then purified by centrifugation (14000 rpm at 6°C for 15 min) and resuspension in 1.0 ml of distilled H2O. *M. pneumoniae* was then inactivated with 4% formalin.

SERS Measurements and Chemometric Analysis

The SERS spectrum of each virus and the mycoplasma specimens was collected by applying a 1.0 μl droplet of each of the biological samples to an OAD fabricated substrate and allowing it to dry prior to the SERS spectral acquisition. The SERS signal was integrated for 30 s and five spectra were collected from different locations on the substrate and from different substrates.

Spectral preprocessing was required prior to chemometric analysis of the SERS spectra. Two approaches were taken to smooth and baseline-correct the raw spectra. In the first approach, the spectra were smoothed with a binomial function (7 point), baseline-corrected, and zeroed using GRAMS/32 AI version 6.00 (Galactic Industries Corporation). Using the second approach, SERS spectra of the biological samples were imported into The Unscrambler version 9.6 (CAMO Software AS, Woodbridge, NJ) software and the first derivative of each spectrum was computed using a nine-point Savitzky-Golay algorithm. The first derivative eliminated the need for manual and subjective baseline-correction of each spectrum (*32*). Each altered spectrum, using either of the above methods, was then normalized with respect to its most intense band to account for slight differences in enhancement factors from substrate to substrate. The normalized spectra were then imported into MATLAB version 7.2 (The Mathworks Inc., Natick, MA) where the samples were classified with principal component analysis (PCA), using the PLS Toolbox version 4.0 (Eigen Vector Research Inc., Wenatchee, WA).

Results and Discussion

Substrate Characterization

Ag Nanorod Morphology – SEM and TEM Characterization

Ag nanorod substrates prepared by OAD fabrication methods have been primarily characterized with SEM to evaluate the nanorod array structure. Top view and cross-sectional view SEM images of the silver nanorod array are shown in Figures 1a and 1b, respectively. These reveal that the Ag nanorods are aligned parallel to each other at an angle of 71 ± 4° with respect to the substrate surface normal. Under the deposition conditions for this experiment, the Ag nanorods were found to have an average length of 868 ± 95 nm and an average diameter of 99 ± 29 nm with the average gap between two rods ~ 177, resulting in a density of 13.1 ± 0.5 rods/μm2. Varying the deposition time and/or rate has been found to influence the nanorod length, but not the nanorod tilt angle or surface density, which are controlled by the deposition angle. TEM measurements have also been performed to visualize individual, isolated nanorods (*33*).

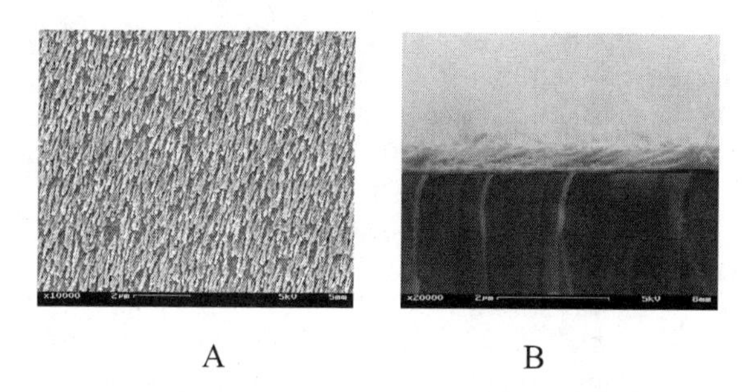

A B

Figure 1. SEM image of Ag nanorods (A) top view and (B) cross-sectional view. (Adapted with permission from (33)).

UV-VIS Characterization

Mounting, indisputable evidence supports the existence of a connection between the surface plasmon resonance of nanostructures and the nanostructures ability to provide surface enhancement of Raman scattering (*34,35*). The surface plasmon resonance is easily measured as the wavelength of maximum extincition; thus, the extinction spectrum of the Ag nanorod array was measured and details were reported in (*36,37*).

SERS Characterization

The SERS response of the nanorod array substrate was evaluated by spreading low concentrations of trans-1,2-bis(4-pyridyl)ethene (BPE) in methanol on the 1 × 1 cm substrate and allowing the solvent to evaporate. The relationship between BPE concentration and the intensity of SERS scattering has been thoroughly investigated for the ~900 nm length Ag nanorod substrates. To determine the concentration dependent range for SERS scattering, increasing amounts of BPE were sequentially deposited onto the Ag nanorod array substrate with Ag film and SERS spectra were collected from 10 spots for each deposition step. The SERS response, defined as the integrated band area of the 1200 cm^{-1} band, was found to be nearly linear over 3 orders of magnitude (10^{-12} – 10^{-9} moles) in the log-log plot, after which the signal approaches a limiting value.

The influence of nanorod aspect ratio on the SERS surface enhancement factor (SEF) was systematically investigated by Murphy and was shown to be a significant parameter for the case of isolated nanorods (*23*). Therefore, the SEF dependence on nanorod length for the Ag nanorod array substrates was probed. Figure 2 highlights the results of this study, while details can be found elsewhere (*28,33*). It is concluded that optimum SERS substrates are those with nanorod lengths of ~900 nm which provide a SEF of ~5 × 10^8 for BPE (*33*).

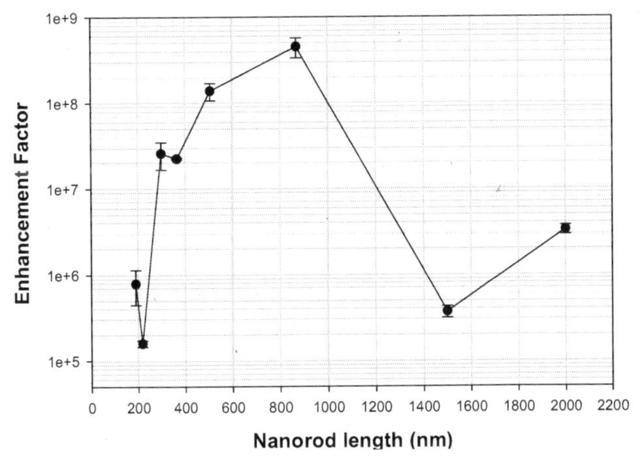

Figure 2. Surface enhancement factors plotted as function of nanorod length from 190–2000 nm. Each data point represents the average SEF taken from 5 different locations and the error bars represent the standard deviation. (Adapted with permission from (33)).

The characterization studies in the preceeding paragraphs were necessary for the production of optimized SERS substrates for sensing applications; however, the fundamental mechanisms by which these Ag nanorod arrays act as highly sensitive SERS substrates has also been examined in polarized SERS measurements. A detailed discussion of these data is provided in references (*38*) and (*36*).

Detection of a Purified Sample

Infectious Agent Classification

The Ag nanorod substrates developed in the previous section were first applied to the detection of purified infectious agents. Initial studies were to determine whether the Ag nanorod substrate could be used to measure a SERS spectrum of a pathogen, and if so, would it be possible to differentiate between different pathogens based on their SERS spectra. SERS spectra of wild type strains of RSV (strain A2), HIV, influenza, and *M. pneumoniae* (strain M129) were were collected from several locations from at least three different substrates Representative spectra (smoothed, baseline-corrected, and normalized) are plotted in Figure 3. Unique spectra, with very distinct SERS signatures, can be easily visualized for each infectious agent.

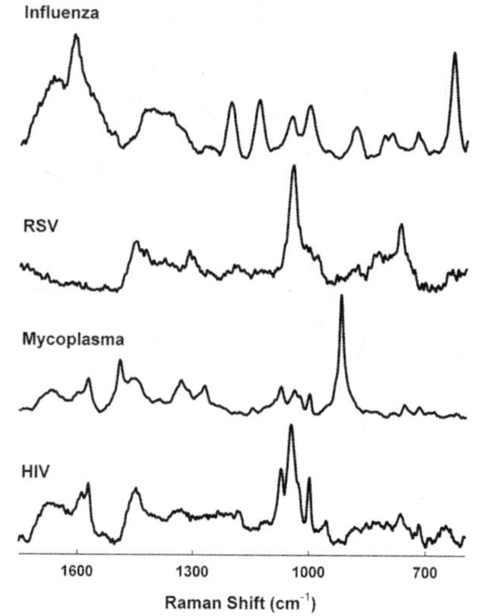

Figure 3. Representative SERS spectra for influenza virus, RSV, M. pneumoniae, and HIV adsorbed on a Ag nanorod substrate. The spectra have been baseline corrected, normalized, and offset for visualization.

Spectral interpretation is possible as many reports have shown (*39-42*). For, example, in the SERS spectrum of purified RSV, the bands at 527 cm^{-1} and 546 cm^{-1} can be assigned to a disulfide stretching mode (*43,44*). The strong band at 837 cm^{-1} corresponds to Tyr. The main feature in the RSV SERS spectrum is a strong band at 1044 cm^{-1} that has been assigned to the C-N stretching vibration in previous SERS studies (*45,46*). Similar analysis of the other infectious agent spectra can be performed to assign the SERS bands; however, this is extremely tedious and little effort was placed on band assignment for the purpose of this study. The chemical composition of these pathogens is very similar (i.e., DNA and protein) and slight variations in genomes, proteins, and adsorption to the sensor will result in measurable differences in the SERS spectra. These issues that cause spectral interpretation to be challenging can be exploited using chemometrics to classify pathogens directly without having to decipher the spectra.

Average spectra from multiple substrates for each biological sample was preprocessed by taking a 9-point Savitsky-Golay first derivative followed by normalizing each spectrum with respect to its most intense band. Prinicipal component analysis was performed on the postprocessed infectious agent spectra and the resulting two-dimensional scores plot is presented in Figure 4. The spectra for each pathogen are clearly separated into individual clusters, and the random distribution of the spectra within a single cluster is evidence that substrate heterogeneity and sampling errors are not a confounding factor in this analysis and do not contribute to the spectral classification.

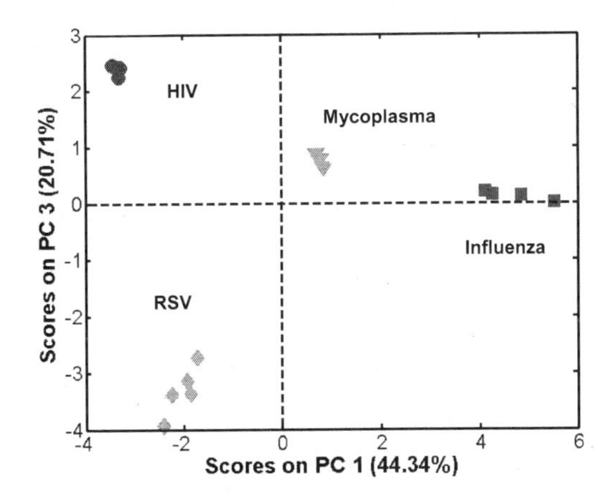

Figure 4. PCA scores plot of PC3 vs. PC1 computed from the SERS spectra of the infectious agent samples: HIV (●), influenza (■), RSV (◆), and M. pneumoniae (▼).

Infectious Agent Subtyping

After measuring the SERS spectra for several infectious agents and successfully classifying samples based on their unique spectra, the sensitivity of the SERS signature to slight differences in the biological sample was explored. The SERS spectra for four strains of RSV, including RSV strain A2, RSV strain A/Long, and RSV strain B1, along with a recombinant wild type strain A2 (6340) with a single G gene deletion (ΔG) were collected. Figure 5 shows overlaid spectra for each strain collected from many different locations and from at least three different substrates. The overlaid spectra demonstrate the high degree of spectral reproducibility from spot-to-spot and from substrate-to-substrate. While slight differences in the spectra can be observed for these strains, such as a peak shift from 1045 cm^{-1} to 1055 cm^{-1} for the A/Long strain, the differences are much less obvious than the differences in Figure 3. To exploit these minor variations, PCA and HCA were performed.

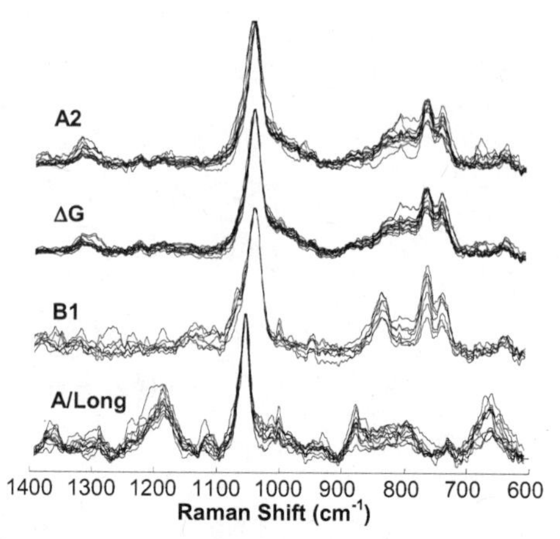

Figure 5. SERS spectra of the RSV strains A/Long, B1, A2 with a G gene deletion (ΔG), and A2 , collected from several spots on multiple substrates and normalized to the peak intensity of the most intense band (1045 cm-1) and overlaid to illustrate the reproducibility on the Ag nanorod substrate. (Adapted with permission from(49))

The spectra in Figure 5 were analyzed with PCA and a scores plot is presented in Figure 6. All three RSV strains were clearly separated into individual clusters, i.e. A/Long, A2 and B1 viral strains. However, significant overlap in the ΔG and the A2 viruses clusters are observed in the 2-D scores plot reflecting the extremely close biochemical similarity between these two virus strains. To further identify statistical differences between the virus spectra, the HCA can be performed. Details of the HCA classification of the RSV strains into their respective classes can be found in (*47*).

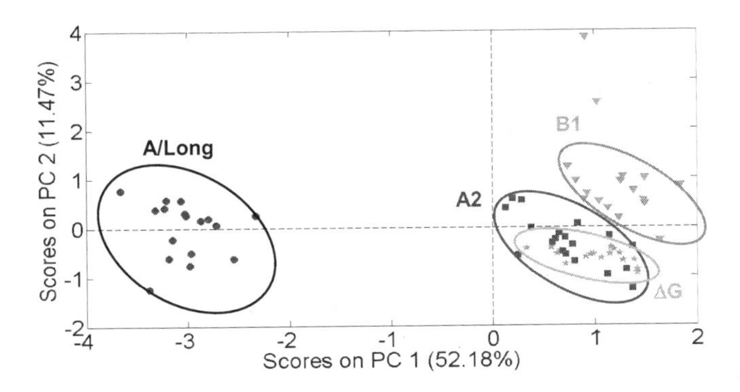

Figure 6. PCA scores plot of PC2 vs. PC1 computed from the SERS spectra of the RSV strains A/Long (●), B1 (▼), A2 (■), and the recombinant strain A2 G gene deletion mutant (ΔG) (*).

Similar studies were performed on *M. pneumoniae* to differentiate between three strains, M129, II-3, and FH. The SERS spectra for each purified strain, adsorbed on the Ag nanorod substrate, is shown in Figure 7. The M129 and II-3 spectra are very similar while that of FH is quite different. The PCA scores plot (Figure 8) emphasizes this finding, featuring a well-isolated clustering of data for the FH strain and overlapping clusters for the M129 and II-3. If the spectral differences in the M129 and II-3 spectra are statistically significant, a more rigorous chemometric analysis of the data will be necessary to exploit these minute differences for classification. It is noteworthy, however, that these two strains are thought to differ at the nucleotide level by a single basepair deletion, and at the protein level by the loss of the P30 protein and reduced stability of the P65 protein (*48*). Further studies are in progress to classify other *Mycoplasma* species and to isolate the components of the bacteria that provide the SERS signature.

Detection in a Biological Matrix

We have demonstrated that Ag nanorod-based SERS is not only sensitive to purified virus, but also is able to sense the presence of virus after infection in biological media (*49*). To demonstrate this, we compared the SERS spectra of uninfected Vero cell lysate, RSV-infected cell lysate and purified RSV. The results show that major Raman bands can be assigned to different constituents of the cell lysate and the virus, such as nucleic acids, proteins, protein secondary structure units and amino acid residues present in the side chains and the backbone. However, our most significant result was that vibrational modes due to the virus could be unambiguously identified in the SERS spectrum of the Vero cell lysate after infection (*49*).

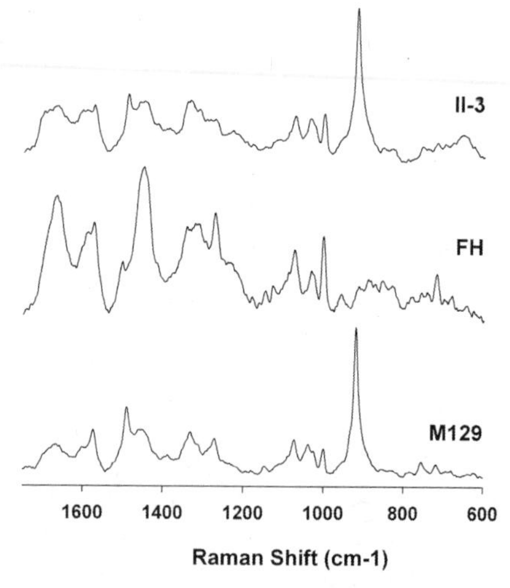

Figure 7. Average SERS spectra of the M. pneumoniea strains II-3, FH, and M129, collected from 5 locations from 3 different substrates for each strain

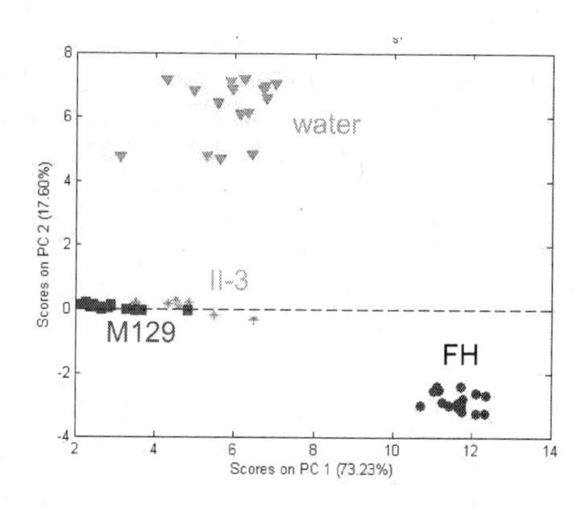

Figure 8. PCA scores plot of PC2 vs. PC1 computed from the SERS spectra of the M. pneumoniea strains FH (●), M129, (■), II-3 (*), and water/background (▼).

Antibody-based Detection

The specificity of the SERS substrate can be tuned via chemical modification of the Ag nanorod surface. To demonstrate, a self-assembled multilayer nanorod-antibody-virus immunoassay was developed for binding of RSV. In this system, thiol-derivatized IgG2a monoclonal antibody was immobilized to the Ag nanorod substrate. After 1 hour the excess unreacted antibody was washed off with saline solution. SERS spectra of the Ag nanorod/antibody complex were collected at 5 different spots on the substrate using the same spectrograph and similar spectral data collection conditions to those described above: 785 nm excitation wavelength, ~20 mW power, 10 s exposure time.

Virus binding was accomplished by exposing the IgG-coated Ag nanorod to a solution of RSV virus. After 1 hour incubation, the excess RSV on the surface of the substrate was removed. SERS spectra of the RSV-IgG complex on the Ag nanorods were then collected using the same spectral conditions as before.

The results of these experiments are shown in Figure 9. Of the spectral features apparent in the IgG2a antibody spectrum, the most intense band at ~1000 cm^{-1} likely arises from the in-plane ring deformation mode of Phe in IgG (*50,51*). The amide III protein mode at ~1260 cm^{-1} may be observed in both the IgG and RSV+IgG spectra (*52*), however, unique, prominent bands are observed in the 1400 – 1600 cm^{-1} region, presumably due to selectively enhanced nucleic acid and/or side-chain vibrations (*53,54*).

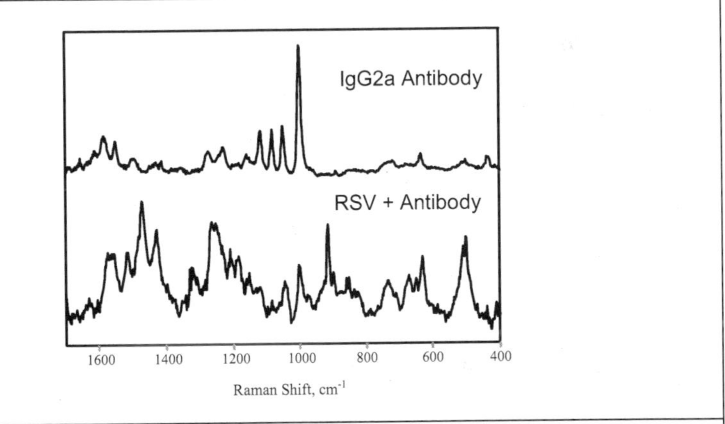

Figure 9. SERS spectra of Ag nanorod substrate modified with a thiol-derivatized anti-RSV antibody and the antibody- modified substrate with bound RSV.

This study illustrates several potential advantages SERS possesses over other widely used biomedical spectroscopic tools such as fluorescence. For example, no fluorescent reporter molecule is needed for SERS, eliminating an assay step, thereby reducing assay time and cost. Additionally, the analyte-specific SERS spectrum may allow discrimination between specific binding to the assay substrate and nonspecific binding, which often plagues sandwich-type immunoassays.

Conclusions

There is a critical, yet currently unmet, need for a rapid and sensitive means of diagnosing infectious diseases that inflict significant burdens on human health. Current diagnostic assays used to dictate disease intervention strategies are limited in sensitivity, have a poor predictive value, or are time-consuming. Among the potential bioanalytical methods available for diagnostic assays, SERS stands out as a method that has exciting potential when applied to the field of biomedical sensing. The trace analytical capabilities of SERS have reached the single molecule level, thus potentially eliminating the amplification steps (*e.g.* PCR) currently needed in many cases. The single largest limitation to SERS, however, is the requirement that the target molecules must be attached to specially engineered substrates that can provide the proper electromagnetic enhancements. We demonstrated that a nanofabrication technique based on vapor deposition produces nanorod substrates that exhibit extremely high SERS activity. Compared with existing SERS substrate preparation techniques, the OAD method offers several strategic advantages for nanofabrication, namely the combination of large enhancements, reproducibility, high throughput, ease of preparation, and low-cost instrumentation. The fundamental aspects of how these nanorods act as SERS substrates has been investigated which led to the development of substrates with SEFs exceeding 5×10^8.

These nanorod substrates have also been evaluated as potential bioanalytical sensors for infectous agent detection. We have shown that SERS can easily distinguish between virus types and among individual virus strains bound to the surface for the case of purified samples. We have also demonstrated that the SERS substrates can be used to analyze more complex samples, i.e., cell lysate, for the presence of viruses. In a final example of the use of these substrates, the Ag nanorods were modified with an antibody selective for a targeted RSV strain. It is currently unclear whether bare or surface modified substrates will emerge as the more useful format, but it is clear that SERS detection and OAD fabricated substrates offer many advantages over current diagnostic techniques.

References

(1) Carey, P. R. *Biochemical Applications of Raman and Resonance Raman Spectroscopies*; Academic Press: New York, 1982.

(2) Kneipp, K.; Kneipp, H.; Itzkan, I.; Dasari, R.; Feld, M. S. *J. Phys.: Condens. Matter* **2002**, *14*, R597-R624.

(3) Kneipp, K.; Wang, Y.; Kneipp, H.; Perelman, L. T.; Itzkan, I.; Dasari, R.; Feld, M. S. *Phys. Rev. Lett.* **1997**, *78*, 1667-1670.

(4) Xu, H.; Bjerneld, J.; Käll, M.; Börjesson, L. *Phys. Rev. Lett.* **1999**, *83*, 4357-4360.

(5) Kneipp, K.; Kneipp, H. *Appl. Spectrosc.* **2006**, *60*, 322A-334A.

(6) Driskell, J. D.; Kwarta, K. M.; Lipert, R. J.; Porter, M. D.; Neill, J. D.; Ridpath, J. F. *Anal. Chem.* **2005**, *77*, 6147-6154.

(7) Mulvaney, S. P.; Musick, M. D.; Keating, C. D.; Natan, M. J. *Langmuir* **2003**, *19*, 4784-4790.

(8) Tian, Z. Q.; Ren, B.; Wu, D. Y. *J. Phys. Chem. B* **2002**, *106*, 9463-9483.
(9) Fleischmann, M.; Hendra, P. J.; McQuillan, A. J. *Chem. Phys. Lett.* **1974**, *26*, 163-166.
(10) Ahern, A. M.; Garrell, R. L. *Anal. Chem.* **1987**, *59*, 2813-2816.
(11) Neddersen, J.; Chumanov, G.; Cotton, T. M. *Appl. Spectrosc.* **1993**, *47*, 1959-1964.
(12) Xue, G.; Dong, J.; Zhang, M. S. *Appl. Spectrosc.* **1991**, *45*, 756-759.
(13) Norrod, K. L.; Sudnik, L. M.; Rousell, D.; Rowlen, K. L. *Appl. Spectrosc.* **1997**, *51*, 994-1001.
(14) Schlegel, V. L.; Cotton, T. M. *Anal. Chem.* **1991**, *63*, 241-247.
(15) Van Duyne, R. P.; Hulteen, J. C.; Treichel, D. A. *J. Chem. Phys.* **1993**, *99*, 2101-2115.
(16) Semin, D. J.; Rowlen, K. L. *Anal. Chem.* **1994**, *66*, 4324-4331.
(17) Roark, S. E.; Rowlen, K. L. *Anal. Chem.* **1994**, *66*, 261-270.
(18) Wang, D. S.; Kerker, M. *Phys. Rev. B* **1981**, *24*, 1777-1790.
(19) Nikoobakht, B.; El-Sayed, M. A. *J. Phys. Chem. A* **2003**, *107*, 3372-3378.
(20) Tao, A.; Kim, F.; Hess, C.; Goldberger, J.; He, R. R.; Sun, Y. G.; Xia, Y. N.; Yang, P. D. *Nano Lett.* **2003**, *3*, 1229-1233.
(21) Moskovits, M.; Jeong, D. H. *Chem. Phys. Lett.* **2004**, *397*, 91-95.
(22) Maxwell, D. J.; Emory, S. R.; Nie, S. M. *Chem. Mater.* **2001**, *13*, 1082-1088.
(23) Orendorff, C. J.; Gearheart, L.; Jana, N. R.; Murphy, C. J. *Phys. Chem. Chem. Phys.* **2006**, *8*, 165-170.
(24) GarciaVidal, F. J.; Pendry, J. B. *Phys. Rev. Lett.* **1996**, *77*, 1163-1166.
(25) Zou, S. L.; Schatz, G. C. *Chem. Phys. Lett.* **2005**, *403*, 62-67.
(26) Hankus, M. E.; Li, H.; Gibson, G. J.; Cullum, B. M. *Anal. Chem.* **2006**, *78*, 7535-7546.
(27) Steele, J. J.; Brett, M. J. *J. Mater. Sci.: Mater. Electron.* **2007**, *18*, 367-379.
(28) Chaney, S. B.; Shanmukh, S.; Dluhy, R. A.; Zhao, Y. P. *Appl. Phys. Lett.* **2005**, *87*, 031908.
(29) Tripp, R. A.; Moore, D.; Jones, L.; Sullender, W.; Winter, J.; Anderson, L. J. *J. Virol.* **1999**, *73*, 7099-7107.
(30) Tully, J. G., R.F.Whitcomb, H.F.Clark , and D.L.Williamson *Science* **1977**, *233*, 241-246.
(31) Cousin-Allery, A. *Epidemiol. Infect.* **2000**, *124*, 103-111.
(32) Beebe, K. R.; Pell, R. J.; Seasholtz, M. B. *Chemometrics: A Practical Guide*; John Wiley & Sons: New York, 1998.
(33) Driskell, J. D.; Shanmukh, S.; Liu, Y.; Chaney, S. B.; Tang, X.-J.; Zhao, Y.-P.; Dluhy, R. A. *J. Phys. Chem. C* **2007**, *In press*.
(34) Driskell, J. D.; Lipert, R. J.; Porter, M. D. *J. Phys. Chem. B* **2006**, *110*, 17444-17451.
(35) Haynes, C. L.; VanDuyne, R. P. *J. Phys. Chem. B* **2003**, *107*, 7426-7433.
(36) Chaney, S. B.; Zhang, Z.-Y.; Zhao, Y.-P. *Appl. Phys. Lett.* **2006**, *89*, 053117
(37) Zhao, Y. P.; Chaney, S. B.; Zhang, Z. Y. *J. Appl. Phys.* **2006**, *100*.
(38) Zhao, Y. P., Chaney, S.B., Shanmukh, S., Dluhy, R.A. *J. Phys. Chem. B* **2006**, *110*, 3153-3157.
(39) Jarvis, R. M.; Goodacre, R. *Anal. Chem.* **2004**, *76*, 40-47.

(40) Laucks, M. L.; Sengupta, A.; Junge, K.; Davis, E. J.; Swanson, B. D. *Appl. Spectrosc.* **2005**, *29*, 1222-1228.
(41) Premasiri, W. R.; Moir, D. T.; Klempner, M. S.; Krieger, N.; Jones, G.; Ziegler, L. K. *J. Phys. Chem. B* **2005**, *109*, 312-320.
(42) Sengupta, A.; Laucks, M. L.; Davis, E. J. *Appl. Spectrosc.* **2005**, *59*, 1016-1023.
(43) Edwards, H. G. M.; Hunt, D. E.; Sibley, M. G. *Spectrochim. Acta A* **1998**, *54*, 745-757.
(44) Qian, W. L.; Krimm, S. *J. Raman Spectrosc.* **1992**, *23*, 517-521.
(45) Bao, P.-D.; Huang, T.-Q.; Liu, X.-M.; Wu, T.-Q. *J. Raman Spectrosc.* **2001**, *32*, 227-230.
(46) Stewart, S.; Fredericks, P. M. *Spectrochim. Acta A* **1999**, *55*, 1615-1640.
(47) Shanmukh, S.; Jones, L.; Zhao, Y.-P.; Driskell, J. D.; Tripp, R. A.; Dluhy, R. A. *Biosens. Bioelectron.* **2007**, *submitted.*
(48) Balish, M. F.; Krause, D. C. *J Mol Microbiol Biotechnol* **2006**, *11*, 244-55.
(49) Shanmukh, S.; Jones, L.; Driskell, J.; Zhao, Y.; Dluhy, R.; Tripp, R. *Nano Lett.* **2006**, *6*, 2630-2636.
(50) Ljunglof, A.; Larsson, M.; Knuuttila, K.-G.; Lindgren, J. *J. Chromatog. A* **2000**, *893*, 235-244.
(51) Picquart, M.; Haro-Poniatowski, E.; Morhange, J. F.; Jouanne, M.; Kanehisa, M. *Biopolymers* **2000**, *53*, 342-349.
(52) Kitagawa, T.; Hirota, S. In *Handbook of Vibrational Spectroscopy*; Chalmers, J. M., Griffiths, P. R., Eds.; J. Wiley & Sons, Ltd.: Chichester, 2002; Vol. 5, p 3426-3446.
(53) Taillandier, E.; Liquier, J. In *Handbook of Vibrational Spectroscopy*; Chalmers, J. M., Griffiths, P. R., Eds.; J. Wiley & Sons, Ltd.: Chichester, 2002; Vol. 5, p 3465-3480.
(54) Tuma, R.; Thomas Jr., G. J. In *Handbook of Vibrational Spectroscopy*; Chalmers, J. M., Griffiths, P. R., Eds.; John Wiley & Sons: Chichester, 2002; Vol. 5.

Chapter 9

Gold Nanoparticle Based Surface Energy Transfer Probe for Accurate Identification of Biological Agents DNA

Paresh Chandra Ray*, Gopala Krishna Darbha, Oleg Tovmachenko, Uma Shanker Rai , Jelani Griffin, William Hardy and Ana Balarezo

Department of Chemistry, Jackson State University, Jackson, MS, USA

Rapid differentiation and accurate identification of bioagents are crucial to planning timely and appropriate measures for public safety. Driven by the need to detect bioagents selectively, the development of a miniaturized, inexpensive and battery operated ultra-sensitive gold nanoparticle-based surface energy transfer (NSET) probe for screening of the bioagents DNA with excellent sensitivity (800 femto-molar) and selectivity (single base pair mismatch) is reported here. In the presence of various target sequences, detection of sequence specific DNA is possible via independent hybridization process. As proof of concept, multiplexed detection of two target sequences, 1) oligonucleotide sequence associated with the anthrax lethal factor and 2) sequence related to positions 1027–1057 of the *E.coli* DNA which codes for the 23S rRNA, have been demonstrated with high sensitivity and specificity. The quenching efficiency as a function of distance is investigated. The mechanism of distant dependent fluorescence quenching has been discussed.

Introduction

The biological weapon has been around at least since the Middle Ages when soldiers catapulted the bodies of dead smallpox victims over fortress walls in the hope of infecting their enemies or at least demoralizing them. Lately, biological agents have been appearing in the news with increasing frequency. Biological agents are of concern because of the lag time between a biological attack and the appearance of symptoms in those exposed, biological weapons could be devastating. Many biological agents are contagious, and during this lag time, infected persons could continue to spread the disease, further increasing its reach. As a result, the availability of rapid, sensitive and cost-effective diagnostic methods is paramount to the success of a comprehensive national health security system in the USA *(1,2)* . Ideally, detection systems should be capable of rapidly detecting and confirming biothreat agents, including modified or previously uncharacterized agents, directly from complex matrix samples, with no false results. Furthermore, the instrument should be portable, user-friendly, and capable of testing for multiple agents simultaneously. Many local and state authorities are inadequately prepared to deal with biological-based incidents, and first responders to such incidents will face considerable risk. In addition, since public health personnel rarely encounter any of the 30 or so pathogens on various agency threat lists, the ability to rapidly identify their infection is waning. So sensors must not only be sensitive and specific, but must also be able to accurately detect a variety of pathogens. Driven by the need, current efforts have been focused on the development of miniaturized, inexpensive and battery operated ultra-sensitive gold nanoparticle based surface energy transfer (NSET) probe for screening bioagents DNA which has excellent sensitivity and selectivity.

Conventional analytical methods *(3-5)* used to detect biological agents, such as high-performance liquid chromatography, gas chromatography, and mass spectroscopy, require expensive equipments that may be difficult to field-deploy. Standard microbiological methods *(6,7)*, such as culturing and microscopic examination, are time-consuming and labor-intensive. The increasing availability of nanostructures with highly controlled optical properties in the nanometer size range has created widespread interest in their use in biotechnological system for diagnostic application and biological imaging (*8-33*). Merging biotechnology with nano science will allow us not only to take advantage of the improved evolutionary biological components to generate new smart sensors but also to apply today's advanced characterization and fabrication techniques to solve biodefense problems. Although still in its infancy, the application of surface-functionalized nanomaterials such as nanoparticles in sequence recognition schemes has shown *(8-34)* great promise in achieving high sensitivity and specificity, which are difficult to achieve by conventional methods. In this chapter, we have discussed our recent research on gold nanoparticle (Au/NP) based fluorescence resonance energy transfer (FRET) assay to detect multiple bioagenst DNA together. We also described the fabrication of a miniaturized, inexpensive and battery operated ultra-sensitive laser-induced fluorescence (LIF) optical fiber sensor, based on quenching of the LIF signal by gold nanoparticle, to detect single base-mismatch DNA.

Recently, there have been many reports for the development of fluorescence-based assays for bioagents DNA detection *(35-37)*. These assays are based on Föster resonance energy transfer (FRET) *(38)* or non-FRET quenching mechanisms. Although FRET technology is very convenient and can be applied routinely at the single molecule detection limit, the length scale for the detection using FRET-based method is limited by the nature of the dipole-dipole mechanism, which effectively constrains the length scales to distances on the order of <100 Å ($R_0 \approx 60$ Å) and FRET is observed when the donor and an acceptor are placed 20 bases apart on an oligonucleotide. The limitations of FRET can be overcome with a dynamic molecular ruler based on the distance-dependent plasmon coupling of metal nanoparticles. In 1970s, Drehage showed that a nearby metal mirror can alter lifetime of a fluorophore and the lifetime depends on the distance from the metal. After this discovery in last thirty years there has been several theoretical and experimental *(6-54)* contributions about the behavior of fluorophores near metals. This interest has now been extended to nanoparticles that display collective oscillations of electrons known as surface plasmons *(8-34)*. Fluorophores are known to display the optical properties of a oscillating point dipoles. The rates of radiation and the spatial distribution of the radiation can be dramatically altered by near-field interactions of the fluorophore with the metal. These interactions can increase the rates of fluorophore excitation and/or emission. Recently several groups including ours *(19,22-34)* have reported that NSET is a technique capable of measuring distances nearly twice as far as FRET in which energy transfer from a donor molecule to a nanoparticle surface follows predictable distance dependence. In this chapter we have discussed size dependent NSET properties of gold nanoparticles for recognizing bioagents DNA sequence selectively (single-base mutations) in a homogeneous format.

Experimental Methods

Hydrogen tetrachloroaurate ($HAuCl_4 \cdot 3H_2O$), $NaBH_4$, Rhodamine B, mercaptopropionic acid, homocystine, 2,6-pyridinedicarboxylic acid, buffer solution, NaCl, and sodium citrate were purchased from Sigma-Aldrich and used without further purification. 5'-Cy3 modified GTAACTTCCATTTCTTTGG-3', 5'-Texas Red modified GGATTATTGTTAAATATTGATAAGGAT-3', 5'-JOE modified AAACGATGTGGGAAGGCCCAGACAGCCAGG-3' and complementary oligonucleotides were purchased from MWG Biotech. The oligos were incubated with gold nanoparticles overnight. The oligo-particle conjugates were gradually exposed to 0.1 M NaCl in a PBS buffer over a 16-h period according to a procedure we have reported recently *(19,23-29)*.

Gold Nanoparticle Synthesis

Gold nanoparticles with diameters of 15 nm or above were synthesized using reported method *(19,23-34)*. Gold nanoparticles of different sizes and shapes were synthesized by controlling the ratio of $HAuCl_4 \cdot 3H_2O$ and sodium citrate concentration as we reported recently *(19,23-29)*. For smaller gold nanoparticles, we have used sodium borohydride method as reported before. Briefly, 0.5 mL of 0.01 M $HAuCl_4$ trihydrate in water and 0.5 mL of 0.01 M sodium citrate in water were added to 18 mL of deionized H_2O and stirred. Next, 0.5 mL of freshly prepared 0.1 M $NaBH_4$ was added and the solution color changed from colorless to orange. Stirring was stopped and the solution was left undisturbed for 2h. The resulting spherical gold nanoparticles were 4 nm in diameter. Transmission electron microscope (TEM) and UV-visible absorption spectrum were used to characterize the nanoparticles. The particle concentration was measured by UV-visible spectroscopy using the molar extinction coefficients at the wavelength of the maximum absorption of each gold colloid as reported recently [$\varepsilon_{(15)\ 528nm} = 3.6 \times 10^8\ cm^{-1}\ M^{-1}$, $\varepsilon_{(30)\ 530nm} = 3.0 \times 10^9\ cm^{-1}\ M^{-1}$, $\varepsilon_{(40)\ 533nm} = 6.7 \times 10^9\ cm^{-1}\ M^{-1}$, $\varepsilon_{(50)\ 535nm} = 1.5 \times 10^{10}\ cm^{-1}\ M^{-1}$, $\varepsilon_{(60)\ 540nm} = 2.9 \times 10^{10}\ cm^{-1}\ M^{-1}$, and $\varepsilon_{(80)\ 550nm} = 6.9 \times 10^{10}\ cm^{-1}\ M^{-1}$] *(19,23-29)*.

Preparation of Gold Nanoparticle adsorbed DNA Probes

Probe DNA was incubated with gold particles for about 2 hours at a ratio of 30 to 40 RNA molecules per particle. This ratio was used to ensure that each particle was conjugated to at least few DNA molecules. The DNA-gold nanoparticle conjugates were gradually exposed to 0.1 M NaCl in a PBS buffer over a 1-h period, according to a procedure reported by us. Unadsorbed nanoparticle probes were removed from the solution by centrifugation at 12 000 rpm for 20 min. The final product was a red precipitate, which was resuspended and stored in 10 mM phosphate buffer. To quantify the number fo DNA adsorbed in each gold nanoparticle, after hybridization, we centrifuged the solution at 13 000 rpm for 20 minutes, the precipitate was redispersed into 2 mL of the buffer solution for LIF experiment. The amount of ds-DNA was measured by fluorescence, and the gold particle concentration was measured by optical absorbance. By dividing the total number of oligo molecules by the total number of nanoparticles, we estimated that there were about 25-30 DNA molecules per particle

Hybridization Assays

Anthrax toxin comprises three nontoxic monomeric proteins, comprise edema factor (EF), lethal factor (LF), and protective antigen (PA), which assemble at the mammalian cell surface to form toxic complexes *(39-40)*. EF and lethal factor LF are enzymes that are delivered to the cytosol of mammalian cells by the third protein, PA. EF is an adenylyl cyclase that elicits edema when coinjected with PA into animals and likely serves to impair the immune response to infection. LF is a zinc protease that cleaves certain mitogen-activated protein kinases. In combination with PA, LF causes death of animals. So lethal toxin does the most damage within the cell. For the proof-of-concept experiment an oligonucleotide sequence associated with the anthrax lethal factor(5' GGATTATTGTTAAATATTGATAAGGAT 3') was chosen as an initial target.

This sequence is important for bioterrorism and biowarfare applications, and it has been well studied in the literature. Hybridization of the probe and the target DNA was conducted in phosphate buffer solution with 0.3-0.5 M NaCl for half an hour at room temperature. Fluorescence studies showed that the DNA-gold nanoparticle probes are stable under high salt conditions (e.g., 0.5 M NaCl). A 4-fold molar excess of a complementary strand was added and was allowed to hybridize for 30 min at room temperature.

Results and discussion

NSET Probe Design

Scheme 1 shows a schematic diagram of the nanoparticle probes and their operating principles for single DNA hybridization detection process. Dye tagged ss-DNA is adsorbed on the gold nanoparticle and as a result the fluorescence from the dye is completely quenched by gold nanopartcile *(19.23-29,41-42)*. When a dye-labeled oligonucleotide molecule is adsorbed on a nanoparticle, the fluorophore at the distal end can loop back and adsorb on the same particle. Our experimental data showed a quenching efficiency of nearly 100% when the fluorophore was statically adsorbed on the particle (static quenching). An important insight came from our recently reported *(25)* surface-enhanced Raman scattering (SERS) studies, which showed that fluorescent dyes could reversibly adsorb on the surface of colloidal silver and gold nanoparticles.

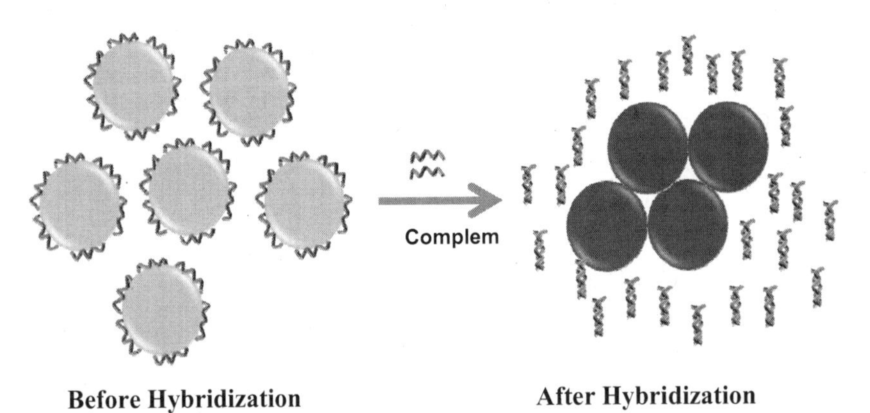

Scheme 1: Schemetic representation of the DNA hybridization process

Thus, when an oligonucleotide molecule is firmly tethered to a particle, the fluorophore at the distal end can loop back and adsorb on the same particle. Because the backbone of single- stranded DNA is conformationally flexible, a favorable conformation for the adsorbed oligos is an arch-like structure, in which both the 3'- and 5'-ends are attached to the particle but the DNA chain does not contact the surface.

DNA Hybridization Detection

Upon target binding, due to the duplex structure, the double-strand (ds) DNA does not adsorb onto gold and the fluorescence persists (as shown in Scheme 1). This structural change generates a fluorescence signal that is highly sensitive and specific to the target DNA.

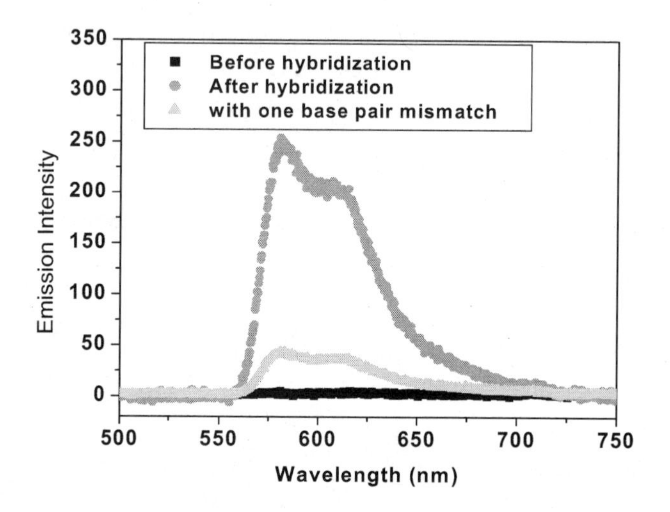

Figure 1: Plot of fluorescence intensity vs. wavelength for 120 pM 5'-Cy3 modified GTAACTTCCATTTCTTTGG--3'oligonucleotides

As shown in Figure 1, we observed a very distinct LIF intensity change after hybridization even at 120-picomolar concentration of probe fluorophore tagged ss-DNA. Since the surface adsorption energies of organic dyes on gold are usually in the range of 8-16 kcal/mol, which are much smaller than the energies involved in DNA hybridization (80-100 kcal/mol), after hybridization, the constrained conformation is opened and the fluorophore is separated from the particle surface. To demonstrate that after hybridization the ds-DNA was not adsorbed onto gold nanoparticle, we centrifuged the solution at 13 000 rpm for 20 minutes, the precipitate was redispersed into 2 mL of the buffer solution for LIF experiment. We have not observed any fluorescence signal from the gold nanoparticles after it was centrifuged, whereas the supernatant shows strong fluorescence and the intensity is about same as we observed after hybridization. This indicates that ss DNA is not able to adsorb onto gold nanoparticle after hybridization. The amount of ds-DNA was measured by fluorescence, and the gold particle concentration was measured by optical absorbance. By dividing the total number of oligo molecules by the total number of nanoparticles, we estimated that there were about 25-30 DNA molecules per particle. This high fluorescence enhancement clearly demonstrates that nanoparticle based LIF assay can be used as a highly sensitive probe for monitoring DNA hybridization. Figure 1 also illustrates single-mismatch detection capability. Our result indicates that our NSET probes are highly specific in discriminating against non-complementary DNA sequences and single-base mismatches.

The addition of noncomplementary nucleic acids had no effect on the fluorescence, and a single-base mismatch reduced the fluorescence intensity by 90% (in comparison with the fluorescence intensity of perfectly matched targets). So our NSET probes will be applicable to rapid detection of single-nucleotide polymorphisms (SNPs) in genomic DNA, an exciting prospect for eliminating time-consuming and expensive gel sequencing procedures that are currently the standard protocol.

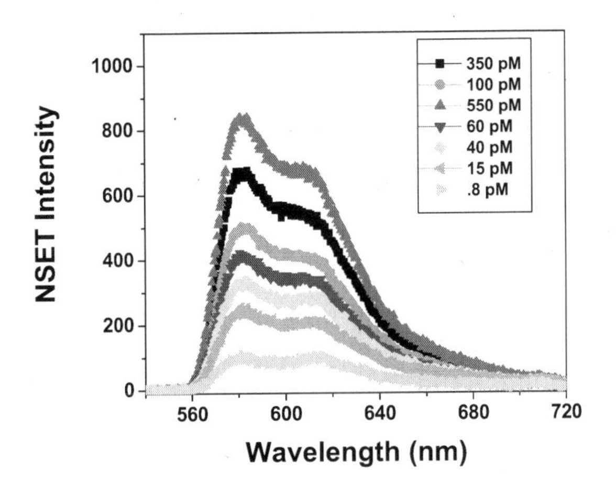

Figure 2: Fluorescence response upon addition different concentration of target anthrax DNA on 150 nm probe DNA

Quantitative Measurement of Sequence Specific Target Anthrax DNA Concentration

To evaluate whether our NSET probe is capable of measuring target anthrax DNA concentration quantitatively, we performed NSET intensity measurement at different concentrations of target DNA. As shown in Figure 2, the NSET emission intensity is highly sensitive to the concentration of target DNA and the intensity increased linearly with concentration. Our data indicate that NSET with 45 nm gold nanoparticle, exhibit sensitivity to detect DNA as low as 800 fM. So our NSET probe can provide a quantitative measurement of anthrax DNA concentration in a sample up to 800 fM.

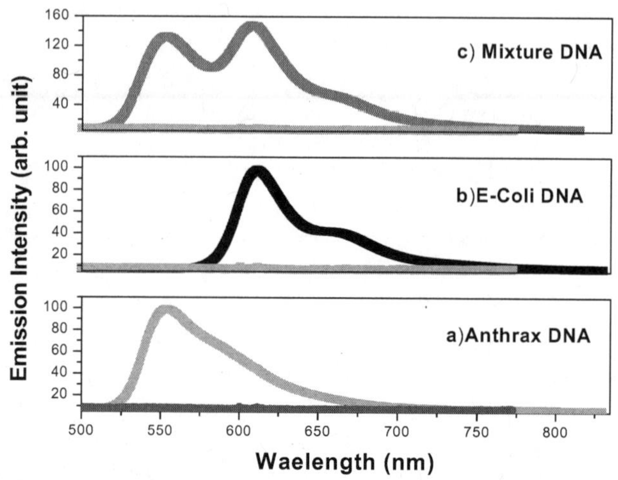

Figure. 3.. Plot of fluorescence intensity vs. wavelength a) for 5'-Texas Red modified GGATTATTGTTAAATATTGATAAGGAT-3' oligonucleotides, b) 5'-JOE modified AAACGATGTGGGAAGGCCCAGACAGCCAGG -3, c) Mixture of both (Reprinted from Nanotechnology, Paresh Chandra Ray, Gopala K Darbha, Anandhi Ray, William Hardy and Joshua Walker, Nanotechnology, 2007, 18, 375504).

Multiplex DNA Detection

To demonstrate that our FRET probe can detect multiple DNA hybridization we have used anthrax DNA as well as E-coli DNA. *Escherichia coli* O157:H7 is important food pathogens in public health, causing severe illnesses associated with contamination of meat, poultry or dairy foods *(44-45)*. Rapid and sensitive detection of these food pathogens is required for food safety surveillance. The sequence of the target DNA which codes for the 23S rRNA (positions 1027–1057 of the *E.coli* 23S rRNA *(46)*) was 5'-AAACGATGTGGGAAGGCCCAGACAGCCAGG-3'. Figure 3 demonstrates that our gold nano-particle based FRET probe can detect multiple DNA very easily. To detect multiple DNA we tried single size nano-particle as well as mixture and our results indicate that FRET intensity become maximum when we used 1:1 mixture of 13 nm and 60 nm gold nano-particle. After hybridization of both DNA, gold nanoparticles undergo aggregation due to the presence of NaCl (as shown in Figure 4).

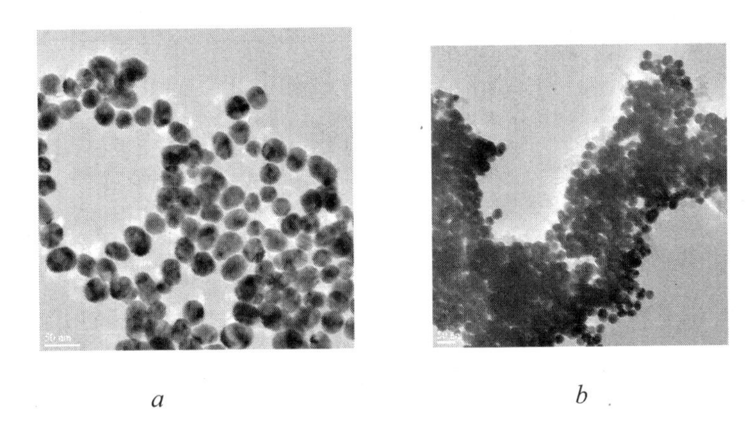

Figure 4. TEM images of gold nanoparticle-DNA solution a) before and b) after hybridization

Distance Dependent NSET

To understand the limitation of out nano-FRET probe in terms of length of DNA and also to understand the mechanism of fluorescence quenching, we have used different length *E.coli* DNA which codes for the 23S rRNA (*46*). We have used positions 1027-1047, 1027–1057, 1027-1077, 1027-1097, 1027-1117, 1027-1137 of the *E.coli* DNA which codes for the 23S rRNA. 3'-thiol (-SH) group and a 5'-fluorophore containing ss-DNA were purchased from MWG Biotech. The oligo-particle conjugates were gradually exposed to 0.1 M NaCl in a PBS buffer over a 16-h period, according to a procedure reported by Mirkin and co-worker (*7*). Scheme 2 shows a schematic diagram of the nanoparticle probes and their operating principles for DNA hybridization process when the ss-DNA is attached with gold nanoparticle through –S- linkage.

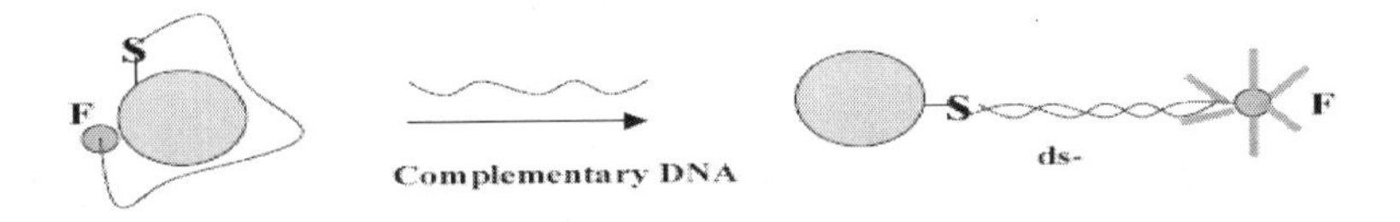

Scheme 2: Schematic illustration of DNA hybridization process when DNA is attached with gold nanoparticle through -SH. The circle represents the gold nanoparticle and F represent the fluorophore

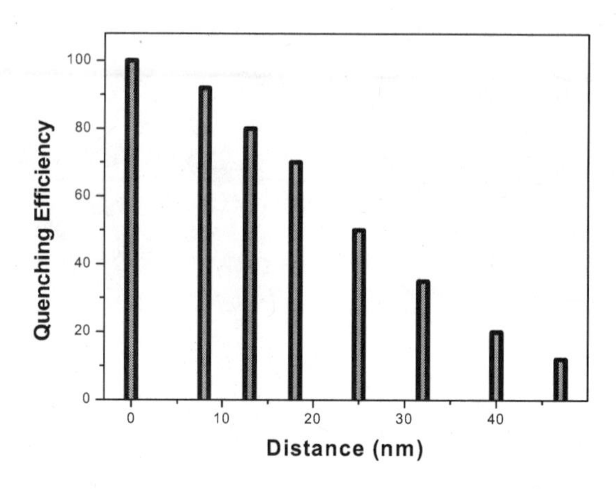

Figure 5. Variation of the quenching efficiency with distance between gold nano-particle and JOE dye (Reprinted from Nanotechnology, Paresh Chandra Ray, Gopala K Darbha, Anandhi Ray, William Hardy and Joshua Walker, Nanotechnology, 2007, 18, 375504).

After hybridization, by varying the DNA lengths, the separation distance between gold nanoparticle and JOE dye can be systematically varied between 8 nm and 40 nm, by varying the number of base pairs, for example, 8.2 nm for 20 bp DNA, 13.04 nm for 30bp DNA, and 23.24 nm for 60bp DNA. The distance from the center of the molecule to the metal surface is estimated by taking into account size of the fluorescent dye, .31 nm for each base pair *(47-48)* , Au-S distance, end base pair to dye distance. We have assumed a linear ds-DNA strand configuration because ds-DNA is known to be rigid having a persistence length 90 nm *(47)*. Figure 5 shows how the quenching efficiency varies with the increase in the distance between gold nano-particle and JOE dye. Our result indicates that our assay can be used conveniently for longer DNA till 110 base pairs.

Our experiment indicates 50% quenching efficiency even at 25 nm separation between donor and acceptor. This type of fluorescence signal change due to energy transfer is typically addressed by optical methods based on FRET between molecular donors and acceptors. Förster formula *(38)* is very convenient and can be applied routinely at the single molecule detection limit. However, the length scale for detection using Förster method is limited by the nature of the dipole-dipole mechanism, which effectively constrains the length scales to distances on the order of <100 Å ($R_0 \approx$ 60 Å). Our FRET probe measuring distance is indeed much higher than the general FRET distance. Because the efficiency general FRET process scales as $1/(1 + (r/r_0)^6)$, where r is the distance between donor and acceptor, r_0 is the distance between donor and acceptor at which the energy transfer efficiency is 50%. FRET efficiency is significant over a very short distance range of r_0 ~ 4-8 nm for typical fluorophores. Although in general the interaction between nanoparticle and dye are quite complex when considering all illumination polarizations, distance ranges, and particle sizes, the situation for very small particles can be understood as follows. 1) Since FRET physically originates from the weak electromagnetic coupling of two dipoles, one can imagine that introducing additional dipoles and thus providing more coupling interactions can circumvent the FRET limit. Light induces oscillating dipole moments in each gold particle, and their instantaneous $(1/r)^3$ coupling results in a repulsive or attractive interaction, modifying the plasmon resonance of the system. The softer dependence of the interaction strength on the particle separation r results in a much longer interaction range compared to FRET. 2) The fluorescence quantum yield is determined by the radiative rate constant, k_r, and its nonradiative counterpart, k_{nr} : $\tau = (k_r + k_{nr})^{-1}$. At small distances (1-2 nm), the large fluorescence quenching efficiency of 99.8% is due to two effects: a) gold nanoparticles increase the nonradiative rate R_{nonrad} of the molecules due to energy transfer; b) the radiative rate R_{rad} of the molecules is decreases because the molecular dipole and the dipole induced on the gold nanoparticles radiate out of phase if the molecules are oriented tangentially to the gold nanoparticle's surface. At higher distance, the distance dependent quantum efficiency is almost exclusively governed by the radiative rate as reported recently. In the case of $\tau_{nr} << \tau_r$, the quantum yield is primarily governed by the nonradiative lifetime and can be approximated as QY $(\tau_{nr}) \approx \tau_{nr} / \tau_r$. Very recently it has been proposed that in which the strength of fluorescence quenching and the accompanying change in the fluorescence lifetime act as measures for the separation between a fluorophore and metal clusters.

Portable NSET Probe

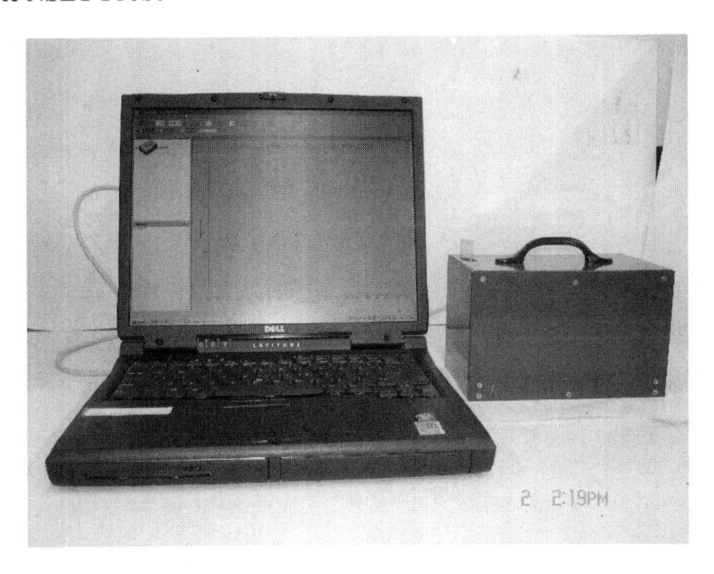

Figure 6: Photograph of the minimized NSET probe box, connected with labtop computer for data acquisition

The photograph of the minimized system components for the NSET sensor configuration to detect biological and chemical toxin is shown in the Figure 6. For fluorescence excitation, we have used a continuous wavelength Melles-Griot green laser pointer (18 Lab 181) operating at 532 nm, as an excitation light source. The laser pointer can maintain 10~13 hours with two AAA size batteries, and its maximum output power is ~5mW. This light source has a capability that can minimize whole sensor configuration. The total size of the sensor configuration was 5'' x 8`` x 5`` including the laser pointer, optical fiber and OOI spectrometer in the aluminum box. This probe consists of total seven optical fibers, each having 200 μm core diameters with one launching fiber and six surrounding collecting fibers (as shown figure 6). Excitation light source was first attenuated using appropriate neutral density (ND) filter and coupled to the excitation arm of the Y-shaped reflection probe through a plano-convex lens (*f*:~4.5mm) as shown in Figure 7. A typical laser energy at the sample was adjusted to ~1.3mW with 0.3 ND filter. The diverging output beam from the fiber provided nearly uniform illumination onto sample.

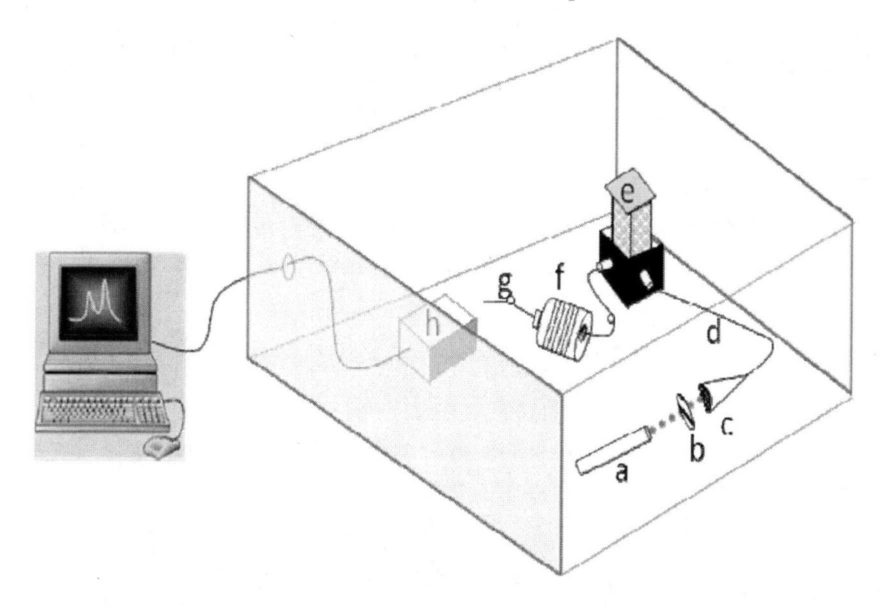

Figure 7.. Schematic diagram of our FRET nano probes. It consists of several components. a) Melles-Griot green laser pointer (18 Lab 181), b) neutral density filter, c) plano-convex lens, d) optical fiber, e) sample holder, f) 532 nm cut-off filter, g) optical fiber, h) ocean optics QE65000 spectrometer. (Reprinted from Nanotechnology, Paresh Chandra Ray, Gopala K Darbha, Anandhi Ray, William Hardy and Joshua Walker, Nanotechnology, 2007, 18, 375504).

In the Coherent laser configuration, 2.0 ND filter was used to attenuate excitation beam to ~1.3mW. NSET signal from sample was simultaneously collected in the backward direction (180° collection geometry) by a ring of six optical fibers around the single illumination channel. The collected emission signal was transmitted through an online filter module (containing a 570 nm cutoff filter to suppress the detection of the back-scattered excitation light) to a 600 μm core diameter UV grade fused silica auxiliary fiber prior to feeding it to an Ocean Optics spectrometer. A low resolution OOI spectrometer with 600/1mm grating (cover 200-850nm) was used in this work. OOI spectrometer was interfaced with a note book computer though USB port. The LIF spectrum was collected with Ocean Optics data acquisition software. All measurements performed with 5ms integration time were instantaneously averaged with 5 spectra using the software. The averaged data were processed using MS excel program. The ON/OFF (Push /Pull) switch of the laser pointer is located at the other end of laser pointer and it is fixed such a way one can operate it from outside of the aluminum box.

Conclusions

In conclusion, in this chapter we have reported a miniaturized, inexpensive and battery operated ultra-sensitive gold nanoparticle based FRET probe for screening bio-agents DNA in femto-molar level. Our method has high sensitivity and convenience. We have demonstrated a multiplexed hybridization detection method in a homogeneous, separation-free format based on multicolor oligonucleotide-functionalized dyes with gold nanoprobes. As proof of concept, multiplexed detection of two target sequences, 1) oligonucleotide sequence associated with the anthrax lethal factor and 2) sequence related to positions 1027–1057 of the *E.coli* DNA which codes for the 23S rRNA, have been demonstrated with high sensitivity and specificity. Our experiment indicates 50% quenching efficiency even at 25 nm separation between donor and acceptor. Our experimental observation paradigm for design of optical based molecular ruler strategies at distances more than double the distances achievable using traditional dipole-dipole Columbic energy transfer based methods. This long-range feature will allow the development of biosensors and homogeneous bioassays that are not possible by using general FRET process e.g. catalytic activities of many enzymes such as HIV proteases and matrix metalloproteases (MMP) as well as the proteins involved in cellular apoptosis. Given the simplicity, speed, and sensitivity of this approach, the described methodology could easily be extended to a high throughput format and become a new method of choice in all applications that require an assay for bio-agent DNA detection. Looking into the future, we expect these compact sensor developments will have important implications in the development of better biosensors and bioassay for application to pathogen detection, clinical analysis and biomedical research. Our observations also point toward the exciting possibility to perform spatially confined detection on array formats of biological recognition, for example, RNA/DNA hybridization or antibody-antigen recognition.

Acknowledgements

Dr. Ray thanks ARO grant # W911NF-06-1-0512, NSF-PREM grant # DMR-0611539, and NSF-CRIFMU grant # 0443547 for their generous funding.

References

1. Walt, D. R.; Franz, D. R. *Anal. Chem.* **2000**, 72, 738A-746A.
2. Chin, J. E. Control of Communicable Diseases Manual; American Public Health Association: Washington, D.C., 2000; pp 20-25.
3. Beltyukova, S. V.; Poluektov, N. S.; Tochidlovskaya, T. L.; Kucher, A. A. *Dokl. Akad. Nauk.* **1986**, 291, 1392-1395
4. Yolken, R. H.; Wee, S. B. *J. Clin. Microbiol.* **1984**, 19, 356-360.
5. De Wit, M. Y.; Faber, W. R.; Krieg, S. R.; Douglas, J. T.; Lucas, S. B.; Montreewasuwat, N.; Pattyn, S. R.; Hussain, R.; Ponnighaus, J. M.; Hartskeerl, R. A. *J. Clin. Microbiol.* **1991**, 29, 906-910
6. Pellegrino, P. M.; Fell, N. F.; Gillespie, J. *B. Anal. Chim. Acta* **2002**, 455, 167.
7. Fasanella, A.; Losito, S.; Adone, R.; Ciuchini, F.; Trotta, T.; Altamura, S. A.; Chiocco, D.; Ippolito, G. *J. Clin. Microbiol.* **2003**, 41, 896-899.
8. Storhoff, J. J.; Lucas, A. D.; Garimella, V.; Bao, Y. P.; Muller, U. R. *Nat. Biotechnol.* **2004**, 22, 883-887.
9. Park, S.-J.; Taton, T. A.; Mirkin, C. A. *Science* **2002**, 295, 1503-1506.
10. Taton, T. A.; Lu, G.; Mirkin, C. A. *J. Am. Chem. Soc.* **2001**, 123, 5164-5165.
11. Cao, Y. C.; Jin, R.; Mirkin, C. A. *Science* **2002**, 297, 1536-1540.
12. Nicewarner-Pena, S. R.; Freeman, R. G.; Reiss, B. D.; He, L.; Pena, D. J.; Walton, I. D.; Cromer, R.; Keating, C. D.; Natan, M. J. *Science* **2001**, 294, 137-141.
13. Bruchez, M.; Moronne, M.; Gin, P.; Weiss, S.; Alivisatos, A. P. *Science* **1998**, 281, 2013-2016.
14. Chan, W. C. W.; Nie, S. M. *Science* **1998**, 281, 2016-2018.
15. Xiaoyu Z, Matthew A. Y, Olga L, and Van Duyne R. P., *J. Am. Chem. Soc.*, **2005**, 127, 4484.
16. Bailey, R C.; Kwong, G A.; Radu, C G.; Witte, O N.; Heath, J R., *J. Am. Chem. Soc.* **2007**, 129, 1959-1967.
17. J-S Lee, S. I. Stoeva, and C. A. Mirkin, *J. Am. Chem. Soc.* **2006**, 128, 8899.
18. L. Yougen, Y. T. Hong Cu, D. Luo, *Nature Biotechnology* **2005**, 23, 885
19. P. C. Ray, *Angew Chem.* **2006**, 45, 1151-1154.
20. N. L. Rosi and C. A. Mirkin, *Chem. Rev.* **2005**, 105, 154.
21. K. Kinbara and T Aida, *Chem. Rev.* **2005**, 105, 1377.
22. J. A. Hansen; R. Mukhopadhyay; J. O. Hansen and K. V. Gothelf, *J. Am. Chem. Soc.* **2006**, 128, 3860.
23. Darbha G K, Ray A and Ray P. C. *ACS Nano*, **2007,** 3, 208-214.
24. Ray PC , Darbha GK, Ray A, Hardy W and Walker J, *Nanotechnology*, **2007**, 18, 375504
25. Tiwari V S, Tovmachenko O, Darbha G K, Hardy W, Singh J P and Ray P C, **Chemical Physics Letters**, **2007**, 446, 77-82
26. Ray, P. C.; Fortner, A and Gopala Krishna Darbha , *J. Phys. Chem. B*, **2006**, 110 , 20745 –20748.
27. Kim, C. K.; Kalluru, R. R.; Singh, J. P.; Fortner, A.; Griffin, J.; Darbha, G. K.; Ray, P. C. *Nanotechnology* **2006**, 17, 3085,

28. Jennings, T. L.; Singh, M. P.; Strouse, G. F. *J. Am. Chem. Soc.* **2006**, 128, 5462,
29. Ray, P. C.; Fortner, A.; Griffith, J.; Kim, C. K.; Singh, J. P.; Yu, H. *Chem. Phys. Lett.* **2005**, 414, 259,
30. Li, H.; Rothberg, L. J. *Anal. Chem.* **2004**, 76, 5414,
31. Maxwell, D. J.; Taylor, J. R.; Nie, S. *J. Am. Chem. Soc.* **2002**, 124, 9602.
32. Shi, L.; De Paoli, V.; Rosenzweig, N.; Rosenzweig, Z., *J. Am. Chem. Soc.* **2006**, 128, 10378.
33. Seelig, J.; Leslie, K.; Renn, A.; Kuhn, S.; Jacobsen, V.; van de Corput, M.; Wyman, C.; Sandoghdar, V., *Nano. Lett.* **2007**, 7, 685.
34. Shi, L.; Rosenzweig, N.; Rosenzweig, Z., *Anal. Chem.* **2007**, 79, 208
35. M. A. Stoff-Khalili, P. Dall, D. T. Curiel, *Can. Gene. Ther.*, **2006**, 13, 633.
36. P. Swiderek, *Angew. Chem. Int. Ed,* 2006, 45, 4056.
37. S. Ptasińska, S. Denifl, S. Gohlke, P. Scheier, E. Illenberger, T. D. Märk, *Angew. Chem. Int. Ed*, **2006**, 45, 1893.
38. T. Foster, *Ann. Phys.* **1948**, 2, 55.
39. Neumeyer, T.; Tonello, F.; Dal Molin, F.; Schiffler, B.; Orlik, F.; Benz, R. *Biochemistry*, **2006**, 45, 3060.
40. Cooksey, B. A.; Sampey, G. C.; Pierre, J. L.; Zhang, X.; Karwoski, J. D.; Choi, G. H.; Laird, M. W.. *Biotechnol. Prog*, **2004**, 20, 1651.
41. Lakowicz, J. R. Principles of Fluorescence Spectroscopy, 2nd ed.; Plenum Press: New York, 1999; pp 237-265.
42. Fan, C; Wang, S; Hong, J W.; Bazan, G C.; Plaxco, K W.; Heeger, A J. *Proc. of the National Academy of Sciences of USA*, **2003**, 100, 6297-6301.
43. You, C-G; Miranda, O R.; Gider, B; Ghosh, P S.; Kim, Ik-Bum; Erdogan, B; Krovi, S A ; Bunz, U H. F.; Rotello, V M. *Nature Nanotechnology*, **2007**, 2, 318-323.
44. Parkes, R. J. In Ecology of microbial communities; Fletcher M., Gray T. R. G., Jones, J. G., Eds.; Cambridge University Press: Cambridge, 1987; pp 147-177.
45. Bauer-Kreisel, P.; Eisenbeis, M.; Scholz-Muramatsu, H. *Appl. Environ. Microbiol.* **1996**, 62, 3050-3052.
46. Fuchs B. M., Syutsubo K, Ludwing W, Amann R, *Appl. Environ, Microbiol.* **2001**, 67, 961.
47. Hays J B, Magar M E and Zimm B H *Bioplymers* **1969**, 8 531.
48. Lu Y, Weers B and Stellwagen N C *Bioplymers* **2002**, 61, 261.

Chapter 10

Enhanced Raman Detection using Spray-On Nanoparticles/Remote Sensed Raman Spectroscopy

Michael L. Ramirez, Leonardo C. Pacheco, Marcos A. Barreto and Samuel P. Hernández-Rivera

Chemical Imaging Center / Center for Chemical Sensors Development Department of Chemistry, University of Puerto Rico-Mayagüez, P.O. Box 9019, Mayagüez, PR 00681

This report presents current research directed towards the use of Surface Enhanced Raman Spectroscopy for *in situ* analysis and remote Raman sensing applications. The Raman detection of trace amounts of explosives and other hazardous materials and test compounds deposited on surfaces was improved by ten to thousand fold (10-1000x) with the addition of colloidal metallic nanoparticles to contaminated substrate areas. In this contribution application of different deposition techniques such as direct transfer and smearing, and spraying techniques, which included ultrasonic spraying and pneumatically assisted nebulization via an electrospray needle to prepare SERS active substrates is reported. Colloidal Ag nanoparticles sols were deposited on glass, silicon and stainless steel in order to accomplish the proposed objective. The spectroscopic evaluation of samples was made by Raman microscopy. Distribution and morphology of nanoparticles and aggregates was compared for the various deposition techniques used. Also, the use of nanoparticles sols in combination with telescope assisted Remote Raman detection showed promising results for the detection of trace amounts of threat compounds.

The enhancement of Raman scattering signals when a molecule is adsorbed on a roughened metal surface termed Surface Enhanced Raman Scattering (SERS) has been extensively studied since the late seventies when research groups such as Fleishmann's *(1)*, Van Duyne's *(2)* and Albrecht's *(3)* presented their observations on this phenomenon. This enhancement is the result of an increase in the magnitude of the induced transition moment therefore increasing the intensity of the inelastic scattering cross section and has both a physical (electromagnetic) and a chemical component. A related modality, Surface Enhanced Raman Resonance Spectroscopy (SERRS), consists on matching the incident photon energy of the incident light to the difference in energy of an analyte electronic transition (resonance Raman). Both approaches increase intensity, reduce fluorescence and lower the detection limits of analytes, thus overcoming conventional Raman Spectroscopy (RS) limitations.

Silver, gold and copper are the most commonly used metals for SERS and the variety of substrates include, among others, roughened surfaces, nanoparticle colloidal suspensions, island films and nanostructured surfaces *(4)*. Current trends in SERS and Raman spectroscopy include the development of new substrates which includes immobilized particles, functionalized nanoparticles and nanolayered films. These recent development have created new opportunities for chemical and biological sensing.

The transfer of colloidal nanoparticles onto surfaces to take advantage of the surface enhancement has been reported previously. Recent studies mention the addition of silver colloids to the surface of fluorescent materials such as lipsticks *(5)* and inks *(6)*. The fluorescence quenching allowed characterizing and differentiating similar materials. Many researchers agree that critical aspects of the deposition of SERS materials on a surface are the distribution and coverage of the sampling area *(7-8)*. Other important parameters are size, shape, and interparticle spacing of the material as well as the dielectric environment.

In this work different deposition techniques were evaluated for the addition of metal nanoparticles to a contaminated area for *in situ* surface enhanced Raman spectroscopy. Also metallic colloids were added to solutions for Raman characterization at 7 meters of standoff distance. The goal was to combine SERS and a Raman telescope for remote detection of materials at trace levels.

Direct Deposition and Sample Smearing

Direct deposition is probably the most common technique to deposit a colloidal material to a surface for SERS experiments. It consists of using a micropipette *(10)* or syringe *(11)* to deposit a drop of 10-50 μL to the area. For example, silver colloids were applied to a thin layer chromatography (TLC) silica plate after the separation for *in situ* characterization of the separated species *(12)*. Another related method of transferring analytes onto surfaces is by sample smearing *(13-15)*. To carry out these experiments, samples were deposited on thin stainless steel plates by applying a known amount of the compound on the surface. Taking into account the surface area covered by the target analyte, the amount of compounds was weighted and diluted in methanol in order to have loading concentrations of 5 $\mu g/cm^2$ and 10 $\mu g/cm^2$. Aliquots of 20 μL were deposited in one end of the plate. Using a Teflon sheet, the samples were smeared from a plate end *(13-14)*. Finally, the samples were allowed to dry for few seconds and then followed with the spectroscopic measurements.

A Renishaw Raman Microspectrometer, model RM-2000 equipped with Leica microscope, 10x magnification objective and CCD detector was used to study the signal enhancement of analytes deposited on stainless steel (ss) plates by the direct application of silver colloids. The spectra were obtained in the range of 600-1800 cm^{-1} using one acquisition and 3 s of integration time (Fig. 1). Spontaneous Raman spectra were obtained from pentaerythritol tetranitrate (PETN) high explosive (HE) and from SEMTEX, a PETN based explosive formulation. PETN was synthesized according to the method described by Urbanski *(16)*.

The normal or spontaneous Raman spectra of neat recrystallized samples of PETN and SEMTEX are shown in Fig 1(a) and 1(b), respectively. SEMTEX samples were smeared over stainless steel plates as previously described. Figure 1(b) shows that the amount of SEMTEX in the field of view of the objective is so little that only one signal (1291 cm^{-1}, NO_2 symmetric stretch) is clearly detected by spontaneous Raman spectroscopy (which has a limit of detection of several picograms, depending on the analyte). It is estimated that there were 15-30 picograms (pg) in the interrogation area, if it is assumed that the analyte was uniformly distributed on the metal surface.

Silver colloids prepared according to the Lee and Meisel method *(17)* were dissolved in methanol in a 1:10 colloidal suspension to transfer solvent ratio. Then, 10 μL of this suspension was deposited over the smeared area. When the solvent evaporated an optical microscope was used to locate the metal colloid deposits on the sample. The resulting Surface Enhanced Raman spectrum is shown in Fig. 1(c) shows obtained by focusing the laser over the particle nanoaggregates. The SERS spectrum of SEMTEX containing metallic nanoaggregates showed an enhancement of about ~ 13 fold with respect to the normal Raman spectrum of SEMTEX at the same position on the test surface. The signal enhancement was calculated by comparing the signal intensities (baseline to peak) of the 1291 cm^{-1} PETN vibrational signature (NO_2 symmetric stretch. The normal or spontaneous Raman spectrum of PETN showing the principal vibrational features corresponding to the important HE is also included for comparison purposes.

Ultrasonic Sprayer

Ultrasonic nozzles employ high frequency sound waves, and convert that into vibratory motion at the same frequency. The liquid passing through the atomizing surface absorbs some of the vibrational energy, setting up wave motion in the liquid on the surface. If the vibrational amplitude is correct atomized drops will be produced. The control of drop size was achieved by controlling the frequency. A photo of the ultrasonic nozzle used for these experiments is included in Figure 2. The flow of liquid into the atomizing chamber was controlled with an electronic syringe pump. A mechanical belt conveyor was used to spray the surface area at a continuous rate.

The ultrasonic sprayer can be used to deposit a layer of nanoparticles over a surface to detect and characterize contaminants. Contaminated silicon plates were generated by depositing 10 μL of a dilute solution of test analytes

(Rhodamine 6G or adenine) in methanol. Silver colloids prepared according to the Lee and Meisel method *(17)* and diluted with methanol tenfold. Two milliliters of the suspension were transferred to the syringe. The liquid was dispensed at a rate of 50 μL/min. The frequency of the sprayer was set to 1.6 Hz and the belt speed was set to 1ft/min.

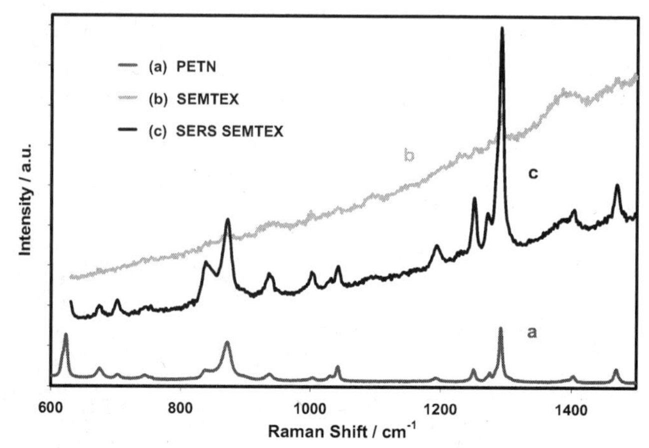

Figure 1. Raman spectra on stainless steel plate of: (a) neat PETN; (b) SEMTEX smeared on plate; (c) after covering with directly deposited Ag sol.

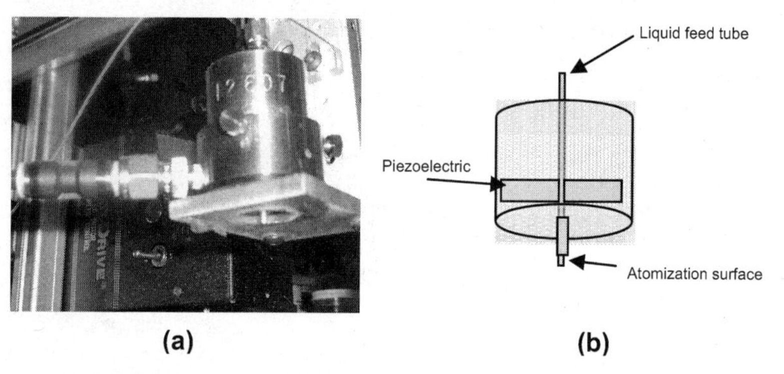

Figure 2. (a) Photograph and (b) schematic diagram of an ultrasonic sprayer based deposition system.

Figure 3 shows white light images obtained with a 50x objective of the silver islands formed over silicon slides. The substrate was exposed 10 times to the fine mist of the aerosol generated in the ultrasonic sprayer. A Jobin/Yvon Raman Microspectrometer, model LabRam-HR equipped with an Olympus BX52 microscope and 532 nm Spectra Physics laser was used to collect spectra of the contaminated area coated with the silver metal deposits. Spontaneous Raman and SERS spectra were collected before and after depositing silver nanoparticles, respectively, on the test surfaces at the same positions on the substrate. Figure 4 shows a comparison between normal Raman spectrum of Rhodamine 6G (Rh6G) and the subsequent surface enhanced Raman spectrum obtained at the same location on the metal plate. The normal Raman spectrum (Fig. 4(a), multiplied by 10x) was obtained by a single 20s integration scan. A 2 s integration time SERS trace is shown in Fig. 4(b). The spontaneous and enhanced Raman spectra of adenine are presented in Figure 5. The enhancement was estimated to be larger than 1000 fold. Raman signal enhancements for this and other analytes were calculated by comparing the intensities of persistent vibrational signatures of the test compounds. In the case of adenine the marker peak can be observed at 728 cm^{-1} (729 cm^{-1}, SERS). Enhancement of substrate bands (Si) is also observed as well as several, typically weak, bands of adenine.

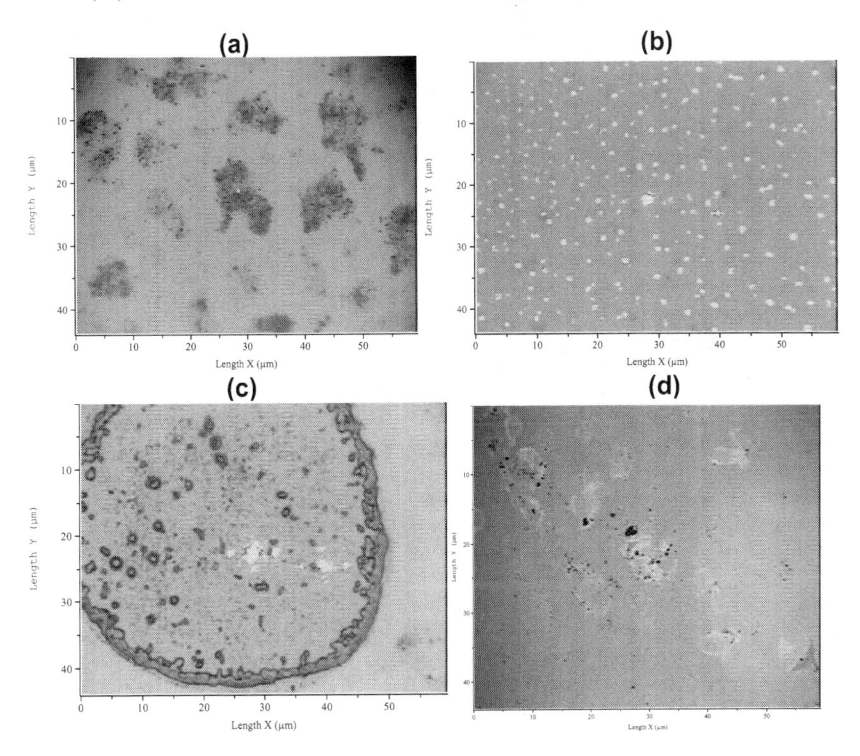

Figure 3. Optical micrographs of glass slides showing the formation of silver islands after multiple depositions of Ag sol with ultrasonic sprayer: (a) and (b); Substrates with Ag nanoparticles sprayed-on: (c) adenine; (d) Rhodamine 6G.

Pneumatically Assisted Nebulization

The electrospray nebulizer consists of an extra fine stainless steel tube or needle with a diameter of 0.92 mm that is confined into a stainless steel body. The nebulizing gas (nitrogen) is fed into the needle through a port positioned at 65° with respect to the sample inlet port. The gas travels from the head of the nebulizer tangentially to the fine needle toward the exit of the needle or nozzle. A closer view of the electrospray head containing the needle and other parts is shown in Figure 6.

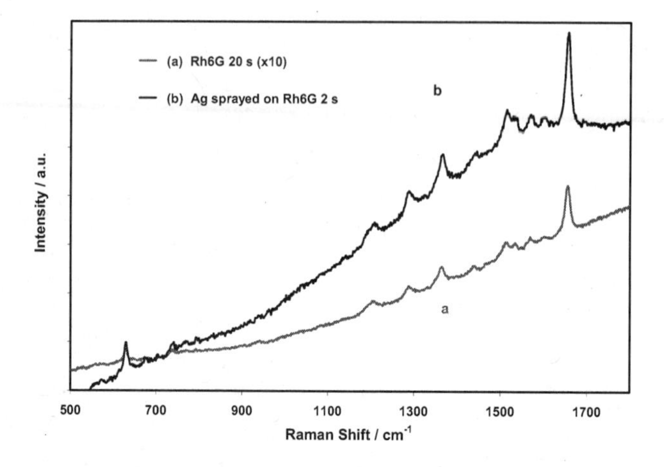

Figure 4: (a) Raman spectra of R6G (x10) deposited on glass substrate; (b) SERS spectra of R6G after spraying with Ag sol. Signal enhancements were of the order of 250x.

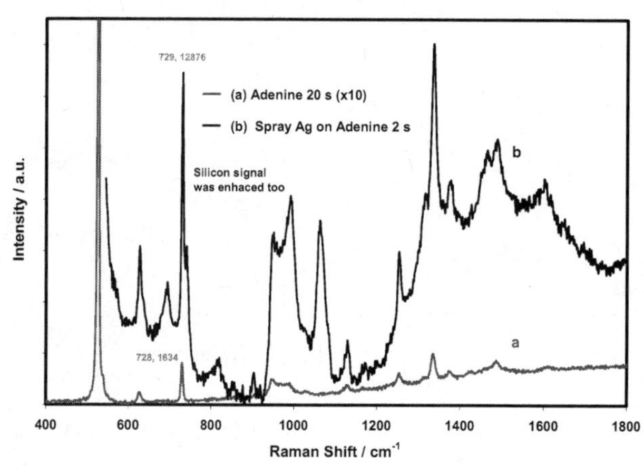

Figure 5. (a) Normal Raman spectrum of adenine on Si substrate; (b) SERS spectrum obtained at same location on metal substrate after spraying Ag sol. Raman signal enhancement larger than 1000 were observed. Si bands from substrate as well as other analyte bands are also observed in enhanced mode.

The particles deposited with this method were characterized with Scanning Electron Microscopy. SEM measurements were performed on a Philips FEG-SEM XL-30 system. The SEM micrographs are presented in Figure 7. These images show a homogeneous distribution of the nanoparticles on the target area. The sizes of the deposited particle are in the range of 40 to 90 nanometers, which is a rather narrow size distribution of particles for the Ag sols.

The use of pneumatically assisted nebulization (PAN) was explored by depositing the colloidal silver solution over the samples. Contaminated samples were generated by depositing 10 μL of a 10^{-3} M solution of adenine in methanol over a silicon plate. The colloid was sprayed for 30 s at a distance of approximately 1 cm. The Raman spectra of the samples prepared with the PAN method showed no enhancement. It is believed that more nanoparticle agglomeration is required for SERS signal enhancement. Also, isolated nanoparticles deposited on the surface may not be effective in reaching the analyte-nanoparticle proximity needed for SERS signal improvement.

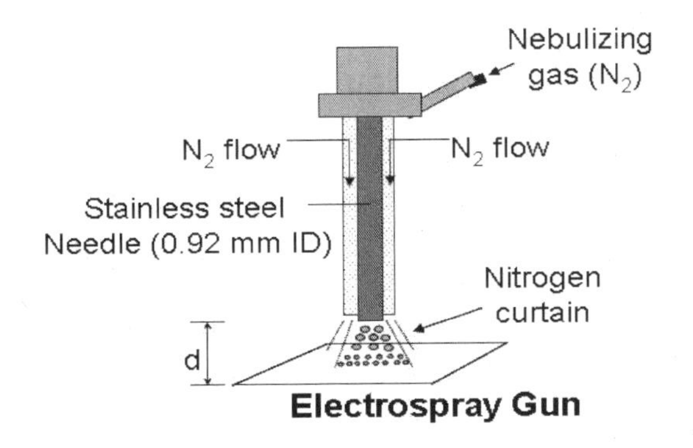

Figure 6. Diagram of PAN system used to spray-on SERS active materials.

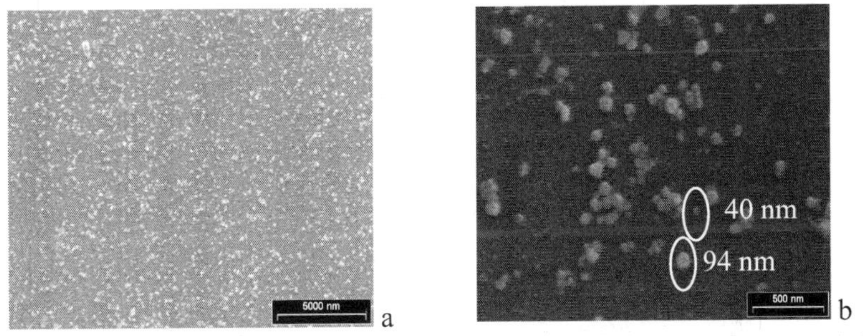

Figure 7. SEM images of distribution of silver particles sprayed on glass slides using PAN (a: low magn); Size of nanoparticles produced (b: high magn).

Nanoparticle Assisted Remote Sensing

Silver nanoparticles were also used to improve the detection of liquid solutions of analytes at a target-collector distance of 7 m. The experimental setup of the prototype instrument used to perform standoff Raman detection is schematically shown in Figure 8. The prototype system consists of an Andor Technologies Shamrock spectrograph equipped with a charge-coupled device detector (CCD), a reflecting telescope, a fiber optic bundle cable, a notch (or edge) filter assembly and a laser source (532 and 488 nm) for active standoff Raman detection.

The telescope used, a MEADE ETX-125 Maksutov-Cassegrain, 125 mm clear aperture, 1900 mm focal length, $f/15$, was coupled to the Raman spectrometer with a non-imaging optical fiber bundle (600 μm diameter, model AL 1217, Ocean Optics, Inc.) containing the appropriate filters for the rejection of the

Rayleigh scattered radiation. Two lenses were used collimate the light from the telescope output, which was directed into the fiber optic from which the focusing objective had been removed. The output of the fiber was directly coupled to the entrance slit of the Andor spectrograph.

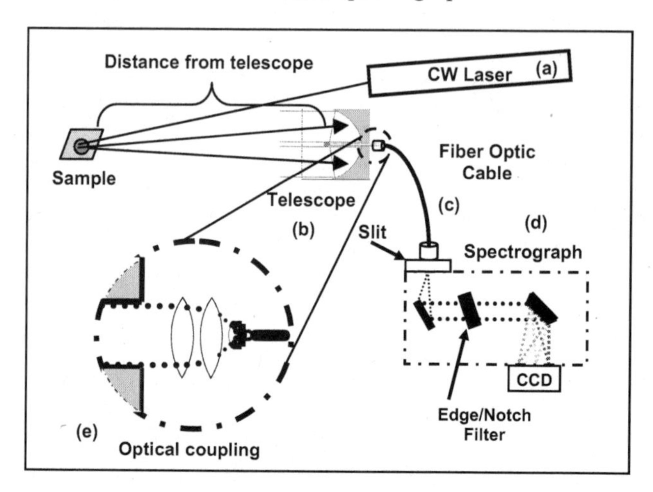

Figure 8. Schematic diagram of standoff Raman spectrometer. Essential parts are: (a) excitation source: laser; (b) collector: telescope; (c) fiber optics coupling; (d) fast spectrograph equipped with edge or notch filters. Details of the optical coupling of the telescope to the fiber optics bundle are shown in the inset (e).

A dilute solution of the purine adenine in slightly acidic ultra pure water with a nominal concentration of $1.0x10^{-4}$ M was prepared and transferred to a clear glass sample vial. The samples were placed 7 m away from the excitation laser source. A 532 nm, Coherent VERDI, solid state diode was used to excite the Raman scattered radiation at the maximum current standoff distance (based on the limitations of the lab room). Figure 9 shows the spectra acquired when 0.1 mL of colloidal silver suspension was added to 10 mL of adenine solution.

A typical Micro Raman SERS spectrum of adenine-Ag sols is also shown in Figure 9 (a) for comparison purposes. The intensities of the Micro Raman spectrum have been divided by 20 to fit both traces on the same scale. Integration times for Micro Raman and remote Raman spectra were 2 s. It is worth mentioning that it was not possible to acquire a remote Raman spectrum of adenine solutions without the use of the silver sol nanoparticles.

The diameter for the uncollimated laser beam was approximately 0.5 cm at the sample or 0.2 cm^3 interrogation volume in the glass vial used. A laser power at sample of 500 mW (max) was used to excite the Remote Raman spectra in the sample. This resulted in a power density value of approximately 2.5 W/cm^3. In contrast, typical Micro Raman SERS experiments used power densities in the order of 20,000 W/cm^3 to excite the Raman Shift spectra of samples contained in capillary tubes with an interrogation volume of $5x10^{-6}$ cm^3. This represents 8,000-fold increase in energy density for the microscope experiments than the telescope based Raman experiments.

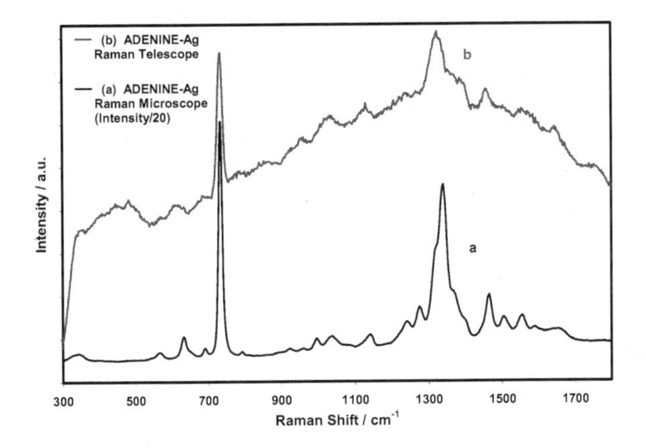

Figure 9. SERS spectra of adenine collected with: (a) Raman Microscope; (b) Raman Telescope collected at 20 ft standoff distance.

Conclusions

Silver nanoparticles from colloidal suspensions prepared via the citrate mediated reduction of Ag^+ salts (modified Lee and Meisel procedure) were successfully used in spray-on SERS experiments. The sols were directly applied on test surfaces containing analytes of interest to evaluate the nanoparticles for SERS enhancement with the spraying-on and direct deposition methods. In the initial experiments, adenine and Rhodamine 6G, which are very strong SERS scatterers, were used as test compounds. However, the results of direct deposition and remote sensing approaches suggest that spraying-on metallic nanoparticles can be used for a wide variety of *in situ* SERS sensing scenarios and analytes. The results presented suggest that particle size and aggregation may have a critical influence on SERS enhancement of sprayed-on nanoparticles, more even than the particle spatial distribution and surface arrangement. Pneumatically Assisted Nebulization (PAN) deposition and ultrasonic sprayer deposition were successfully used for transferring both colloids and target compounds, although no significant Raman signal enhancements were obtained from PAN delivery of silver sols. The use of silver nanoparticles from colloidal suspensions coupled to telescope collector based Remote Raman detection showed promising results for the detection of trace amounts of test compounds at a distance of 7 m.

This series of experiments represent good examples of this "Proof of Concept" investigation which could lead to an infinite breath of possibilities and opportunities to expand the portfolio of applications of Surface Enhanced Raman Scattering as applied in areas of interest to National Defense and Security.

Acknowledgments

This work was supported by the U.S. Department of Defense, University Research Initiative Multidisciplinary University Research Initiative (URI)-

MURI Program, under grant number **DAAD19-02-1-0257**. The authors also acknowledge contributions from Aaron LaPointe of Night Vision and Electronic Sensors Directorate, Department of Defense and from the Department of Chemistry of the University of Puerto Rico–Mayagüez.

Part of the work on the last stage of the project was supported by the U.S. Department of Homeland Security under Award Number **2008-ST-061-ED0001**. The views and conclusions contained in this document are those of the authors and should not be interpreted as necessarily representing the official policies, either expressed or implied, of the U.S. Department of Homeland Security.

References

1. Fleischmann, H.; Hendra, P. J.; MacQuillan, A. J. *Chem. Phys. Lett.* **1974**, *26*, 163.
2. Jeanmaire, D. L.; Van Duyne, R. P. J. Electroanal. Chem. 1977, 84, 1–20.
3. Albrecht M. G.; Creighton, J. A.; *J. Am. Chem. Soc.* **1977**, *99*, 5215.
4. Haynes, C. L.; McFarland, A. D.; Van Duyne, R. P. J. *Analytical Chemistry*. **2005**, *338*, A-346.
5. Rodger, C.; Rutherford, V.; Broughton, D.; White, P. C.; Smith, W. E. *Analyst*. **1998**, *123*, 1823.
6. White, P. C.; Munro, C. H.; Smith, W. E. *Analyst*, **1996**, *121*, 835.
7. Lewis, I. R.; Edwards, H. G. M. ed., Handbook of Raman Spectroscopy, Marcel Dekker: New York, NY, 2001.
8. Fang, J.; Zhong, C.; Mu, R. Chem. Phys. Lett. **2005**, *401*, 271.
9. Muniz-Miranda, M.; Sbrana, G. *J. Molec. Struct.* **2001**, *565*, 159.
10. Kocsis, L.; Horva´th, E.; Kristo´f, J.; Frost, R. L.; Re´dey, A´.; Mink, J. *J. Chrom.* A. **1999**, *845*, 197.
11. Kho K. W., Z.X., Zeng, H.C., Soo, K.C., Olivo, M; Anal. Chem. **2005**, *77*, 7462.
12. Caudin, J. P.; Beljebbar, A.; Sockalingum, G. D.; Angiboust, J. F.; Manfait, M. *Spectrochimica Acta* Part A. **1995**, *51*, 1977.
13. Hernández-Rivera, S. P.; Pacheco-Londoño, L. C.; Primera-Pedrozo, O. M.; Ruiz, O. Soto-Feliciano Y.; Ortiz, W. *International Journal of High Speed Electronics and Systems*. **2007**, *17*(4): 827.
14. Primera-Pedrozo, O. M.; Pacheco-Londoño, L. C.; De la Torre-Quintana, L. F.; Hernandez-Rivera, S. P.; Chamberlain, R. T.; Lareau, R. T. In *Proc. SPIE*. **2004**, *5403*, 237.
15. Primera-Pedrozo, O. M.; Pacheco-Londoño, L. C.; Ruiz, O.; Ramirez, M.; Soto-Feliciano, Y. M.; De La Torre-Quintana, L. F.; Hernandez-Rivera, S. P. *Proc. SPIE*. **2005**, *5778,* 543.
16. Urbanski, T. *Chemistry and Technology of Explosives;* Macmillan Co.: New York, NY, 1964.
17. Lee P. C. ; Meisel, D. ; *J. Phys. Chem.* **1982**, *86*, 3391.
18. Ayora, M. J.; Bayesteros, L.; Perez, R.; Ruperez, A.; Lacerna, J. J. Anal. Chim. Acta. **1997**, *355*, 15.

Chapter 11

Virus Nanoparticles for Signal Enhancement in Microarray Biosensors

Amy Szuchmacher Blum, Carissa M. Soto, Gary J. Vora, Kim E. Sapsford, and Banahalli R. Ratna

Naval Research Laboratory, Center for Bio/Molecular Science and Engineering, 4555 Overlook Avenue SW, Washington, DC 20375

Although fluorescent dye loading is often accompanied by fluorescence quenching, biosensor sensitivity depends on localizing large numbers of dye molecules in a specific area. Our results demonstrate that organized spatial distribution of fluorescent reporter molecules on a virus capsid eliminates this commonly encountered problem. We have used this enhanced fluorescence for the detection of DNA-DNA hybridization and the detection of protein toxins. When compared with the most often used detection methods in a microarray-based genotyping assay for *Vibrio cholerae* O139, and a sandwich assay for staphylococcal enterotoxin B, these viral nanoparticles markedly increased assay sensitivity thus demonstrating their applicability for existing microarray protocols.

Introduction

Bionanotechnology, which encompasses concepts from biology, chemistry, physics, and engineering, is a newly emerging and exciting area of research.*(1-4)* Its foundation is the controlled molecular self-assembly of simple building blocks into complex structures. In the biosensing arena, bionanotechnology has the potential to greatly improve sensitivity using particles that are inherently compatible with biological fluids and receptors. An example of such a biological building block is Cowpea Mosaic Virus (CPMV), which has been well characterized in the literature.*(5-7)* Previous work has demonstrated that CPMV can be used as a building block for the development of nanoassemblies.*(8-13)* Here we discuss progress in using CPMV labeled with fluorescent dyes and coupled with analyte-specific binding molecules such as antibodies and DNA for multiplexed biosensor applications.

In many biosensor configurations, fluorescent organic dyes are an extremely useful reporting agent.*(14-16)* In all such dye-based techniques, the sensitivity can be limited by quenching of the fluorescent signal when large numbers of dye molecules are in close proximity on the antibody or other binding agent.*(16)* For example, antibodies labeled with more than six Cy5 dye molecules are virtually non-fluorescent due to the formation of non-fluorescent dye dimers that are also efficient quenchers.*(14)* If the formation of these non-fluorescent quenching dimers is in fact the primary source of quenching for Cy5 molecules attached to IgGs, then it should be possible to load many more dye molecules onto a protein if the dye-dye distances can be controlled to suppress the formation of quenching dimers. Using an icosahedral viral capsid such as CPMV as a scaffold should enable us to increase the number of dye molecules in a localized volume by controlled positioning of the dye molecules. When used as a tracer in microarray-based sensors, these dye-decorated CPMV capsids may result in improved sensitivity relative to plain dye labeled antibodies.

Cowpea Mosaic Virus

CPMV is a member of the comovirus group of plant viruses, infecting legumes with yields reaching 1-2g/kg of leaves.*(17)* The CPMV capsid is an icosahedron with a spherical average diameter of 28.4 nm, formed by 60 identical copies of an asymmetric subunit. The 3-D structure of wild-type CPMV has been determined by X-ray crystallography to a resolution of 2.8 Å. *(6)* The virus particles are stable to a wide variety of conditions, including temperatures up to 60 °C, the presence of up to 30% organic solvent such as DMSO, and pH values from 3.5 to 9 indefinitely at room temperature.*(7)* These properties make them suitable for further reaction under many different conditions. In its natural state, CPMV displays no solvent accessible cysteine residues on the exterior capsid surface. Thus, there are no naturally occurring thiol groups on the virus surface, and thiol reactive species, such as gold nanoparticles, do not bind to the wild-type virus. By engineering in cysteines at selected points on the capsid exterior, we can control the locations of thiol-binding moieties such as dyes*(7)* and 5 nm gold colloidal particles.*(18, 19)*

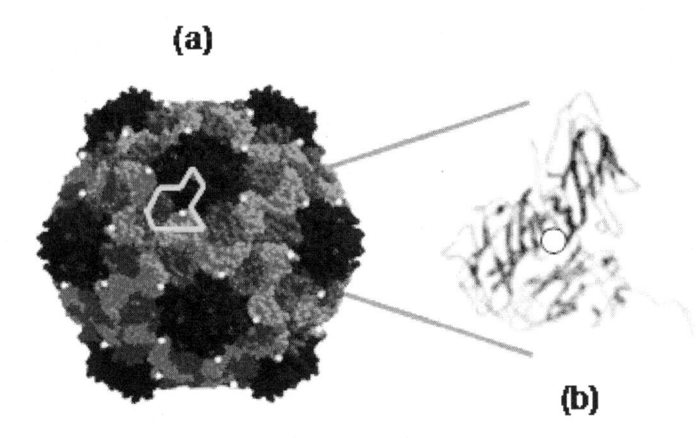

Figure 1. Cysteine-mutant of CPMV. PDB file available for wild type virus (1NY7.pdb). (a) A 30 nm diameter icosahedral virus particle, EF-CPMV, made of 60 identical protein subunits containing a total of 60 cysteines (thiol-containing group shown as white circles). (b) EF-CPMV protein subunit to which a single cysteine was incorporated via the addition of a five residue loop (GGCGG) placed between positions 98 and 99(20). (Adapted with permission from reference (21). Copyright 2006, American Chemical Society.)

We used the EF-CPMV mutant,*(20)* which contains a single cysteine introduced as a GGCGG loop placed between positions 98 and 99 of the large protein resulting in a total of 60 cysteines, represented as white circles on the capsid (Figure 1). The EF-CPMV mutant is an excellent candidate as a scaffold for this work since the distance between thiol groups is 5.3 nm for neighbors around the same 5-fold axis (icosahedral 5-fold symmetry axis) and 6.5 nm for the next nearest neighbor on adjacent 5-fold axes.*(18)* These distances are large enough to prevent non-fluorescent dimer formation and quenching.*(20)*

Enhancing Fluorescence Intensity

Using standard maleimide coupling chemistry, we used the EF-CPMV as a scaffold for both Cy5 and NeutrAvidin (NA) to produce NA-Cy5-CPMV bionanoparticles.*(21, 22)* These NA-Cy5-CPMV bionanoparticles can act as both a fluorescent signal generating element (via adducted Cy5 molecules) and as a recognition element (via adducted NeutrAvidin proteins). The fluorescence yield of NA-Cy5-CPMV (EF-dye, 42 dye/virus; Figure 2a) was compared with two experimental controls: wild type (WT)-dye mix (dye not bound to the WT viral protein scaffold) and free Cy5 in solution. The concentrations of both Cy5 and virus, as determined via Absorbance spectroscopy, were held constant for all samples. Figure 2a shows the fluorescence intensities (relative quantum yield) under excitation at 605 nm for all three samples, normalized to the absorbance at 605 nm. 605 nm is used for excitation instead of the absorbance maximum at 650 nm in these experiments so that scattering at the excitation wavelength does not interfere with the emission at ~660 nm. Fluorescence intensities are on the same order of magnitude for all three samples, however, the covalently coupled NA-Cy5-CPMV sample shows two significant differences from the other two control samples. A red shift (2 nm) of the fluorescence maximum of EF-dye sample in comparison to the free dye is similar to that observed in the absorption spectrum, and is indicative of dye bound to protein.*(23)* Furthermore, while the normalized fluorescence intensity for the free dye and the WT-dye mix is identical within the limits of the experiment, there is a significant increase in the normalized intensity for the NA-Cy5-CPMV. This suggests that covalent coupling the Cy5 dye to the viral scaffold not only eliminates quenching, but also enhances the fluorescence quantum yield of Cy5.

In order to examine the effect of the number of dye molecules per virus on the absolute fluorescence intensity more closely, a series of NA-Cy5-CPMV samples containing different average ratios of dye molecules per virus were prepared. The fluorescence signal was plotted vs. dye to virus ratio in Figure 2b. The data shows fluorescence intensity increasing linearly over a wide range of dye per virus ratios, suggesting negligible quenching in the virus conjugate even at dye loads of 42 dye/virus.

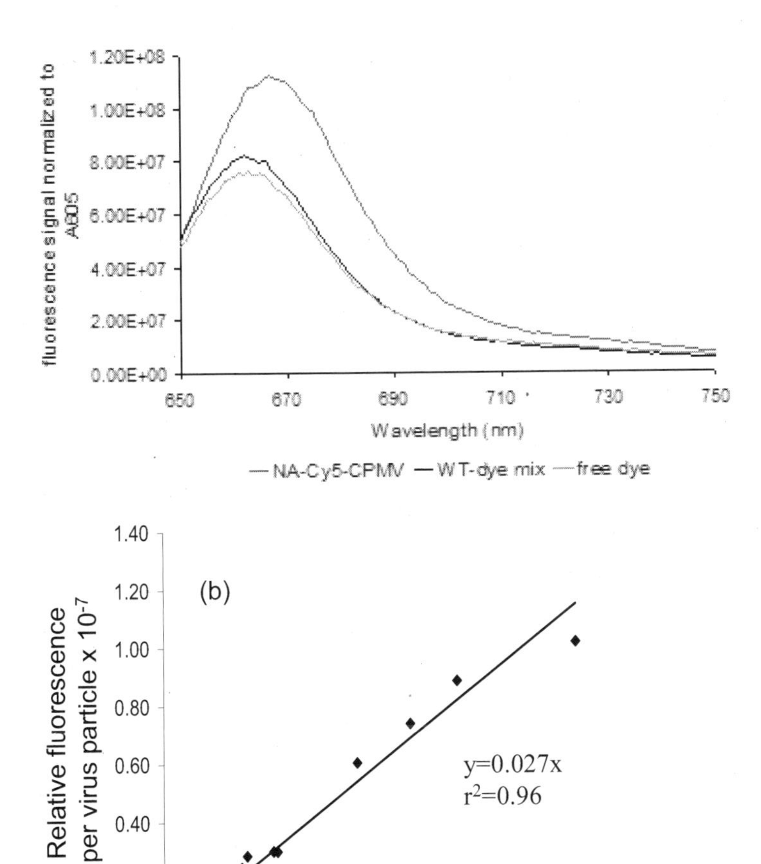

Figure 2. (a) Fluorescence spectra of NA-Cy5-CPMV conjugates (dark gray), WT-dye mix (black) and free dye (light gray) excited at 605 nm, normalized to Cy5 absorbance at 605 nm. Concentration of virus is comparable for NA-Cy5-CPMV and WT-dye mix. (b) Fluorescence intensity vs. number of dye per virus in NA-Cy5-CPMV. Relative fluorescence per virus particle is defined as the ratio: fluorescence at 666 nm/Absorbance at 260 nm. The solid line corresponds to the linear regression from the data (black diamonds). (Adapted with permission from reference (21). Copyright 2006, American Chemical Society.)

DNA Microarrays

The use of DNA microarray technology enables the rapid and simultaneous interrogation of thousands of genetic elements. In recent years, it has quickly become a preferred tool for applications in DNA and RNA sequence analysis, gene expression profiling, genotyping of single-nucleotide polymorphisms and the molecular detection of pathogenic organisms.*(24, 25)* Despite the tremendous utility of DNA microarray technology, one of the major recurrent problems has been assay sensitivity and the need for target or signal amplification.*(26)* In the context of pathogen detection, assay sensitivity is one of the most important issues to consider in the development of microarray-based tools and protocols. As a result, most attempts to improve DNA microarray detection sensitivity utilize nucleic acid amplification strategies to amplify the amount of specific target present prior to microarray hybridization (target amplification).*(27, 28)* As a more recent complementary strategy, attempts to increase the signal generated by each label or hybridized DNA molecule (signal amplification) also have been developed.*(29, 30)*

Enhancing sensitivity in a DNA sensor platform.

Having demonstrated successful coupling of Cy5 dye to CPMV capsid particles and confirmed the absence of fluorophore quenching on such bionanoparticles, we demonstrated the utility of the NA-Cy5-CPMV bionanoparticles in a DNA microarray detection platform. DNA microarrays consist of an array of spatially immobilized cDNA or DNA oligonucleotide elements (probes). The signal intensity, and thus the detection sensitivity, for each element of the array is determined by the number of reporters that can be localized at the hybridization reaction site. Hence, a practical signal amplification strategy would be to increase the amount of fluorophores per reaction site post-hybridization to improve detection sensitivity. As discussed previously, the limitation of such a strategy is the possible self-quenching of fluorophores when they are not adequately separated from each other. Thus, we tested CPMV-dye complexes as a highly fluorescent recognition element for current DNA microarray applications to provide a proof-of-principle demonstration of sensitivity enhancement in a sensor platform. Our NA-Cy5-CPMV bionanoparticles represent an inexpensive and efficient way to boost sensitivity in pathogen detection without adding additional target amplification steps.

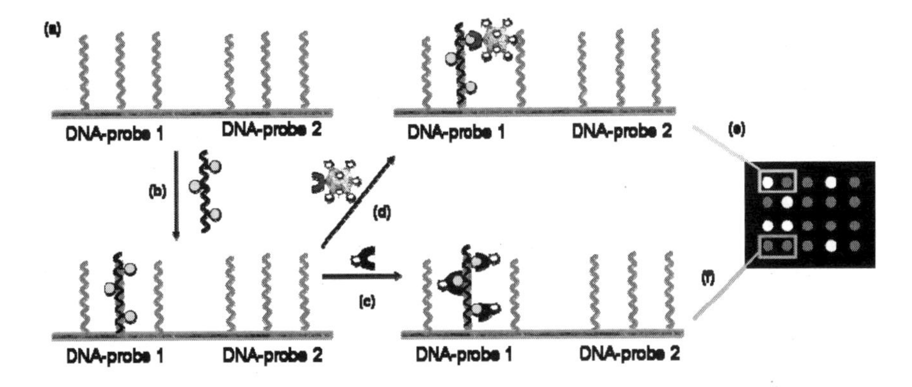

Figure 3. DNA microarray detection scheme. (a) DNA oligonucleotides 1 and 2 (probes) are immobilized in a microarray format on glass slides. (b) DNA probe-1 hybridizes with a previously amplified and biotinylated target DNA molecule. This hybridization event is detected using (c) streptavidin-Cy5 or (d) NA-Cy5-CPMV. (e) Quantification post-detection indicates a true positive signal for the NA-Cy5-CPMV detection method (white spot, hybridization with DNA-probe 1) or a true negative signal (gray spot, non-hybridization with DNA-probe 2). (f) A false negative signal (gray spot) is observed for streptavidin-Cy5 as the total number of fluorophores at the DNA-probe 1 spot results in the generation of a signal that is below the detection threshold (or background). (Adapted with permission from reference (21). Copyright 2006, American Chemical Society.)

Figure 3 shows the pathogen detection procedure schematically, demonstrating how NA-Cy5-CPMV can enhance detection sensitivity without increasing the number of false positives by increasing the number of dyes per biotin binding event. Amplification is provided by increasing the local concentration of reporter dye molecules vs. standard techniques. The NA-Cy5-CPMV bionanoparticles act as both fluorescent reporters (via the attached Cy5 molecules) and as recognition elements for target DNA molecules containing biotin (via the attached NeutrAvidin proteins).

Detection of genes from *Vibrio cholerae*

Vibrio cholerae O139 genomic DNA was used as the template for all amplification reactions using protocols, gene targets, probes and primer sequences that have been described elsewhere.*(21)* To create the microarray, oligonucleotide probes were designed and synthesized with a 5' amino modifier and 12 carbon spacer (Qiagen), and spotted onto 3-aminopropyltriethoxysilane (silanization) + 1,4-phenylene diisothiocyanate (crosslinker)-modified glass slides for covalent probe immobilization as previously described.*(31)* The fluorescent signal from each microarray element was considered positive only when its quantified intensity was >2X that of known internal negative control elements. Three independent amplification and hybridization experiments were performed from each of the DNA template dilutions (10^5 – 10^0) for the hybridization detection methods interrogated.

In order to demonstrate the usefulness of NA-Cy5-CPMV bionanoparticles in a microarray platform three detection methods were compared: i) direct enzymatic incorporation of Cy5-dCTP during DNA amplification (Figure 4a), ii) direct enzymatic incorporation of biotin-14-dCTP during DNA amplification followed by detection with Cy5-labeled streptavidin (Cy5-streptavidin, Figure 4b), and iii) direct enzymatic incorporation of biotin-14-dCTP during DNA amplification followed by detection with NA-Cy5-CPMV (Figure 4c). As anticipated, when the amount of template DNA was not limiting (i.e. 10^5 genomic copies, Figure 4d) each of the three methods tested reliably detected 100% of the targeted genes. As the amount of template DNA was decreased via serial dilution, and sensitivity became more of an issue, the viral nanoparticles outperformed the other two detection methods with respect to the percentage of targeted genes detected (Figure 4d). This is clearly highlighted at the lowest template concentrations of 10^1 and 10^0 (Figure 4a-d). Thus, in direct comparison with the two most often used methods of microarray hybridization detection, the NA-Cy5-CPMV bionanoparticles provided the greatest overall detection sensitivity. It is important to note that while a case could be made for polymerase-mediated label incorporation bias (Cy5-dCTP versus biotin-14-dCTP) when comparing the detection sensitivities of the Cy5-dCTP direct incorporation method with the NA-Cy5-CPMV, that case cannot be made when comparing Cy5-streptavidin and NA-Cy5-CPMV as the biotin-14-dCTP labeling method was used for both. Therefore, any increase in detection sensitivity can be attributed solely to the larger numbers of reporter fluorophores and the absence of fluorophore quenching. The results suggest that these bionanoparticles have immediate utility for the detection of trace amounts of pathogen nucleic acids as the enhancement in detection sensitivity improves the lower limits of detection (genomic copy number or organisms)

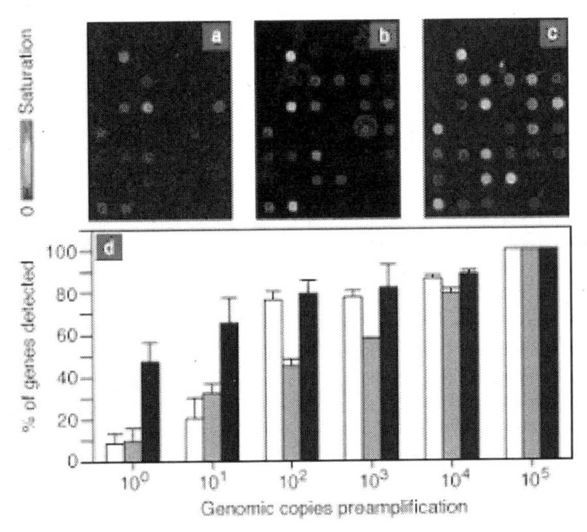

Figure 4. Comparison of NA-Cy5-CPMV with commonly utilized microarray hybridization detection methods. Fluorescent microarray hybridization profiles of Vibrio cholerae 0139 amplified DNA (101 genomic copies starting material) detected via (a) direct incorporation of Cy5-dCTP; (b) Cy5-streptavidin; (c) NA-Cy5-CPMV, 42 dye/virus. (d) Graphical representation of the percentage of targeted genes detected as a function of preamplification genomic DNA copy number. Cy5-dCTP (white), Cy5-streptvidin (gray) and NA-Cy5-CPMV (black). The data shown represent means + standard deviation (SD) of three independent amplification and hybridization experiments. (Adapted with permission from reference (21). Copyright 2006, American Chemical Society.)

and reduces the risk of false negative determinations at low pathogen concentrations.

Antibody Microarrays

Antibody microarrays are a useful tool for rapid detection of multiple biohazard targets. A typical antibody microarray consists of a modified microscope slide on which an array of antibody capture molecules is immobilized. Multiple samples can be passed over different portions of the slide for simultaneous monitoring. If a target biohazard is present, the capture molecule binds it in proportion to the concentration in the sample. After capture, a solution of fluorescently labeled tracer molecules (usually antibodies labeled with Cy5 or AlexaFluor dye) is passed over the array and binds to the captured targets.*(32-35)* As in DNA microarrays, the sensitivity of antibody microarrays is limited by the number of fluorescent molecules that can be bound to the captured biohazard. Dye labeled CPMV bionanoparticles, bound to antibodies, can be effective tracers, increasing the sensitivity of antibody based microarrays. The virus nanoscaffold allows labeling of the capsid surface with a large number of dyes, which, when used as a tracer may result in improved sensitivity relative to plain dye labeled antibodies.

Initial experiments were performed to demonstrate the functionalization of CPMV with both functional antibodies (to bind to the captured target) and a signal transduction species, in this case a dye, to demonstrate its potential application as a tracer for immunoassays. Two mutants of CPMV were used in this study (provided by The Scripps Research Institute), EF-CMPV which has a single cysteine group inserted (at residue 98) per subunit generating a total of 60 thiols on the surface of the capsid*(7)* and DM-CPMV in which two cysteines are inserted (at residues 228 and 2102) per subunit giving a total of 120 surface thiols.*(18)* In both mutants described there are four lysines per subunit, each with varying reactivity,*(36)* generating a total of 240 per capsid. Initial studies to optimize the antibody-CPMV coupling chemistry were carried out with IgG proteins which are inexpensive, well characterized, and structurally equivalent to antibodies. Following immobilization of functional antibodies on the CPMV surface optimization of the dye-to-virus ratio, to improve immunoassay sensitivity, was investigated. Samples were prepared as previously described.*(22)*

Initial experiments to optimize protein-virus labeling involved functionalizing EF-CPMV with dye and IgG protein. A total of three modifications of the Alexa-EF-CPMV were prepared: one modified with chicken IgG, one with mouse IgG and the final with both chicken and mouse IgG. This scheme takes advantage of the native lysines and inserted cysteines present on the CPMV to controllably functionize its surface. The final EF-CPMV, modified with AlexaFluor 647 dye and both chicken and mouse IgG, was characterized by UV-visible spectroscopy giving the final average ratio of 13.7 dyes per virus. Functionalization of the virus with dye and IgG was confirmed with a direct immunoassay using a rabbit anti-chicken IgG/goat anti-mouse IgG modified waveguide. The Alexa-EF-CPMV-chicken IgG was only captured by the rabbit-anti-chicken IgG and likewise the Alexa-EF-CPMV-mouse IgG by goat anti-mouse IgG, as demonstrated by the increased signal intensity observed in these regions from the dual dye and IgG labeled virus. The Alexa-EF-CPMV modified with both the chicken and mouse IgG, as expected, was captured by both the rabbit anti-chicken IgG and the goat anti-mouse IgG with little effect on the final intensity reached relative to Alexa-EF-CPMV modified with a single protein species. Thus, the dye-labeled virus, functionalized with up to two different species of IgG can be used as a tracer.*(22)*

Sandwich Assays

After demonstrating the effectiveness of modified CPMV bionanoparticles as tracers in a direct assay, the next step was to immobilize functional antibodies to the surface of a dye-labeled virus and use the resulting complex as a tracer in sandwich immunoassays. For these studies Alexa-EF-CPMV was functionalized with antibodies to either SEB, botulinim toxin or *Campylobacter jejuni*.*(22)* Experiments carried out using antibody pre-labeled with Cy3 fluorescent dye suggest that typically 3-5 antibodies are attached to the virus surface using this conjugation procedure. As expected, the Alexa-EF-CPMV-antibody complex is only captured in the regions of the slide functionalized first with the appropriate capture antibody and second exposed to the correct analyte. Array slides were analyzed using a custom software application written in LabWindows/CVI; the program creates a mask consisting of data squares (where the capture antibody is patterned) and background rectangles located on either side of a data square. The net intensities obtained by subtracting the average background signal from the average data signal for each Alexa-EF-CPMV-antibody complex are shown in Table 1. The Alexa-EF-CPMV-anti-SEB was found to bind non-specifically to the biotinylated rabbit anti-*Campylobacter jejuni*, but not to the biotinylated rabbit anti-Botulinim toxin columns. This non-specific binding occurs both in the presence and absence of SEB, suggesting it is the Alexa-EF-CPMV-anti-SEB bionanoparticle that is non-specifically binding and not the SEB target. Non-specific binding of rabbit anti-SEB has previously been observed for Alexa-labeled rabbit anti-SEB, suggesting it is an antibody issue and not an effect of the CPMV scaffold. These studies clearly demonstrate that once immobilized to the surface of the virus, the antibodies remain functional and bind to their specific analyte. In direct mole-to-mole comparisons between standard Alexa-labeled antibodies and the Alexa-EF-CPMV-antibody complexes used as tracers in immunoassays, however, we found that the plain antibodies gave a lower limit of detection. This is likely a combined effect of the fairly low improvement in the number of dyes per virus which was 13.7, relative to 4.5 dyes per plain antibody, and the overall larger size of the antibody-virus-complex compared to the antibody alone.

Table 1. Sandwich assay results

Sample	Intensity
Bot. toxin Blank	1118
Bot. toxin 1μg/ml	5430
SEB Blank	-15
SEB 50 ng/ml	11200
Camp Jejuni Blank	-102
Camp Jejuni $1x10^5$ cfu/ml	2616

Standard deviation 10%. Tracer is Alexa-EF-CPMV-antibody complex. Data reproduced from reference *(22)*.

In order to improve the limit of detection obtained by the Alexa-EF-CPMV-antibody complex, it is necessary to further increase the overall dye-to-virus ratio. In order to achieve this we developed an alternative scheme, which used a double CPMV mutant (DM-CPMV: 228/2102), containing 120 surface thiols, to increase the number of dyes-per-virus. In this scheme the DM-CPMV is first modified with the GMBS functionalized antibody, in this case sheep-anti-SEB, and then the Alexa-dye. However, unlike in the initial scheme, both the antibody and the Alexa dye target the surface thiol groups. The final ratio of

AlexaFluor 647 per virus, characterized by UV-visible spectroscopy, was calculated to be 60:1, a factor of 4.4 improvement over the initial scheme.

Table 2. Signal intensity as a function of SEB concentration

SEB (μg/mL)	Antibody Intensity	Virus Intensity
100	12210	20113
10	2062	6695
1	213	635
0.1	54	38

Standard deviation 10%. Tracer is Alexa-DM-CPMV-anti-SEB. Data reproduced from reference *(22)*.

Table 3. Signal intensity as a function of capture antibody concentration

Anti SEB (μg/mL)	Antibody intensity	Virus intensity
10	12753	20219
1	12210	20113
0.1	3262	6911
0.01	127	-232
0	141	-234

Standard deviation 10%. Tracer is Alexa-DM-CPMV-anti-SEB. Data reproduced from reference *(22)*.

Sandwich immunoassays were run to confirm successful modification of the DM-CPMV with both dye and anti-SEB antibodies and also to investigate potential improvements in the limit of detection. Tables 2 and 3 compare mole equivalent amounts of the Alexa-DM-CPMV-anti-SEB complex to the Alexa-anti-SEB when used as tracers in sandwich immunoassays for SEB, respectively. As in the previous experiment, the Alexa-DM-CPMV-anti-SEB complex only binds to the surface of the waveguide in the regions functionalized with both the capture anti-SEB antibody and the SEB analyte. The signal intensity is found to decrease as the concentration of the SEB in solution decreases. The same holds true when Alexa-anti-SEB is used as a tracer. The images were analyzed and the signal intensity calculated as a function of the SEB concentration (Table 2) or the concentration of the biotinylated rabbit anti-SEB exposed to the surface (Table 3). Clearly in both cases the Alexa-DM-CPMV-anti-SEB complex produces a stronger signal than the mole equivalent of Alexa-anti-SEB when used as a tracer, demonstrating the advantage of CPMV as a nanoscaffold to couple active biomolecules and a larger number of reporter dye molecules on the same capsid.

Conclusions

This chapter reviews current work in improving biosensor sensitivity using viral nanoparticles as scaffolds to produce bright fluorescent bionanoparticle tracers. By choosing appropriate reactive groups to covalently couple the fluorescent dye molecules, it is possible to control dye-dye distances in order to prevent the formation of non-fluorescent quenching dimers, a major limitation when overloading other carrier proteins, such as IgG. In addition to the lack of quenching in NA-Cy5-CPMV, we also observed an enhancement of fluorescence output in comparison to similar molar amounts of dye in solution. Further research is underway to understand the fluorescence of the dye-virus combined system. Beyond allowing for the controlled distribution of dye molecules, the virus also offers the additional advantage that a variety of reactive groups are available to couple other proteins which may serve as a recognition element for the detection of pathogenic organisms and toxins, making it versatile for several applications. This strategy may be transferable to other viral particles or protein scaffolds that may offer similar control of reactive group positioning. Finally, although we have specifically addressed the use of the CPMV bionanoparticle tracers for microarray-based pathogen detection, toxin detection and genotyping assays, this tool may also have utility in any number of other microarray formats. The engineered CPMV mutants can be readily manipulated permitting the incorporation of a variety of molecular recognition elements and reporters, including dyes and nanoparticles. These bionanoparticles may be applied towards other biotechnology applications in which high throughput reporter molecules are needed for multiplexing or sensitivity enhancement.

References

1. Storhoff, J.J., R.C. Mucic, and C.A. Mirkin, *J. Cluster Sci.*, **1997**, 8, 179-216.
2. Choi, J.W., Y.S. Nam, and M. Fujihira, *Biotechnol. Bioprocess Eng.*, **2004**, 9, 76-85.
3. Oakley, B.A. and D.M. Hanna, *IEEE Trans. Nanobiosci.*, **2004**, 3, 74-84.
4. Clark, J., et al., *BioTechniques*, **2004**. 36, 992-1001.
5. Lomonossoff, G.P. and J.E. Johnson, *Curr. Opin. Struct. Biol.*, **1996**, 6, 176-182.
6. Lin, T., et al.,. *Virology*, **1999**, 265, 20-35.
7. Wang, Q., et al.,. *Angew. Chem. Int. Ed.*, **2002**, 41, 459-462.
8. Blum, A.S., et al., *Small*, **2005**, 1, 702-706.
9. Blum, A.S., et al., *Nanotechnology*, **2006**, 17, 5073.
10. Medintz, I.L., et al., *Langmuir*, **2005**, 21, 5501-5510.
11. Strable, E., J.E. Johnson, and M.G. Finn, *Nano Lett.*, **2004**, 4, 1385-1389.
12. Steinmetz, N.F., G.P. Lomonossoff, and D.J. Evans, *Langmuir*, **2006,** 22, 3488-3490.
13. Steinmetz, N.F., et al., *Langmuir*, **2006**, 22, 10032-10037.

14. Gruber, H.J., et al., *Bioconj. Chem.*, **2000**, 11, 696-704.
15. Hoen, P.A.C.t., et al., *Nuc. Acids Res.*, **2003**, 31, e20 1-8.
16. Anderson, G.P. and N.L. Nerurkar, *J. Immuno. Meth.*, **2002**, **271,** 17-24.
17. Johnson, J., T. Lin, and G. Lomonossoff, *Annu. Rev. Phytopathol.*, **1997**, 35, 67-86.
18. Blum, A.S., et al., *Nano Lett.*, **2004,** 4, 867-870.
19. Soto, C.M., et al., *Electrophoresis*, **2004,** 25, 2901-2906.
20. Wang, Q., et al., *Chem. Biol.*, **2002**, 9, 813.
21. Soto, C.M., et al., *J. Am. Chem. Soc.*, **2006,** 128, 5184-5189.
22. Sapsford, K.E., et al., *Biosens. Bioelectron*, **2006,** 21, 1668-1673.
23. Schobel, U., et al., *Bioconjugate Chem.*, **1999,** 10, 1107-1114.
24. Wang, D., et al., *Proc. Natl. Acad. Sci. U S A*, **2002,** 99, 15687-15692.
25. Duggan, D.J., et al., *Nat Genet.*, **1999**, 21, 10-14.
26. Vora, G.J., et al., *Applied and Environmental Microbiology* **2004**, 70, 3047-3054.
27. Andras, S.C., et al., *Molecular Biotechnology*, **2001**, 19, 29-44.
28. Lisby, G., *Molecular Biotechnology*, **1999,** 12, 75-99.
29. Lee, T.M.H., L.L. Li, and I.M. Hsing, *Langmuir*, **2003**, 19, 4338-4343.
30. Greninger, D.A., et al., *Journal of Nanoscience and Nanotechnology*, **2005**, 5, 409-415.
31. Charles, P.T., et al., *Langmuir*, **2003**, 19, 1586-1591.
32. Shriver-Lake, L.C. and F.S. Ligler, *IEEE Sensors Journal*, **2004**, 1-6.
33. Chris A. Rowe, et al., *Anal. Chem.* , **1999**, 71, 3846-3852.
34. Frances S. Ligler, et al., *Anal. Chem.* , **2002**, 74, 713-719.
35. Kim E. Sapsford, et al., *Anal. Chem.* , **2002**, 74, 1061-1068.
36. Chatterji, A., et al., *Chem. Biol.*, **2004**, 11, 855-863.

Chapter 12

Novel GaN-based Chemical Sensors for Long-range Chemical Threat Detection

K.-A. Son[1], B. H. Yang[1], N. Prokopuk[2], J. S. Moon[3], A. Liao[1], M. Gallegos[1], J. W. Yang[4], T. Katona[4], and M. A. Khan[4]

[1]Jet Propulsion Laboratory, California Institute of Technology, Pasadena,CA
[2]Naval Air Warfare Center, China Lake, CA
[3]HRL Laboratories, LLC, Malibu, CA
[4]University of South Carolina, Columbia, SC

We are developing micro chemical sensor nodes consisting of GaN HEMT (High Electron Mobility Transistor) sensors and a RF communication link for long range chemical threat detection and early warning. In this paper, we discuss our research on (1) high selectivity detection of chemical agents (stimulants) using the GaN HEMT sensors, (2) the effects of the materials and design of the gate electrode on the sensitivity of the sensor, and (3) optimal operating parameters for high sensitivity detection.

Introduction

Early warning capability of chemical or biological threats is crucial for the safety and security of the public as well as military forces, and it has been a driving force for development of stand-off detection technologies. The current stand-off monitoring systems for chemical or biological agents are mostly based on optical spectral analysis and deployed on mobile/fly-by platforms (*1-3*). These systems offer excellent chemical identification capability, but are in general large, expensive, and require long data acquisition times. Functionality of the optical systems can be hampered by bad weather or by air clutter (e.g., smoke or fire) due to undesirable signal attenuation (*1-3*). For defense and security applications, real-time and autonomous monitoring of the continuously changing chemical and biological environment is highly beneficial. Continuous monitoring of the environment is also critical for proper implementation of the decontamination protocol. To this end, we are developing micro chemical sensor nodes that will enable real-time remote detection and monitoring of chemical threats and rapid transmission of the early warning signals. Our sensor node consists of an array of GaN-based micro chemical sensors, a micro controller, and a RF link. In this paper, we will discuss the selectivity and sensitivity of chemical agent (simulants) detection using GaN HEMT (High Electron Mobility Transistor) sensors.

GaN HEMT devices have been developed mostly for RF (radio frequency) transceivers, and their potential for chemical or biological sensors has been recognized only recently. Unlike the ISFET (Ion-Sensitive FET), which is traditionally used for chemical sensing, the GaN HEMT has a two-dimensional electron gas (2DEG) conducting channel near the AlGaN/GaN interface caused by the differences in spontaneous polarizations between AlGaN and GaN layers and piezoelectric polarization of the pseudomorphic AlGaN layer (Fig. 1) (*4-7*). This 2DEG induces surface polarization sheet charges on the sensor surface. While the details of chemical sensing mechanism of GaN HEMT is not clearly known, it is generally understood that chemical species adsorbing on the GaN HEMT sensor can modify the surface polarization charges, modulating the charge density and conductivity of the 2DEG and thus providing a transduction mechanism for chemical sensing (*8*). In GaN HEMT, the AlGaN layer is typically 10-30 nm thick. Therefore the conduction within this 2DEG channel is very sensitive to surface polarization interactions induced by chemical species (on the sensor surface, especially on the gate electrode). The resulting transconductance change is amplified by the gain of the HEMT, offering potential for high sensitivity detection of chemical analytes with a very short response time of a few milliseconds.

To date, GaN HEMT sensors have been investigated primarily for extreme environment sensors with focus on detecting automobile combustion gases (*9*) and hydrogen (*10*) and on measuring pressure or stress (*11-13*). Development toward biological sensors has also been reported recently for GaN HEMT sensors (*14,15*). In our work, GaN HEMT sensors have been studied for chemical agent detection with the ultimate goal of developing a chip-level GaN chemical sensor node (*16*). In this paper, we discuss our research on (1) high selectivity detection of chemical agents (stimulants) using the GaN HEMT sensors, (2) the effects of the materials and design of the gate electrode on the sensitivity of the sensor, and (3) optimal operating parameters for high sensitivity.

Experimental

The GaN HEMT sensors are fabricated with $Al_xGa_{1-x}N/GaN$ (x=0.25-0.3) heterostructures grown on semi-insulating 4H-SiC substrates. The HEMT structures are grown by a combination of pulsed atomic layer epitaxy (PALE) and conventional metalorganic chemical vapor deposition (MOCVD). The AlGaN layer is 20-25 nm thick and with a thin AlN spacer layer above the 2DEG channel. The GaN layer was ~1300 nm thick and delta-doped. Typical sheet charge density is $\sim 1 \times 10^{13}/cm^2$ for these structures, and electron mobility is $\sim 1500\ cm^2/Vs$.

Mesa isolation of devices is carried out with chlorine-based reactive ion

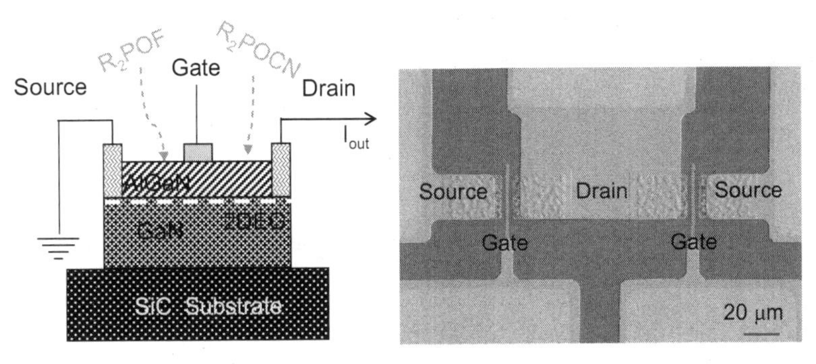

Fig. 1 Schematic of a cross sectional view of a GaN HEMT sensor (left; not drawn to scale) and SEM image of a pair of GaN HEMT sensors (right). (Reproduced from reference 16. Copyright 2007 SPIE)

etching. Ohmic contacts for the source and drain electrodes are made with Ti/Al/Ni/Au layers by rapid thermal annealing at 800°C, and specific contact resistance of ~0.5 ohm-mm is obtained. The excellent ohmic contact resistance results in a lower knee voltage, allowing the sensor operation at source-drain potential difference of less than 3 V. The GaN HEMT sensors are fabricated with various gate electrode designs; various gate length (L), gate width (W), and source-drain distance (D) to investigate potential correlation to sensor sensitivity. For characterization of sensor responses to chemical species, Ni- and Pt-based Schottky contacts are employed as gate electrodes.

The GaN HEMT sensors were tested with DECNP (Diethyl cyanophosphonate, a Tabun simulant), sulfur hexafluoride (SF_6, a nerve agent stimulant), and common chemical solvents including diethyl ether, dichloromethane, ethyl acetate, acetonitrile, acetone, and benzene. To measure sensor responses, drain current (I_{ds}) of a GaN HEMT was monitored at a constant drain voltage (V_{ds}) and a constant gate voltage (V_{gs}) during exposures to various gaseous environments.

Results and discussion

A prototype Ni-gated GaN HEMT sensor with L=0.15 μm, W=200 μm, and D=2 μm was used to investigate sensor responses to DECNP and common chemical solvents. Sensor evaluation was performed at V_{ds}=1.0–2.0 V with I_{ds} in the order of 10 mA, which corresponds to dissipating power of 75 mW/mm. For a background signal, I_{ds} was measured first with a flow of nitrogen. A dilute sample of analyte was then injected to the test chamber while measuring I_{ds}. For samples of low concentration analytes, the nitrogen gas saturated with an organic vapor was diluted with pure N_2 using mass flow controllers.

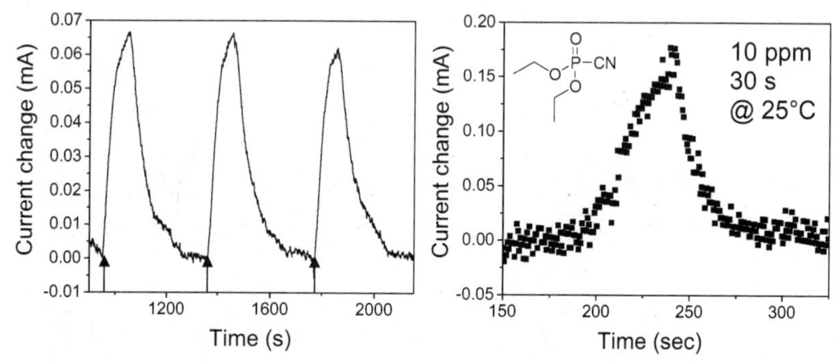

Figure 2 Electrical responses of a prototype Ni-gated GaN HEMT sensor. Source-drain current (I_{ds}) of the GaN HEMT sensor was measured during exposures to diethyl cyanophosphonate (DECNP), a Tabun simulant, at room temperature: (a) 90 s exposures to 0.1% DECNP in N_2, and (b) a 30 s exposure to 10 ppm DECNP. Measurements were made with V_{ds} of 1.5 V and V_{gs} of 0.0 V. The arrows in the plot (a) indicate the exposure initiation times. (Reproduced from reference 16. Copyright 2007 SPIE)

Figure 2 shows sensor response to DECNP at room temperature. The left plot presents sensor's response to repeated exposures of 0.1% DECNP in N_2. Each exposure was 90 s long and the starting point of each exposure is marked with an arrow in the plot. The result indicates that sensor response is immediate and repeatable. As shown in the right plot, the prototype GaN HEMT sensor is capable of detecting 10 ppm DECNP at room temperature, resulting in a clear increase of I_{ds}. The sensitivity of the GaN HEMT sensor can be further improved by optimization of AlGaN/GaN epitaxial structure and gate electrode design.

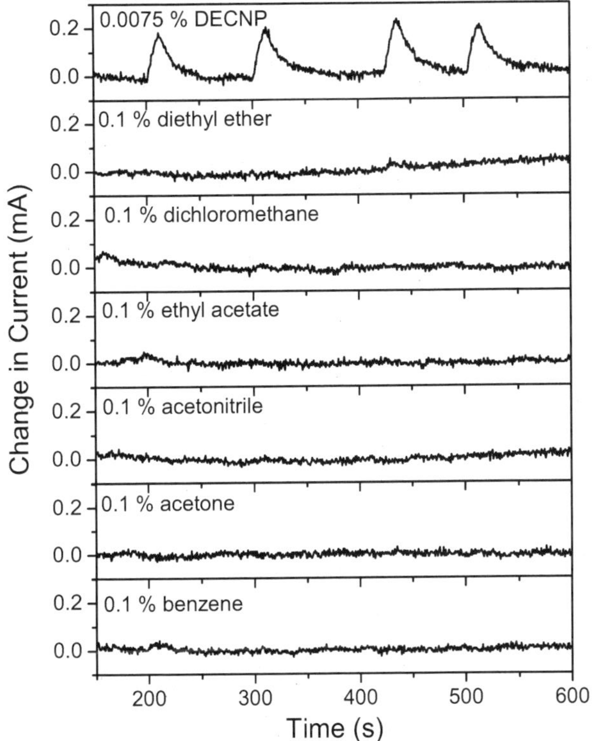

Figure 3 Electrical response of a prototype Ni-gated GaN HEMT sensor to DECNP and common chemical solvents. Among all the chemical species tested, the sensor shows positive responses to only DECNP at the concentrations used for the testing. (Reproduced from reference 16. Copyright 2007 SPIE)

To study selectivity of the sensor, the GaN HEMT has also been investigated with common chemical solvents. Figure 3 shows I_{ds} response, i.e., I_{ds}(analyte exposure)-I_{ds}(background in nitrogen), measured during repeated exposure cycles to DECNP, diethyl ether, dichloromethane, ethyl acetate, acetonitrile, acetone, and benzene at room temperature. The concentrations of the organic vapors used for the test were 0.0075% for DECNP and 0.1% for all other organics. The results show that the GaN HEMT sensor generates positive responses to only DECNP at the given condition, despite its concentration being one order of magnitude lower than that of the other chemicals. Modification of the surface polarization interaction induced by phosphonate and cyano groups is presumed to be a major factor for the high selectivity observed for DECNP, but more detailed study is under progress for better understanding of the selectivity. Positive responses of the sensor have been observed for the other organics tested, except benzene, when the concentration was much higher (~10%). However, such high organic concentration is not relevant for our study since it is not a typical operating condition for the chemical agent sensors. No response to benzene was expected based on the zero dipole moment of the molecule and the absence of a functional group that can induce strong polarization interactions on the sensor (gate electrode) surface at room temperature.

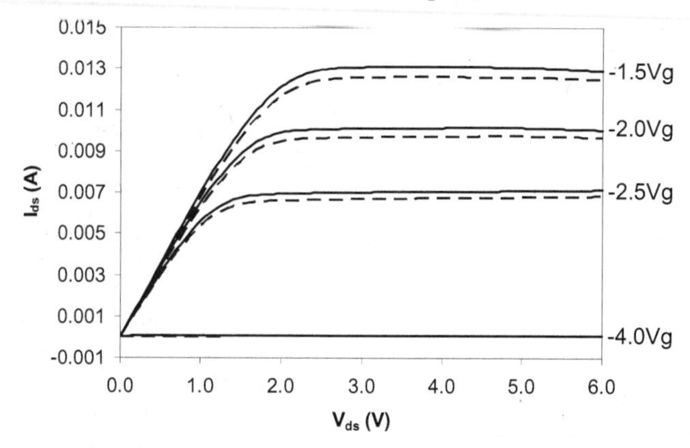

Figure 4 I_{ds}-V_{ds} (source-drain current vs. source-drain voltage) curves of a GaN HEMT sensor measured during exposure to SF_6 in a room ambient condition; solid lines. As a reference, I_{ds}-V_{ds} curves of the sensor measured without SF_6 exposure in the same condition are also shown; dashed lines. The sensor was fabricated with a 2 μm long and 50 μm wide Ni gate electrode with a source-drain distance of 6 μm. Measurements were made at a gate voltage (V_{gs}) of -1.5, -2.0, -2.5 and -4.0 V.

To probe correlation between sensor design and sensitivity, GaN HEMT sensors fabricated with Ni and Pt gate electrodes in dimensions of L=2, 4, 8, 12 μm, W= 5, 10, 25, 50 μm, D=W+4 μm have been studied. Sensor responses were measured in a room ambient condition with repeated exposures to a 25 sccm flow of pure SF_6, which was directed to the sensor at a ~2.5 cm distance. SF_6 was used for the test due to the fact that it is a stimulant of a nerve agent but it is non-toxic and stable in a room environment. While SF_6 has zero dipole moment, it has low-lying and localized unoccupied electronic states and shows a positive electron affinity (rending it to form SF^-_6 readily) (*17,18*). Thus a decent amount of I_{ds} increase resulting from electron transfer from the gate electrode can be expected with SF_6. Figure 4 shows I_{ds}-V_{ds} (source-drain current vs. source-drain voltage) curves of a GaN HEMT sensor measured during exposure to SF_6. Comparison with the curves measured without SF_6 presents clear increase of I_{ds} upon SF_6 exposure, especially in the saturation region of the I_{ds}-V_{ds}. One thing to note in the figure is that the degree of I_{ds} increase is dependent on the gate electrode voltage. This observation will be discussed in more detail in a later section.

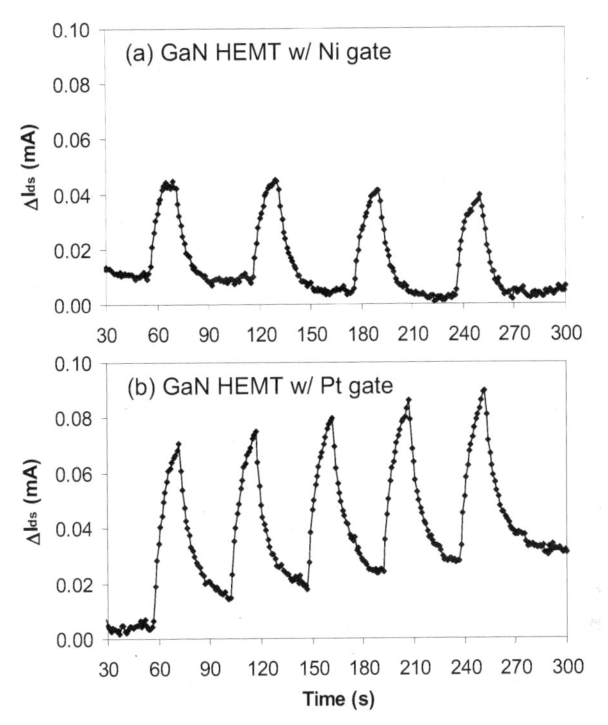

Figure 5 Electrical response of GaN HEMT sensor to SF_6 measured using the sensors fabricated with two different gate electrode metals: (a) Ni and (b) Pt. In both sensors, the gate electrode was 2 μm long and 50 μm wide (L2W50), and the source-drain distance was 6 μm. The drain current (I_{ds}) was monitored at the gate voltage (V_{gs}) of -3 V and the drain voltage (V_{ds}) of 2 V. Each SF_6 exposure was 15 s long with a duty cycle of 60 s for the Ni-gated sensor and 45 s for the Pt gated sensor. SF_6 exposures were carried out in air using a 25 sccm pure SF_6 flow directed from a ~2.5 cm distance.

Figure 5 shows I_{ds} change measured with both Ni- and Pt-gated GaN HEMT sensors during exposure to SF_6. In both cases, the sensors show immediate and repeatable I_{ds} increases upon SF_6 exposures. Based on the similarity of response observed with Ni- and Pt-gated sensors, in the following sections we present only the data acquired with Pt-gated GaN HEMT sensors.

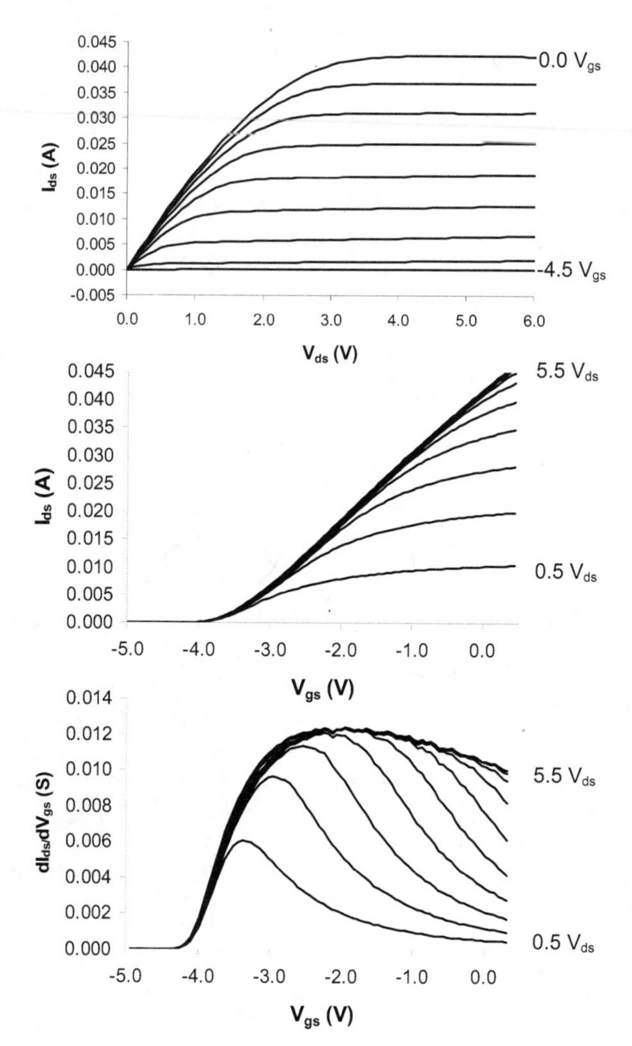

Figure 6 Electrical characteristics of the GaN HEMT sensor used for testing with SF_6. The sensor was fabricated with a 2 μm long and 50 μm wide (L2W50) Pt gate electrode with a source-drain distance of 6 μm. (a) I_{ds}-V_{ds} curves measured at the gate bias V_{gs} of -4.5 V (bottom) to 0.0 V (top); the curves for-4.5V_{gs} and -4.0V_{gs} are overlapping at I_{ds}~0.0 mA. (b) I_{ds}-V_{gs} transfer curves measured at the drain bias V_{ds} of 0.5 V (bottom) to 5.5 V (top) (c) transconductance (dI_{ds}/dV_{gs} vs. V_{gs}) of the GaN HEMT sensor measured at the drain voltage V_{ds} of 0.5 V (bottom) to 5.5 V (top).

Prior to probing factors governing sensor sensitivity, we analyzed electrical characteristics of the GaN HEMT sensors first. Figure 6 presents I_{ds}-V_{ds} (source-drain current vs. source-drain voltage), I_{ds}-V_{gs} (source-drain current vs. source-gate voltage) transfer characteristics, and transconductance (dI_{ds}/dV_{gs} vs. V_{gs}) characteristics of the GaN HEMT sensor used for testing with SF_6. The gate electrode of the sensor is a 2 μm long and 50 μm wide (L2W50) and 40 nm thick Pt-based Schottky contact located in the middle of the 6 μm gap between the source and the drain electrode. The I_{ds}-V_{ds} curves were measured at the V_g of -4.5 V to 0.0 V.

The sensor resistance in the linear region of the I_{ds}-V_{ds} curve is 2.5 ohm-mm, indicating that the ohmic contact resistance is quite small. The ohmic contact resistivity of the sensor that we measured is ~3.5×10^{-6} ohm cm^2. As shown in Figure 4, upon exposure to the analyte gases, the sensors show a greater response in the saturation regime of the I_{ds}-V_{gs} curves than in the linear regime. Clear understanding on the mechanism behind this observation requires theoretical modeling, which will be discussed elsewhere. Responses (I_{ds}-time curves) of the GaN HEMT sensor shown in this paper were measured in the saturation regime. The I_{ds}-V_{gs} transfer curves and the transconductance characteristics of the GaN HEMT sensor shown in the figure were measured at the V_{ds} of 0.5 to 5.5 V. The peak transconductance is 240 mS/mm at V_{ds}=5 V, indicating that the GaN HEMT sensor has excellent transconductance. In short, the GaN HEMT sensors are fabricated with excellent ohmic contact resistivity and small knee voltages (i.e., V_{ds} at the start of I_{ds} saturation), enabling sensor operation at V_{ds}< 3 V and consequently with a small DC power dissipation.

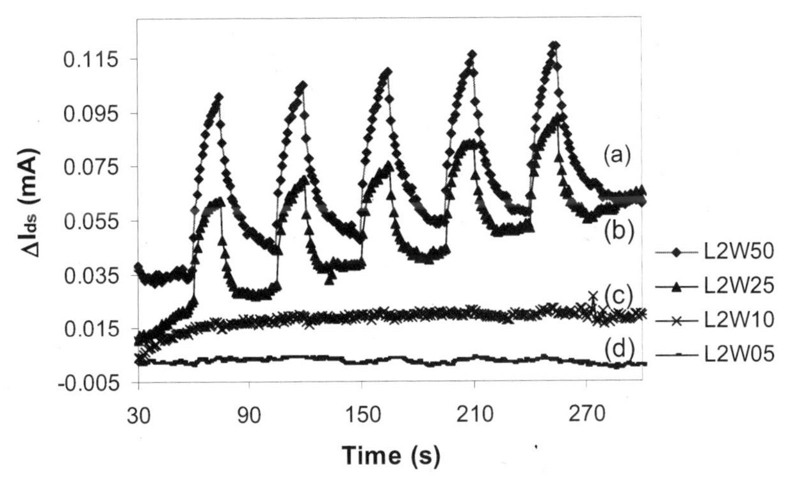

Figure 7 Effects of gate width on the responses of GaN HEMT sensors to SF_6. The GaN sensors were fabricated using Pt gate electrodes with various gate widths (W=5, 10, 25, & 50 μm) at a fixed gate length (L=2 μm). Source-drain current (I_{ds}) of the GaN sensors was monitored at -3V_{gs} and 2V_{ds} during exposures to SF_6 at room temperature. Each SF_6 exposure was 15 s long with a duty cycle of 45 s. The curves shown above correspond to ΔI_{ds}= I_{ds}(measured with SF_6 exposure cycles)- I_{ds}(background measured with no SF_6). For easier comparison between the curves, an offset value was added to each curve.

Effects of a sensor layout on detection sensitivity are investigated with the GaN HEMT sensors fabricated with various gate lengths (L) and gate widths (W). Figure 7 shows I_{ds} response to SF_6 exposures measured using the GaN HEMT sensors fabricated with W=5, 10, 25, 50 μm at L=2 μm. Higher sensitivity with an increasing gate width is clear. Figure 8 shows I_{ds} response measured using GaN HEMT sensors fabricated with L=2, 4, 8 μm at W=50 μm and with L=2 μm at W=25 μm; in these sensors the source-drain distance (D) is D=L+4 μm.

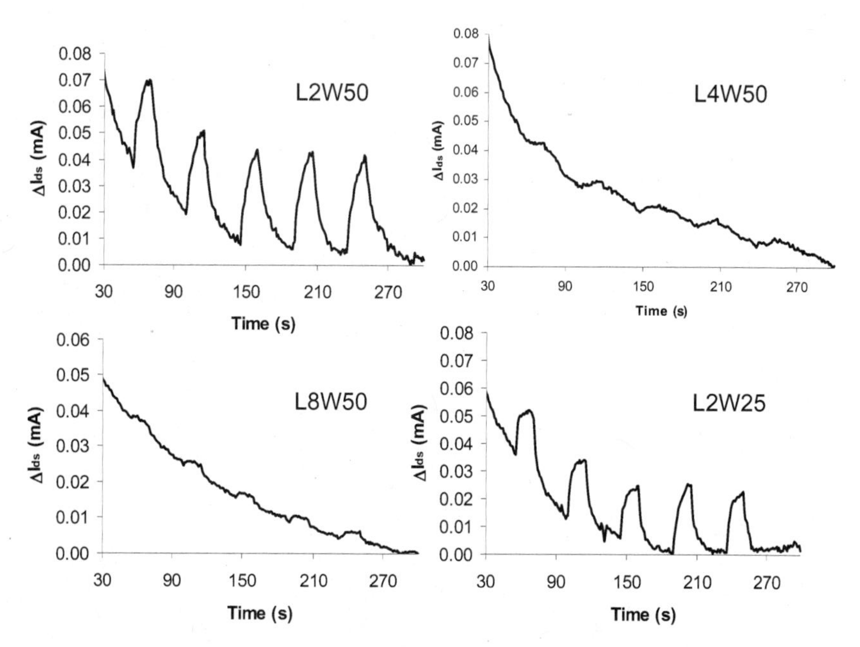

Figure 8 Responses of GaN HEMT sensors to SF_6 measured with various gate electrode designs: (a) L2W50, (b) L4W40 (c) L8W50 (d) L2W25, where L is the length of a gate electrode and W is the width of the gate electrode. The GaN sensors were fabricated using Pt gate electrodes, and the source-drain current (I_{ds}) of the GaN sensors was monitored at $-3V_{gs}$ and $2V_{ds}$ during exposures to SF_6 at room temperature. Each SF_6 exposure was 15 s long with a duty cycle of 45 s. The curves shown above correspond to ΔI_{ds}.= I_{ds}.(measured with SF_6 exposure cycles)- I_{ds}.(background measured with no SF_6). SF_6 exposures were made in air using a 25 sccm pure SF_6 flow directed from a ~2.5 cm distance.

The data clearly show decreasing sensitivity with an increasing gate length at W=50 μm. Comparison between L4W50, L8W50, and L2W25 sensors, which corresponds to D8W50, D12W50, and D6W25, respectively, indicates that the sensor sensitivity is not simply proportional to I_{ds}. The I_{ds} of the GaN HEMT is proportional to the gate width W, and the L4W50 and the L8W50 sensors should have higher I_{ds} than the L2W25 sensor. However, the L2W25 sensor shows a higher sensitivity, indicting that the shorter gate length plays a significant role.

The results shown in Figure 7 and 8 indicate that the sensor fabricated with L2W50 show excellent response compared to the other layouts and that the sensor response is strongly dependent on the gate length. These phenomena can be attributed to the fact that the sensor with L2W50 has the highest electric field near the gate-to-drain edge (in the saturation regime of the I_{ds}-V_{ds}), which is modulated by the adsorbed SF_6 on the surface. The results also suggest that sensor sensitivity is not simply proportional to the size of the gate electrode or the amount of I_{sd} of the sensor and that a short gate length and thus a source-drain distance are important factors in determining the sensitivity of the sensor. In order to find optimal operational parameters for GaN HEMT sensors, sensitivity dependence on the gate electrode voltage (V_g) has been studied.

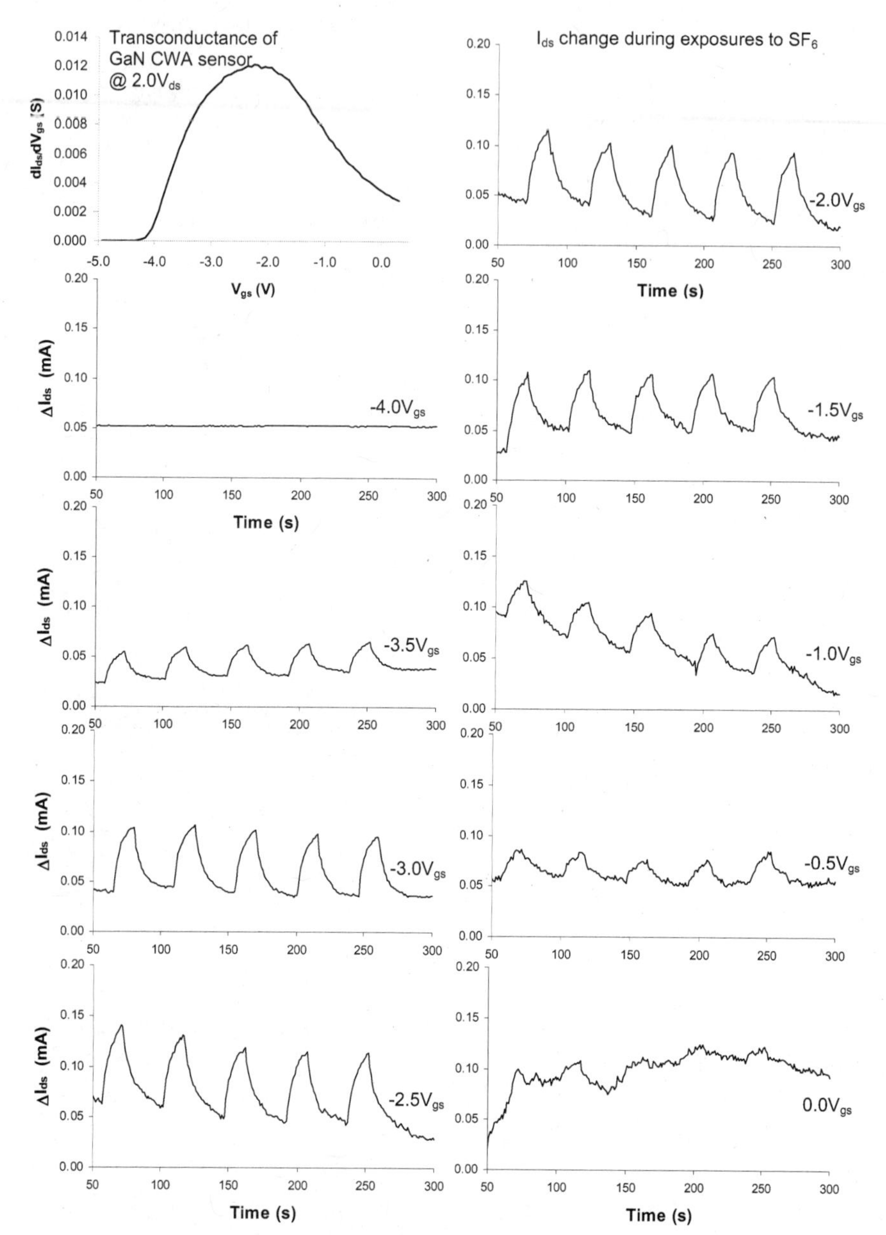

Figure 9 Transfer characteristics of a GaN HEMT sensor fabricated with a L2W50 Pt gate electrode (top left) and I_{ds} responses of the sensor to SF_6 measured at various gate electrode voltages(V_{gs}=-4.0 to 0.0 V). For all the measurements shown above, the drain voltage(V_{gs}) was 2.0 V.

Figure 9 shows I_{ds} responses of a L2W50 sensor to SF_6 exposures measured at various gate voltages V_g at a constant V_{ds}=2.0V. Transconductance curve of the L2W50 sensor measured at V_{ds}=2.0V is also shown in the figure as a reference. The data show that the highest sensitivity or the peak response is obtained when the gate voltage is applied near the peak transconductance. The data indicate a strong correlation between sensor sensitivity and transconductance; sensitivity increases with increasing transconductance. Once the transconductance passes the maximum, the sensitivity and the signal to noise ratio (S/N) seem to drop rapidly. Since the device transconductance is dependent on the gate capacitance, this observation can be attributed to the surface charge accumulated due to the exposure to SF_6 or modulation of gate voltage V_g induced by adsorbing SF_6 molecules.

Summary

We are developing GaN HEMT-based micro chemical sensors and sensor nodes for remotely operated long-range detection of chemical threats. In this paper, our GaN HEMT micro chemical sensor study is described with various device design layouts and wide range of operational parameters using DECNP and SF_6, chemical agent simulants. High sensitivity (10 ppm) and high selectivity detection of DECNP has been demonstrated at room temperature using a prototype Ni-gated GaN HEMT sensor fabricated with L=0.15 μm, W=200 μm, and D=2 μm. The sensitivity study performed with various GaN HEMT sensors (L=2, 4, 8, 12 μm, W= 5, 10, 25, 50 μm,) using SF_6 indicates that GaN HEMT sensor with a shorter gate length and a larger gate width exhibit a higher sensitivity. The GaN sensors also show higher sensitivity in the saturation region of I_{ds}-V_{ds} compared to the linear region. Furthermore, the sensor sensitivity is strongly dependent on transconductance of the sensor; the highest sensitivities are measured near the gate voltage of the peak transconductance. Based on the current observations with SF_6, it is expected that further scaling of the gate-to-channel (2DEG) distance and of the gate length will lead to higher sensitivity detection of chemical agents (simulants) using GaN HEMT sensors.

Acknowledgements

The research described in this paper was carried out by the Jet Propulsion Laboratory, California Institute of Technology, and was sponsored by ARO/DTRA and monitored by Dr. Stephen J. Lee at ARO.

The authors thank Dr. C. J. Whitchurch for valuable input for the project and Prof. P. P. Ruden for helpful discussions.

References

1. Fountain, A. W. III Ed. *Chemical and Biological Sensing VIII*, Proceedings of SPIE, Vol. 6554 (2007).
2. Ewing, K. J.; Gillespie, J. B.; Chu, P. M.; Marinelli, W. J. Eds. *Chemical and Biological Sensors for Industrial and Environment Monitoring III*, Proceedings of SPIE, Vol. 6756 (2007).
3. Jensen, J. O. ; Theriault, J. Eds. *Chemical and Biological Standoff Detection III,* Proceedings of SPIE, Vol. 5995 (2005)
4. Bernardini, F.; Fiorentini, V.; Vanderbilt, D. *Phys. Rev. B*, **1997**, 56, R10024-R10027.
5. Ambacher, O.; Smart, J.; Shealy, J. R.; Weimann, N. G.; Chu, K. et al. *J. Appl. Phys.* **1999**, 85, 3222-3233
6. Ambacher, O.; Foutz, B.; Smart, J.; Shealy, J. R.; Weimann, N.G. et al. *J. Appl. Phys.* **2000**, 87, 334-344.
7. Yu, E. T.; Dang, X. Z.; Asbeck, P. M.; Lau, S. S.; Sullivan, G. J. *J. Vac. Sci. & Tech.* **1999**, 17, 1742-1749
8. Schlwig, J.; Muller, G.; Eickhoff, M.; Ambacher, O.; Stutzmann, M. *Sensors and Actuators* **2002**, B87, 425-430.
9. Schlwig, J.; Muller, G.; Eickhoff, M.; Ambacher, O.; Stutzmann, M. *Mat. Sci. Eng.* **2002**, B93, 207-214
10. Wang, H.; Kang, B. S.; Ren, F.; Fitch, R. C.; Gillespie, J. K. et al. *Appl. Phys. Lett.* **2005**, 87, 172105.
11. Liu, Y. ; Ruden, P.P.; Xie, J.; Morkoç, H.; Son, K.-A. *Appl. Phys. Lett.* **2006**, 88, 013505.
12. Son, K. -A.; Yang, B. H.; Liao, A.; Steinke, I. P.; Liu, Y.; Ruden, et al. *proceedings of NASA Science and Technology Conference,* 2007.
13. Kang, B. S.; Kim, J. Jang, S.; Ren, F.; Johnson, J. W. et al. *Appl. Phys. Lett.* **2005**, 86, 253502.
14. Kang, B. S.; Ren, F.; Wang, L; Lofton, C.; Tan, W. W. et al. *Appl. Phys. Lett.* **2005**, 87, 0253508.
15. Steinhoff, G.; Baur, B.; Wrobel, G.; Ingebrandt, S.; Offenhausser, A.; Dadgar A. et al. *Appl. Phys. Lett.* **2005**, 86, 033901.
16. Son, K.-A.; Yang, B.; Prokopuk, N.; Moon, J.S.; Liao, A.; Gallegos, M.; Yang, J. W.; Khan, M.A. *Proceedings of SPIE, 2007,* vol.6556 *Micro (MEMS) and Nanotechnologies for Defense and Security*, paper # 6556-36
17. Klekamp, A.; Umbach, E. *Surf. Sci.* **1992**, 271, 555-566.
18. Faradzhev, N. S.; Kusmierek, D. O.; Yakshinskiy, B. Y.; Madey, T.E. *Low Temp. Phys.* **2003**, 29, 215-222.

Chapter 13

Solid Supported Polydiacetylene Materials for Detection of Biological Targets

New Material Forms and Detection Examples

Mary A. Reppy[1*] and Bradford A. Pindzola[1,2]

[1]Analytical Biological Services Inc. Wilmington, DE 19801
[2]Current Address: TIAX LLC, Cambridge, MA 02140

The conjugated polymer polydiacetylene (PDA) has shown promise as a sensing material for point-detection of biological agent applications. PDA has environmentally switchable optical properties that make it very attractive for the transduction of chemical information into readily observable optical signals. Historically, most PDA assay materials have been prepared as dispersed liposomes, a format that is not readily incorporated into rugged, field-ready sensors. Two new PDA material formats, coatings on nanoporous membranes and liposomes attached to fiber supports, are used to demonstrate detection of biological targets using the switchable fluorescence of the material.

Reliable and sensitive detection of biowarfare agent (BWA) pathogens and toxins is an important challenge facing both military operations and domestic counter-terrorism efforts. Solutions to this problem would greatly reduce the risk to our soldiers, emergency responders, and the general public; but achieving such solutions has proven extremely challenging. Environmental sampling techniques include direct testing of water supplies, trapping aerosolized microorganisms from air into water, or collecting samples on absorbent materials impregnated with sterile saline or buffer (*1*). The effective detection of BWAs in these aqueous samples requires simple, rapid, portable, and sclcctive detection equipment incorporating sensing materials.

Polydiacetylene (PDA) is a conjugated polymer, formed through the topochemical 1,4-photopolymerization of self-assembled diacetylene monomers (Figure 1) (*2*), with environmentally sensitive optical properties. Self-assembled PDA structures can readily incorporate ligands and substrates for specific detection applications, which along with the built-in optical signal generation arising from changes in its absorption and emission properties make it an advantageous material for detecting biological entities. A recent review of the field, along with references therein, includes a discussion of PDA's optical properties, its various material forms, and its use in detection of biological targets (*3*). In this chapter we will focus on developments in our laboratory in creating new forms of PDA sensing materials.

Figure 1. Photopolymerization of diacetylene amphiphiles to form PDA.

While the blue-to-red color change of PDA was first exploited for detection (*4*), we have focused on using the emission properties. The emissive behavior of PDA parallels the chromic behavior: blue PDA is not fluorescent while red and yellow PDA are. PDA, in its fluorescent form, is excited by wavelengths above 450 nm and emits two broad fluorescent peaks circa 560 nm and 640 nm (Figure 2A) (*5*). Signal generation from PDA emission can be amplified by resonance energy transfer (RET) to appropriate fluorophores (Figure 2), and is combatible with opaque substrates (*6-8*). While changes in PDA absorbance and emission parallel each other, the relative emission changes are larger. This is a particular advantage in assays where the colorimetric response is minor (Figure 3).

We will discuss two new PDA material formats that have demonstrated utility for detection of biological analytes in water. PDA coatings deposited on nanoporous membranes make a solid state sensing platform that, on filtering of analyte solutions, concentrates large targets at the sensing surface and moves detection of smaller analytes from a diffusion controlled regime to a kinetically controlled one (*9*). A second format consists of PDA liposomes covalently

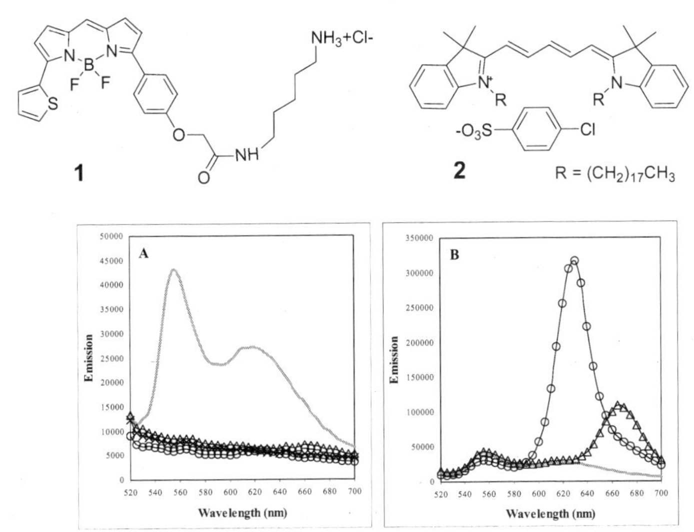

*Figure 2. Fluorophores and emission spectra of poly(10,12-PCDA) liposomes. (A) non-fluorescent with: 0.5% **1** (○), with 0.5% **2** (△), plain (×), and fluorescent (no fluorophore, line); (B) fluorescent: with 0.5% **1** (○), with 0.5% **2** (△) and plain (line).*

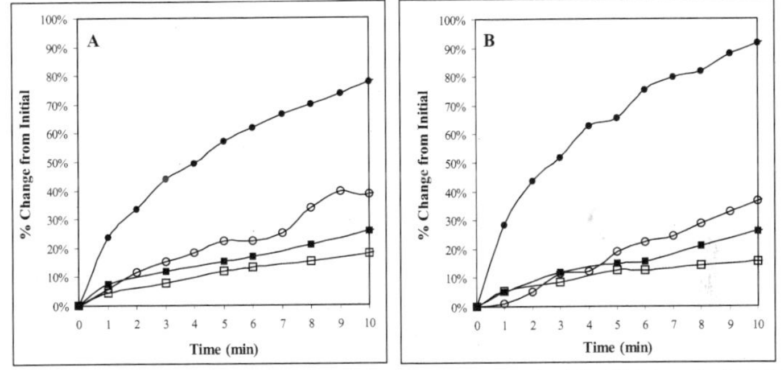

*Figure 3. Phospholipase A_2 (PLA_2) assays with (A) 70% poly(10,12-TRCDA)/30% DMPC liposomes; (B) 70% poly(10,12-TRCDA)/30% DMPC liposomes with 0.5% **1**. Comparison of the emission at 635nm of 6 μM PLA_2 samples (●); emission of buffer control samples (○); CR (calculated from 650 nm and 550 nm absorbance peaks) of PLA_2 samples (■); and the CR of buffer control samples (□).(7)*

attached to glass fiber filter membranes. The attachment stabilizes the particles to conditions common in biological systems that normally lead to aggregation, such as buffers with divalent cations or high protein concentrations, and also provide a portable detection format that only requires addition of target solution. Both of these approaches allow label-free detection of biological targets, which is particularly important in detection devices designed for field use.

These advances in the use of PDA for detection have allowed us to demonstrate the detection of a variety of biologically relevant targets including *Escherichia coli* (*E. coli*), *Bacillus globigii* spores (BG), a simulant for *B. anthracis* and ricin toxin.

Materials & Methods

Trimethoxy-(3- glycidyl)propylsilane was obtained from United Chemical Technologies. BODIPY® TR cadaverine (**1**) and cyanine fluorophore DIC-18(5) (**2**) were obtained from Molecular Probes (now Invitrogen). Ganglioside GM_1, 1-palmitoyl-2-oleoyl-*sn*-glycero-3-phosphocholine (POPC) and 1,2-dipalmitoyl-*sn*-glycero-3-phosphothioethanol (DPPT) were purchased from Avanti Polar Lipids. Asialo-GM_1 was purchased from Matreya LLC. N-hydroxysulfosuccinimide (sulfo-NHS) and bis(sulfosuccinimidyl) suberate (BS3) were purchased from Pierce Biotechnology and EDC from Sigma Aldrich. Ricin toxin was purchased from Vector Laboratories. Anti ricin polyclonal antibodies, anti *Bacillus globigii* (BG) polyclonal antibodies, and BG spores were obtained from the U.S. Department of Defense JPED-CBD Critical Reagents Program. Additional BG spores were generated from *Bacillus atrophaeus var. niger*, obtained from the American Type Culture Collection (ATCC #48337), grown and sporulated according to literature methods (*10*). *Escherichia coli* were purchased from the ATCC (23716) and were propagated in Lennox LB broth. Microorganisms were centrifuged and resuspended in assay buffers or water before assays. Membranes and 96-well filter plates were purchased from Fisher Scientific, with the exception of custom plates manufactured for us by Millipore Inc with 100 μm MCE membranes.

Diacetylene fatty acids were purchased from GFS. N-(2-Hydroxyethyl)-10,12-pentacosadiynamide (**10**) was synthesized by literature methods (*11*). 1-Amino-10,12-pentacosadiyne (**11**) was synthesized in four steps from 10,12-pentacosadiynoic acid: the acid in anhydrous tetrahydrofuran (THF) at 0 °C was reduced to the alcohol through treatment with lithium aluminum hydride (LAH, 2.5 equivalents) in diethyl ether for two hours, the alcohol was then converted to the mesylate by treatment with mesyl chloride (6 equivalents) in methylene chloride with diisopropyl ethyl amine over 30 minutes, the mesylate was displaced by sodium azide (1.5 equivalents) in dimethyl formamide (DMF) at 70 °C over one hour and the azide, in THF, was reduced to the amine with LAH (2 equivalents) in diethyl ether at 0 °C over one hour. N-(10,12-pentacosadiynyl)-glutamic acid (**12**) was prepared from **11** as follows: **11** was reacted with glutaric anhydride (2 equivalents) in DMF, in the presence of diisopropylethylamine (3 equivalents), at 70 °C for 1 hour and the crude product recrystallized from a mixture of chloroform and hexanes. The identity of products were confirmed by ^{1}H and ^{13}C NMR.

PDA coatings on filter membranes were prepared and used in assays as previously described (*9,12*). Antibodies were conjugated to PDA coating surfaces with carboxylic acids incorporated using EDC/sulfo-NHS according to the literature (*13*). Coatings were stored at 4 °C. Attached liposomes were prepared and used in assays as previously described (*14*).

Monomers

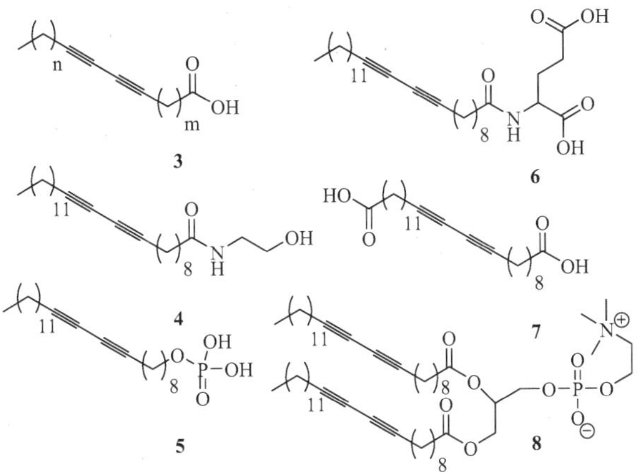

Figure 4. Some common diacetylene amphiphiles.

Self-assembled polydiacetylene materials are formed from diacetylene amphiphiles with polar headgroups and a hydrophobic tails containing the diacetylene moiety. Diacetylenes are usually prepared using the Cadiot-Chodkiewicz reaction consisting of the Cu(I) catalyzed coupling of an acetylene and a halo-acetylene (*15*). A large variety of monomers have been synthesized with different polar headgroups and tail structures; some representative examples are shown in Figure 4 (compounds **3-8**). The molecular structure determines whether the diacetylene will form a self-assembled material and affects the material properties. Headgroup chirality strongly influences colloidal structure; amphiphilic monomers with achiral headgroups usually form spherical liposomes, whereas chiral amphiphiles often form non-spherical structures such as helices and tubules. The length and number of the tails affects the stability of colloidal suspensions and controls the temperature of the melt transition (T_m) from "crystalline" chain packing to "liquid" chain packing. The diacetylene fatty acids (**3**), in particular, 10,12-pentacosadiynoic acid (10,12-PCDA, m,n=8,11), 10,12-tricosadiynoic acid (10,12-TRCDA, m,n=8,9) and their derivatives have been widely used in PDA sensing materials (*3*).

Ligands

An important component of any biosensing material is the ligand, which interacts with and recognizes the target. Ligands range from small molecules to proteins and interact with microorganism and protein targets, usually by simply binding to the target (*16*). Antibodies are one of the most commonly used ligands and have been developed against a wide variety of targets from molecules to microorganisms. Unfortunately antibodies have several major disadvantages: polyclonal antibodies can vary between batches, monoclonal antibodies often bind less well than polyclonal mixtures, and antibodies, as with most proteins, do not tolerate harsh conditions. Antibody fragments and peptide chains are also used as ligands; the phage library approach allows the selection of peptides that selectively and strongly bind to a given target. Enzymatic substrates, such as peptides and lipids can also act as ligands. Cell surfaces are decorated with carbohydrates, both glycolipids and glycoproteins, that bind to both microorganisms and to proteins. For example, asialofetuin (ASF), a glycoprotein with multiple galactose residues, binds to ricin (*17*). These molecules can also be used as ligands, but they are generally less selective than antibodies. RNA and DNA aptamers have also been developed as ligands (*18*).

Liposomes

The majority of PDA sensing materials used by other groups have been prepared as liposome solutions (*3*). We use PDA liposomes both in assays and as precursors for solid supported PDA materials. PDA liposome solutions can be readily prepared by probe sonication of diacetylene amphiphiles followed by polymerization, however, the colloidal solutions are frequently not stable and form aggregates that settle out, either during storage or under common assay conditions. For example, we have shown that the addition of divalent cations to PDA liposomes formed from diacetylene carboxylic acids and the corresponding 2-ethanol amide esters (e.g., **4**) at millimolar concentrations caused swift aggregation (*3*). The liposomes were more stable to monovalent cations but still aggregated upon dilution with 200 mM NaCl and KCl, and 1% BSA. Aggregation affects the optical signal and leads to confounding results in assays.

Sugars, short peptides and nucleic acid sequences, prepared with diacetylene or alkyl tails can be incorporated directly into the diacetylene liposomes at the sonication step, however, most biological substrates and ligands are too large and not stable to probe sonication. Antibodies with conjugated alkyl tails can be inserted into the diacetylene liposome via detergent dialysis prior to polymerization (*9,19*). Alternatively, the surface of the liposomes can be activated and reacted with proteins etc, however, the separation of unreacted species requires size exclusion chromatography or dialysis. Both of these techniques are slow and lead to loss of material. The surfaces of solid PDA are much more easily functionalized as the reagents can be added to the solid and then washed off at the end of the reaction.

Liposomes are also less suitable than films or coatings for use in environmental screening. The use of liposomes generally requires the addition of controlled amount of liquid liposome reagent. Solid supported materials, such as PDA coatings, offer the opportunity for incorporation of the materials in test cartridges, regentless detection, and simple procedures. Simple test procedures are particularly important in cases where the personnel are in biohazard suits or other protective gear. The incorporation of the sensing material in a cartridge further protects the tester from exposure to the sample.

Coatings

Langmuir Blodgett (LB) films have been used for solid supported PDA detection materials, but are relatively difficult to prepare, particularly in volume. Furthermore, LB films need at least three layers to have sufficient material for colorimetric detection (*20,21*). We discovered that it is possible to easily prepare PDA coatings on nanoporous (50 to 450nm) filter membranes by the filtration of diacetylene liposome solutions through the filters (*9*). The liposomes are caught at the surface and merge to cover the filter (Figure 5). Coatings were deposited onto both free-standing membranes and onto membranes in 96-well filter plates; photopolymerization of the deposited diacetylenes usually formed blue, non-fluorescent PDA. The fused liposomes form coatings with thicknesses from 1 - 3 μm with the thickness and morphology dependent on the nature of the diacetylene monomer. Despite the apparent complete covering of the filter surface, it is still possible to pass solutions through the coated filters, thus bringing analytes close to the surface and the recognition elements.

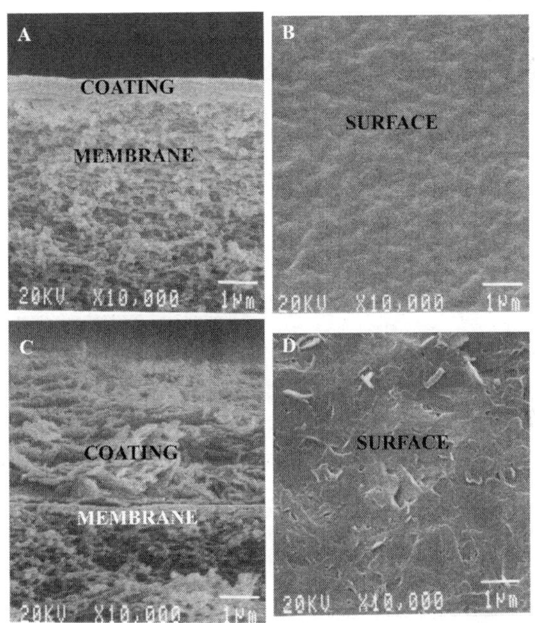

*Figure 5. SEM images of PDA coatings on 50 nm pore MCE filter membranes. (A) and (B) are the cross-section and surface of poly(**6**) coatings; (C) and (D) are the cross-section and surface of poly(**5**) coatings. 1 μm scale bars.*

The fluorescence stability of coatings on different membrane types was monitored over a year and the results showed that stability was dependent on both the monomer tail structure, headgroup and the membrane material (*9*). Typically the emission of coatings would rise slightly over the first few days after deposition and then stabilize. Mixed cellulose ester (MCE), hydrophilic polyvinylidene floride (e.g., Durapore) and hydrophilic polypropylene membranes were used successfully; we also tried polycarbonate membranes, however, many coatings did not adhere well to these membranes. We also investigated the effect of pore size on coating quality and performance using MCE membranes with torturous path pore sizes from 50 nm to 450 nm. Very good coatings formed on the 50 nm membranes, however, the flow through the small pore membranes was not very good and even worse after coating deposition. Better flow was obtained with coatings with 100 nm pores, and the best flow with coatings on 400 or 450 nm pores. There was some difficulty in depositing coatings on the larger pore membranes, as the diacetylene liposomes

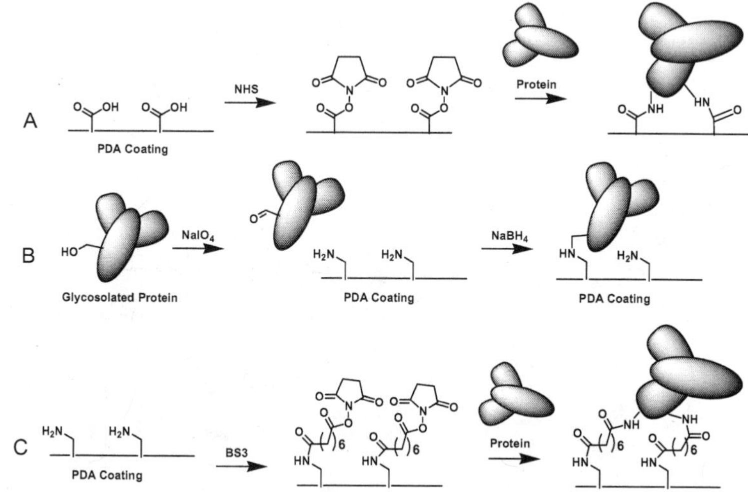

Figure 6. Cartoon of protein conjugation to PDA coating surfaces via (A) NHS activation of the surface and reaction with protein amines, (B) oxidation of the protein to form aldhydes and reaction with amines incorporated in coating and (C) reaction of surface and protein amines with BS3 linker.

or other colloids were more prone to pass through the membranes. We pretreated the filters with polylysine (mw 70,000-150,000) (*9*), this both prevented the liposomes from passing through the filters and improved the adhesion of coatings (*12*).

PDA coatings are easily prepared with ligands, either through incorporation of the ligand in the diacetylene liposome prior to deposition or through reaction of the surface post deposition. The coatings provide a suitable environment for preserving the activity of antibodies, presumably because the polar surfaces remain hydrated even when the coatings are stored "dry". We prepared coatings from liposomes with antibodies incorporated by detergent dialysis; the presence of the antibodies in the coatings was confirmed by immunoassays and by labeling with secondary antibody-gold conjugates and visualization by TEM (*9*). The stability of the antibody presentation was characterized by immunoassay over time and it was seen that the presentation was stable or declined only slowly over the course of more than a year. We have also conjugated ligands, including antibodies, peptides, aptamers, and phage particles, directly to the coating surfaces through standard bioconjugation chemistries, such as

NHS/amine, maleimide/thiol, and aldehyde/amine couplings (Figure 6) (*13*). The advantage of reacting the surface with the ligands is that it avoids the need to pre-functionalize antibodies etc with alkyl tails.

Detection with Coatings

E. coli was used as a model microorganism in our coating development work as they are easily cultured and safe to work with. Detection of pathogenic *E. coli* strains, such as serotype O157:H7, is of additional interest for food safety applications. PDA coatings were prepared from 10,12-PCDA liposomes with 0.5 % **2** incorporated and *E. coli* antibodies inserted via detergent dialysis.(9) A comparison of the response of the coatings to being soaked with solutions of *E. coli* and to having solutions of *E. coli* filtered through is shown in Figure 7. Under the soak conditions, the coatings response 10^5 *E. coli*/mL was only slightly above the buffer while exposure to 10^7/mL gave a more robust response (Figure 7A). Filtration of the samples through coatings improved the sensitivity dramatically. PDA coatings in a 96-well filter plate were exposed to increasing doses of *E. coli* and control buffer (100 μL) with filtration and emission measurement between each dose. The coatings responded to 10^2 *E. coli* per well with emission rises above the buffer samples (Figure 7B). The response increased up to 10^4 *E. coli* per well, where it leveled off. These results demonstrate the advantage of using filtration in overcoming the diffusion limitations of detection at surfaces. The filtering of the solution forces the organisms to come to the surface as the liquid is removed.

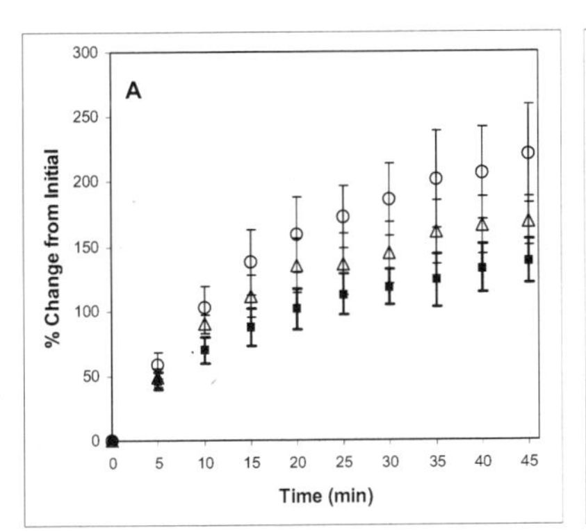

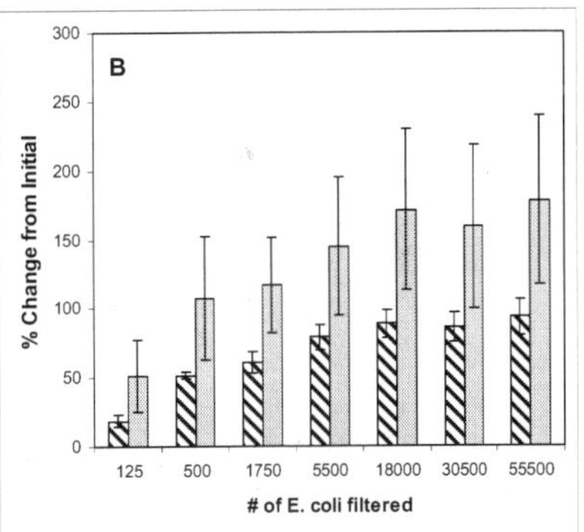

Figure 7. Emission responses of poly(10,12-PCDA) coatings with 0.5% ***2*** *and anti-E. coli antibodies; in 96-well plates caused by (A) Solutions of E. coli: 10^7/mL (○), 10^5/mL (△) and control buffer (■); (B) Iterative filtration of 100 μL of E. coli solutions (grey bars) and control buffer (hatched bars).(9)*

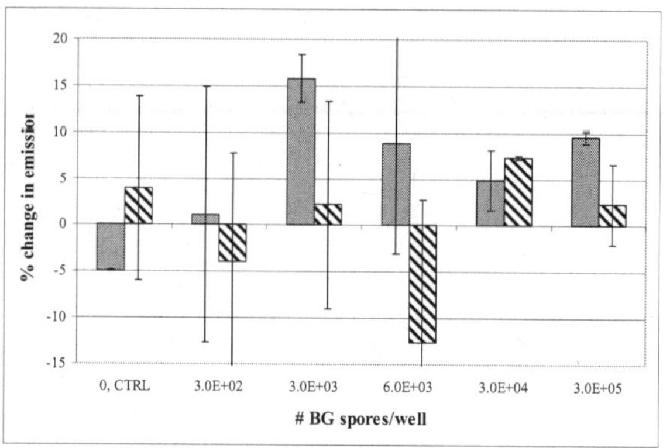

*Figure 8. Chart showing the changes in emission of poly(60% **10**/10% DMPC/30% **12**) coatings with 0.5% **2** and anti-BG antibodies (grey) and anti-Bacillus spore antibodies (hatched) 30 min after filtration of BG spores in buffer. 100 μL/well.*

To improve the coating stability and reduce the non-specific response we moved to preparing coatings from diacetylenes with amide head-groups. The hydrogen bond network formed by amides is stronger than that formed by carboxylic acids; we believed that would increase the stability of the coatings. PDA coatings were deposited from 60% **10**/10% DMPC/30% **12** liposomes with 0.5% **2** and conjugated, using NHS/EDC coupling chemistry, to either anti-BG antibodies or to anti-Bacillus spore antibodies raised against a mixture of Bacillus spores. The emission of the coatings were measured before and after filtration. The coatings with antibodies specific to BG spores detected the BG spores at 10^3 to 10^5 per well, while coatings prepared with more generic anti-Bacillus spores did not detect the BG (Figure 8). The response of the coatings were lower than seen with the 10,12-PCDA coatings and iterative filtration of *E. coli*, and the scatter (as shown by the error bars giving the standard deviation of the averages) was significant. The coatings were more stable than the 10,12-PCDA based coatings to the buffer control, however.

Our work with coatings has led us to identify some technical barriers that remain to be overcome. The steps taken to improve the stability of the coatings led to decreases in sensitivity, and slower responses of the materials. A balance must be struck between generating a strong response to the target and avoiding non-specific responses to other organisms or to assay conditions; investigating different diacetylene monomers can address this issue. Additionally, in our experience, antibodies have significant disadvantages as ligands. We found that polyclonal antibodies tended to cross react, particularly in the case of different Bacillus spore types. For example, coatings prepared with anti-BG antibodies responded to *B. cereus* spores as well as to BG spores. These problems can be addressed by turning to other ligand types, and to combinations of different ligands to improve selectivity.

Despite these issues, the advantage that the PDA coated filters offers in allowing filtration of samples provides impetus for continuted work in this field. Filtration leads to concentration of the bacteria at the coated surface, increasing the sensitivity of the material response and potentially allowing detection of organisms in the larger volumes typical of environmental samples. Even if an analyte is too small to be trapped by the coating, the action of pulling the analyte solution past the sensing surface moves the analyte/ligand interaction from a diffusion-limited to a kinetic-controlled regime (*22*).

Attached Liposomes

In an alternative approach to coating filter membranes, we have attached PDA liposomes through covalent bonds to microporous glass filters (*14*). This approach preserves the liposomes as individual entites, retaining more of their liposome character while still keeping the advantages of working with a solid supported material. The glass filters have much looser structures with larger effective pores than the polymer filters used in the preparation of PDA coatings, however, they still offer the advantage of being able to filter the target solution and can be obtained in 96-well plates. PDA liposomes had previously been attached to glass surfaces via Schiff base formation between amine containing liposomes and aldehyde silane treated slides (*23*), and microcontact printing of fatty acid diacetylene liposomes activated with NHS upon amine-silane treated slides (*24,25*). In our work we chose instead to treat our glass surfaces with trimethoxy-(3-glycidyl)propylsilane to put epoxide groups on the glass and reacted them with DPPT incorporated in our diacetylene liposomes (Figure 9). Epoxides are readily opened by thiols at pHs from 6-8, forming an irreversible covalent attachment (*13*). The filters were then irradiated to form PDA in the liposomes. Cryogenic transmission electron microscopy scans of a attached liposomes/fiber sample show that the liposomes do not open in this process but remain as vesicles (*14*) The immobilized liposomes can be polymerized into the blue form. As expected, immobilization stabilized the liposome response against buffers, including to those containing divalent cations. We previously used emission to detect the action of PLA_2 on free PDA/phospholipid liposomes (Figure 3), and were able to show that PLA_2 activity could also be detected using liposomes attached to the glass fiber membranes (*14*).

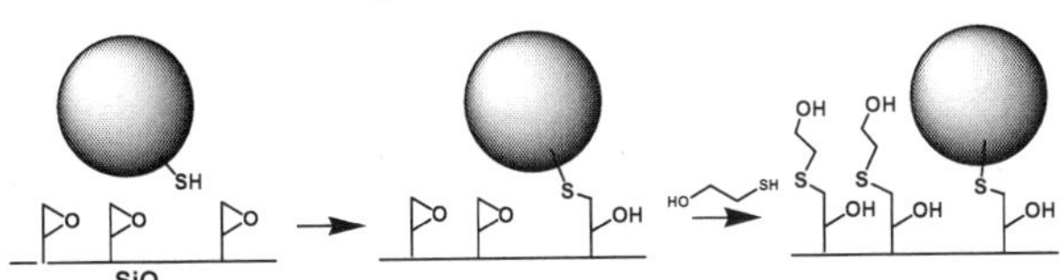

Figure 9. Cartoon showing reaction of diacetylene liposomes with thiols and epoxide groups on glass. Excess epoxides are opened with 2-mercaptoethanol.

Detection with Attached Liposomes

We have investigated using the attached liposomes for detection of ricin toxin and BG spores. This format allowed us to use assay buffers that would aggregate free liposome solutions, for example, buffers containing high concentrations of Ca^{2+}. For the detection of ricin, we screened attached liposome formulations with different ligands, including ganglioside GM_1, AS-GM_1, peptides,(26) asialofetuin (*17*), and anti-ricin antibodies. We also incorporated POPC into the liposomes as it has been shown to be a substrate for the lipase activity of the ricin B chain (*27*). We discovered that combining ligands with POPC in the liposomes improved detection. Figure 10 shows an example of ricin detection with PDA liposomes incorporating both antibodies and POPC. The emission rose as the amount of ricin increased with detection at 0.1 ng of ricin (1 ng/mL), and then dropped at the highest ricin amount. Bovine IgG, a control did not differ from the reference. In general, we have noticed that at higher concentrations of protein targets, the emission response of PDA materials drops off, presumably because the protein is quenching the emission. The range of protein amounts that cause increases in emission versus decreases varies with protein identity and PDA liposome formulation. Proteins that act upon the membrane, such as cholera toxin B(28) or PLA_2, do not show this phenomenon to the same extent, because the perturbation caused by the protein leads to a greater rise in signal, over-riding any quenching effect.

We also explored using the attached liposomes for detection of BG spores. Early experiments suggested that attached liposomes composed of diacetylene fatty acids (**1**) were more responsive to the spores than ones based on diacetylene ethanol amides (**2**). The liposomes were prepared with 20% of **11** incorporated so antibodies could be conjugated via the BS3 linker. We compared the performance of attached fatty acid liposomes in three different buffer systems: phosphate buffers with Na^+ and Cs^+ ions, and MOPS with Ca^{2+} ions. The buffers had a strong effect on the liposome response, with the best responses occurring in the Ca^{2+}/MOPS buffer and detection of 10^3 – 10^5 BG spores (10^4/mL – 10^6/mL) (Figure 11).

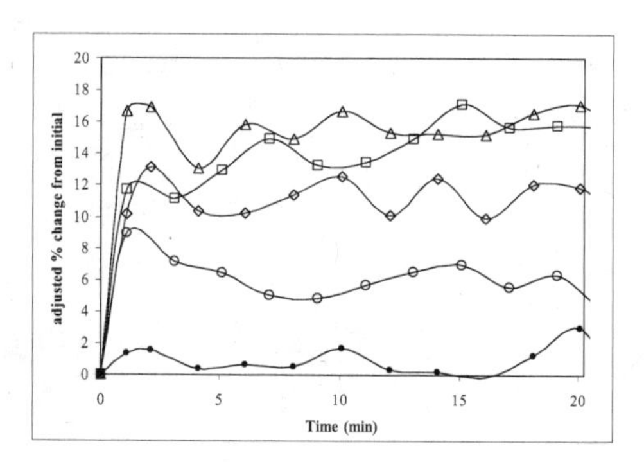

*Figure 10. Ricin detection with poly(60% **10**/20% POPC/10% **11**/10% DPPT) attached liposomes with 0.5% **1** incorporated and anti-ricin antibodies conjugated via BS3; 0.1ng ricin(◇), 1ng ricin (□); 10ng ricin (△); 100ng ricin (○); 100ng bovine IgG (●). 100 μL/well. Control subtracted.*

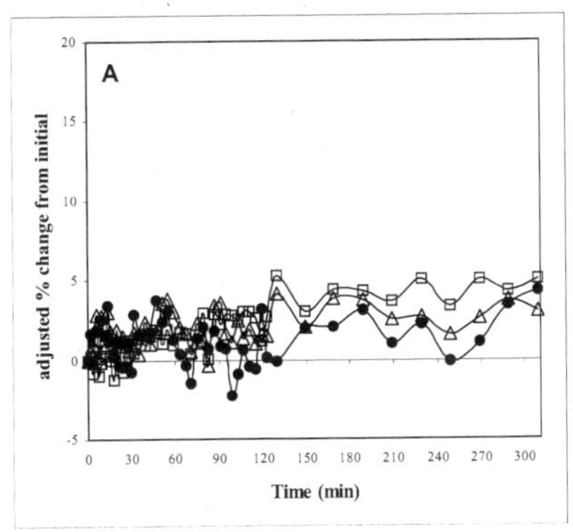

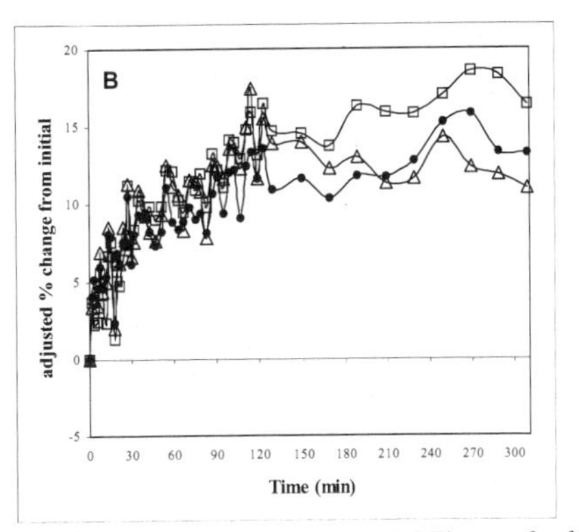

Figure 11. BG detection with poly(70% 6,8-PCDA/20%PCD-NH_2/10% DPPT) attached liposomes with 0.5% ***M*** *incorporated and anti-BG antibodies conjugated via BS3. (A) 10mM NaPO4/150mM NaCl; (B) 10mM)/100mM $CaCl_2$. 10^3spores/well (□); 10^4 spores/well (△); 10^5 spores/well (●). 100 μL/well. Control subtracted.*

The binding of the spores to antibodies at the liposome surface will change the ionic character of the interface at the binding site and may lead to displacement of cations complexed to the surface carboxylates and ultimately to perturbation of the PDA backbone, via the mechanisms proposed by Kew and Hall (*29*). Working with the Ca^{2+}/MOPS buffer system we investigated the effects of varying the concentration of Ca^{2+} in the buffer used to equilibrate the attached liposomes prior to polymerization, and in the assay buffer to see whether we could increase the response to spores and/or decrease the response to the control buffer. The best results were obtained by keeping the Ca^{2+} concentration the same between the equilibration and assay buffers, or by reducing the concentration in the assay buffer relative to the equilibration buffer. Dropping the Ca^{2+} concentration to zero in the assay buffer, or pre-equilibrating at a lower Ca^{2+} concentration and raising it for the assay, both led to indiscriminate responses to the control samples and a failure to detect. These studies reaffirmed the importance of controlling the buffer conditions in testing.

Conclusion

Solid state formats of PDA materials offer significant advantages as sensing materials for biological targets. Self-assembled PDA materials readily incorporate a wide variety of ligands, offering the ability to tune affinity and selectivity and to take advantage of advances in ligand technologies as they develop, and the switchable fluorescence provides an ideal detection signal. The solid-state coating and attached liposome formats presented here provide the added advantages of improved environmental stability and enhanced sensitivity from the ability to force analytes to interact with the sensing material through filtration. Microorganism detection with the coatings on nanoporous membranes was demonstrated with *E. coli* and BG at 10^2 and 10^3 per well respectively. The attached liposomes detected both the protein ricin toxin and BG down to 0.1 ng and 10^3 organisms respectively. These results illustrate the great potential of PDA as a sensing material and the importance of robust solid-state formats.

Acknowledgements

This work was partially funded by a National Science Foundation SBIR Phase II grant (DMI-0239587) and by an Army Research Office STTR Phase II contract (W911NF-O4-C0132). The conclusions are those of the authors alone and not endorsed by any government agency.

References

1. US Centers for Disease Control, National Institute for Occupational Safety and Health, 2002 http://www.cdc.gov/niosh/unp-envsamp.html
2. Odian, G. *Principles of Polymerization*; 3rd ed.; John Wiley & Sons, Inc.: New York, 1991.
3. Reppy, M. A.; Pindzola, B. A. *Chem. Commun.* **2007,** *42,* 4317-4388.
4. Charych, D. H.; Spevak, W.; Nagy, J. O.; Bednarski, M. D. *Mat. Res. Soc. Symp. Proc.* **1993,** *292,* 153-161.
5. Olmsted, J. I.; Strand, M. *J Phys Chem* **1983,** *87,* 4790 - 4792.
6. Reppy, M. A. *J. Fluoresc.* **2007** in press.
7. Reppy, M. A. *Mat. Res. Soc. Symp. Proc.* **2002,** *723,* O5_9_1 - O5_9_6.
8. Reppy, M. A.; Sporn, S. A.; Saller, C. F. US Patent 6,984,528, 2000.
9. Pindzola, B. A.; Nguyen, A. T.; Reppy, M. A. *Chem. Commun.* **2006**, 906-908.
10. Schaeffer, P.; Millet, J.; Aubert, J.-P. *Proc. Nat. Acad. Sci. U.S.A.* **1965,** *54,* 704-711.
11. Spevak, W.; Nagy, J. O.; Charych, D. H. *Adv. Mater.* **1995,** *7,* 85-89.
12. Reppy, M. A.; Pindzola, B. A.; Hussey, S. L. US Patent Application 11/653,260, 2007.
13. Hermanson, G. T. *Bioconjugate Techniques*; Academic Press, Inc.: San Diego, 1996; Vol.
14. Reppy, M. A.; Pindzola, B. A. *Mat. Res. Soc. Symp. Proc.* **2006,** *942,* 0942-W0913-0910.
15. Eglington, G.; McRae, W. In *Advances in Organic Chemistry*, 1963; Vol. 4, pp 252-281.
16. Iqbal, S. S.; Mayo, M. W.; Bruno, J. G.; Bronk, B. V.; Batt, C. A.; Chambers, J. P. *Biosens. Bioelectron.* **2000,** *15,* 549-578.
17. Dawson, R. M.; Paddle, B. M.; Alderton, M. R. *J. Appl. Toxicol.* **1999,** *19,* 307-312.
18. Jayasena, S. D. *Clin. Chem.* **1999,** *45,* 1628-1650.
19. Huang, A.; Huang, L.; Kennel, S. J. *J. Biol. Chem.* **1980,** *255,* 8015 - 8018.
20. Geiger, E.; Hug, P.; Keller, B. A. *Macromol. Chem. Phys.* **2002,** *203,* 2422-2431.
21. Gaboriaud, F.; Volinsky, R.; Berman, A.; Jelinek, R. *J. Colloid Interf. Sci.* **2005,** *287,* 191 - 197.

22. Dai, J.; Baker, G. L.; Bruening, M. L. *Anal. Chem.* **2006,** *78,* 135 - 140.
23. Kim, J.-M.; Ji, E.-K.; Woo, S. M.; Lee, H.; Ahn, D. J. *Adv. Mater.* **2003,** *15,* 1118 - 1121.
24. Shim, H.-Y.; Lee, S. H.; Ahn, D. J.; Ahn, K.-D.; Kim, J.-M. *Mat. Sci. Eng. C - Bio. S.* **2004,** *24,* 157-161.
25. Kim, J.-M.; Lee, Y. B.; Yang, D. H.; Lee, J.-S.; Lee, G. S.; Ahn, D. J. *J. Am. Chem. Soc.* **2005,** *127,* 17580 - 17581.
26. Khan, A. S.; Thompson, R.; Cao, C.; Valdes, J. J. *Biotechnol Lett* **2003,** *25,* 1671-1675.
27. Lombard, S.; Helmy, M. E.; Pieroni, G. *Biochem. J.* **2001,** *358,* 773-781.
28. Pan, J.; Charych, D. H. *Langmuir* **1996,** *13,* 1365-1367.
29. Kew, S. J.; Hall, E. A. H. *Anal. Chem.* **2006,** *78,* 2231-2238.

Chapter 14

Porous Silicon Waveguides for Small Molecule Detection

Sharon M. Weiss and Guoguang Rong

Department of Electrical Engineering and Computer Science, Vanderbilt University, Nashville, TN 37235, USA

A porous silicon resonant waveguide biosensor is demonstrated for the selective detection of DNA and other small molecules. Due to the large surface area of porous silicon and the enhanced interaction of target molecules with the electric field in the waveguide, porous silicon waveguides serve as the basis for efficient biosensing devices. In contrast to traditional evanescent wave sensors for which biomolecules are exposed to an exponentially decaying field, in the porous silicon waveguide sensor, biomolecules are infiltrated directly into the core of the waveguide where the electric field is most strongly concentrated. Experimental demonstrations of the selective detection of complimentary DNA oligos, with negligible response for both non-complementary DNA and buffer controls, have been performed.

The detection of minute quanitities of small molecules, including toxins, poses a significant challenge to many biosensing technologies. Specifically, for optical biosensors, the capture of small molecules causes only a small refractive index change, which is often difficult to detect. Evanescent wave-based sensors, such as surface plasmon resonance sensors (*1*), fiber optic sensors (*2*), and planar waveguide sensors (*3*) rely on the interaction of biomolecules with an exponentially decaying field. Porous materials are an attractive option for biosensors due to their potential to expose biomolecules to a propagating wave instead of an evanescent wave, and to increase the surface area available for molecular binding. This chapter explores the current and future capabilities of porous silicon waveguide sensors, including discussions on the importance of the fraction of the electric field interacting with biomolecules and the benefits of a large active sensing surface area.

Materials and Optical Properties of Porous Silicon

While first reported in 1956 (*4*), strong interest and meticulous characterization of porous silicon did not begin until several years later (*5, 6*). The most common method of fabricating porous silicon is by electrochemical etching of a crystalline silicon wafer in a hydrofluoric acid-based solution, although stain etching has also been performed (*7*). Through manipulation of the electrolyte constituents and their concentrations (e.g., hydrofluoric acid, water, ethanol, and organic solvents), the magnitude of applied current or voltage, the duration of applied current or voltage, and the substrate doping, the porous silicon morphology can be precisely controlled. For example, the porosity can be tuned from below 30% to more than 90%, the pore size can be adjusted between a few nanometers and several microns, and the thickness can be controlled from less than 10 nm to several hundred microns (*8*). Figure 1 shows an example of a typical porous silicon sample with randomly arranged nanoscale pores.

For mesoporous silicon, with pore sizes much smaller than the wavelength of light, the refractive index of the material can be modulated by simply changing the porosity of the material. Several effective medium approximations (EMAs) have been applied to porous silicon to determine the quantitative relationship between porosity and refractive index (*9*). The choice of the most appropriate approximation depends on the porous silicon morphology, and some customized EMAs have been proposed that incorporate birefringence or the presence of an oxide layer (*10, 11*). Figure 2 illustrates the relationship between porous silicon porosity and refractive index for the three most common EMAs. Given the flexibility in refractive index that can be achieved, multilayer porous silicon Bragg mirrors (*12*), rugate filters (*13*), microcavities (*14*), and waveguides (*15*) have been fabricated. Since electrochemical etching proceeds primarily at the pore tips where the electric field is concentrated and sufficient charge carriers are present, multilayer porous silicon structures can be fabricated without degradation of the upper layers that are formed first (*9*). These porous silicon photonic structures have served as the basis for several proposed sensing devices (*16-22*).

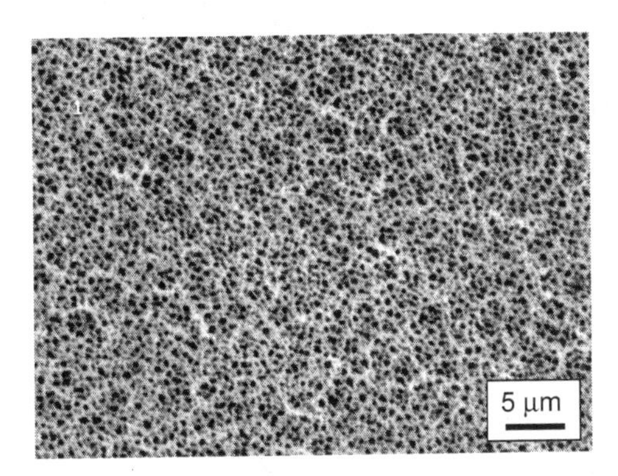

Figure 1. Plan view scanning electron microscopy image of porous silicon. The dark, circular regions are the air pores and the interconnected brighter region is the silicon matrix. The pore size and porosity can be changed by modifying the formation conditions and silicon doping.

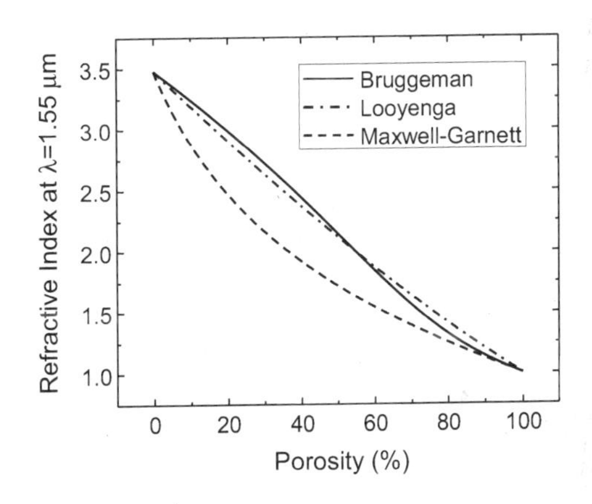

Figure 2. Effective medium approximations relating porous silicon porosity to refractive index for a wavelength of 1.55 μm. The differences arise from different assumptions made about the porous silicon morphology.

Porous Silicon Waveguide Biosensor

Design and Principle of Operation

Here we focus on the capabilities of a porous silicon waveguide biosensor for the detection of small molecules. The porous silicon waveguide, shown schematically in Figure 3a, consists of two porous silicon layers. Light is trapped in the top, high refractive index layer, based on total internal reflection at the interfaces with air and the bottom, low refractive index porous silicon layer. Light is coupled into the waveguide via a prism, in a manner similar to surface plasmon resonance sensor instruments. At a particular incident angle, the wavevector of the incident wave matches that of a propagating waveguide mode, and light is coupled into the waveguide. Figure 3b shows a typical waveguide spectrum; the resonance angle is the angle at which light couples into the waveguide. This resonance angle depends on the effective index of the porous silicon waveguide. Consequently, as illustrated in Figure 3b, if biomolecules are attached inside the pores of the waveguide, the effective index changes and the resonance angle shifts. The amount of material present in the pores can be quantified by measuring the magnitude of the resonance shift.

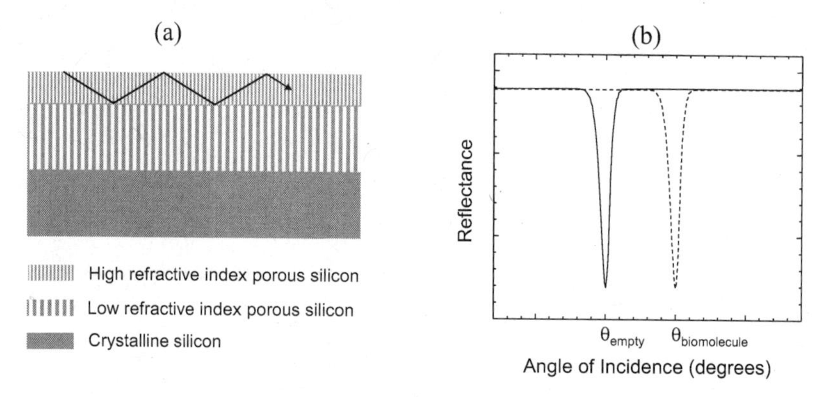

Figure 3. (a) Schematic of porous silicon waveguide. Due to total internal reflection, light coupled into the waveguide remains guided in the high refractive index porous silicon waveguide layer. (b) Representative prism coupling spectra of porous silicon waveguide with (dotted line) and without (solid line) biomolecules attached. The resonance angle corresponds to the incident angle at which light couples into the waveguide. Changing the effective refractive index of the porous silicon waveguide, due to the immobilization of biomolecules, changes the resonance angle.

Electric Field-Biomolecule Interaction: Relationship to Biosensor Performance

The sensitivity of biosensors using optical detection techniques is determined by how the optical properties of the sensor respond to changes in refractive index caused by biomolecule attachment. The magnitude by which the optical properties change is directly correlated to the fraction of the electric field in the region of the sensor where the biomolecules are present. Understanding the fundamental relationship between electric field distribution and biomolecule detection sensitivity is essential for designing highly sensitive optical biosensors.

Figure 4 shows the relationship between the porous silicon waveguide resonance shift and electric field interaction with biomolecules, in this case 24-base pair DNA, inside the porous silicon waveguide. In this first order analysis (*23, 24*), the total power of the field and its distribution remains constant while the number of DNA oligos inside the porous silicon is changed. The slope of the line, which gives the sensitivity of the device, depends on biomolecule size and the electric field distribution in the sensor (*25*). As the electric field-biomolecule interaction increases, the magnitude of the resonance shift also increases. The inset of Figure 4 shows the optical field distribution in the waveguide, which is obtained by solving eigenmode equations for the TE waveguide mode (*24*). The peak of the electric field falls within the top porous silicon waveguiding layer, where the biomolecules are immobilized. Depending on the functionalization conditions and target biomolecule incubation time, biomolecules may also be present in the lower porous silicon layer, where the field is exponentially decaying. In comparison to a planar waveguide biosensor, for which the electric field-biomolecule interaction only occurs in the air region, the porous silicon waveguide biosensor clearly has a larger field-molecule interaction region and hence is more sensitive to the presence of a few small molecules. Recent advances in the field of evanescent wave planar waveguides have led to design modifications that reduce the thickness of the waveguiding layer and consequently create a leaky mode that resides more prominently in the air cladding (*26*).

Calculations have been performed to estimate the detection limit of the porous silicon waveguide biosensor based on the percent of optical power interacting with biomolecules in the waveguide (*25*). The sensor is capable of detecting pictogram quantities of small molecules in a 1 mm^2 surface area. The detection limit depends on the number of probe molecules immobilized on the pore walls, the size of the biomolecules relative to the size of the pores, and the efficiency at which the probe molecules capture the target species.

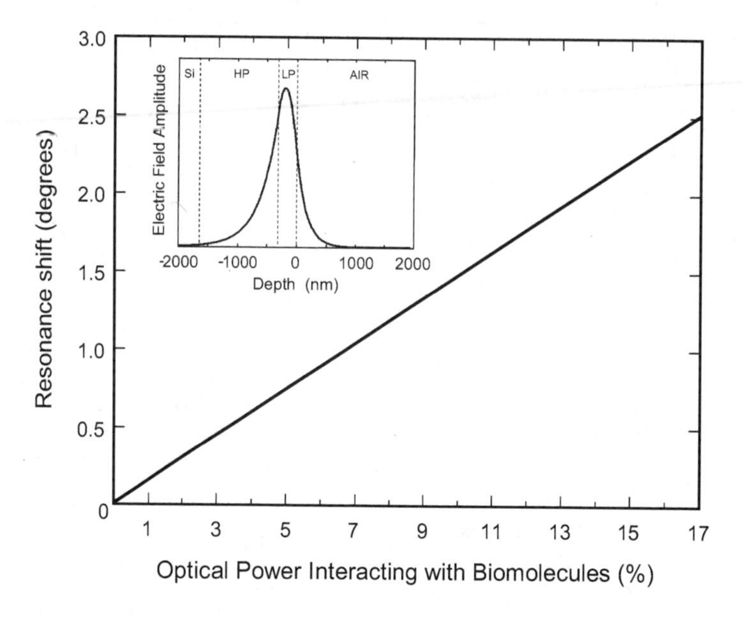

Figure 4. The response of the porous silicon waveguide sensor upon exposure to biomolecules depends on the percent of the optical field interacting with the molecules. The inset shows the field distribution in the waveguide with the majority of the field inside the porous silicon (LP = low porosity layer, HP = high porosity layer) where biomolecules may be captured. Therefore, even if only a few biomolecules are present in the waveguide, they can be detected.

Selective DNA Detection

Experimental verification of the porous silicon waveguide sensor capabilities has been performed. The porous silicon waveguides are formed by electrochemical etching of p-type (0.01 Ω·cm) silicon substrates in 15% hydrofluoric acid in ethanol. With these preparation conditions, the average porous silicon pore diameter is approximately 20 nm. The specifications of the waveguide are as follows: the high refractive index porous silicon layer (n ~ 2.08) has a thickness of 310 nm and a porosity of 56% while the low refractive index layer (n ~ 1.41) has a thickness of 1330 nm and a porosity of 84%. The waveguide is optimized for 1550 nm light since silicon absorption losses are significantly reduced in the near-infrared region (*27*). In order to immobilize probe molecules on the pore walls, the porous silicon must be prepared using a series of functionalization steps. The porous silicon was first oxidized by heating in a tube furnace at 900°C for 10 minutes. Standard aminosilane-glutaraldehyde chemistry was then followed (*28*). The porous silicon waveguide was exposed to 4% 3-aminopropyltriethoxysilane in water, followed by 2.5% glutaraldehyde in HEPES buffer. Amine terminated probe DNA was then attached to the glutaraldehyde. In order to stabilize the Schiff base formed during the reaction of aldehydes (from glutaraldehyde) with amines (from aminosilane and probe DNA), 5 M sodium cyanoborohydride was included in the functionalization protocol. Finally, 3 M ethanolamine was added to close any open glutaraldehyde sites that did not have DNA attached, in order to reduce the chance of non-specific binding. As shown in Figure 5, the porous silicon waveguide resonance was measured after oxidation, after exposure to 3-aminopropyltriethoxysilane, and after exposure to glutaraldehyde and probe DNA. All measurements were performed after the waveguide was dried in nitrogen. The resonance shift after each step confirms that the functionalization was successful and molecules were attached to the pore walls.

The prepared porous silicon waveguide sensor was tested by exposing it to complimentary DNA, non-complimentary DNA, and buffer solution. The DNA used in these experiments were 24 base pair sequences. The non-complimentary DNA had an 18 base mismatch. The melting temperature of both the complimentary and non-complimentary DNA sequences in buffer was approximately 63°. For ease of comparison, three different porous silicon waveguides were spotted with 100 μL of 50 μM complimentary DNA in buffer, 100 μL of 50 μM non-complimentary DNA in buffer, and 100 μL of buffer solution without DNA. Future experiments will be performed to test the reusability of the porous silicon waveguide sensor. The waveguides were incubated in the various solutions at room temperature in a humid environment for 1 hour. Before measurement, the wavguides were rinsed with buffer and dried with nitrogen to remove any species that were not specifically bound to the pore walls. The results of the experiments are shown in Figure 5 (a)-(c) and summarized in (d). The porous silicon waveguide sensor can selectively discriminate between complimentary (shift = 0.046°) and non-complimentary (shift = 0.008°) DNA as the resonance shift after exposure to complimentary DNA is nearly six times greater than the resonance shift after exposure to non-complimentary DNA, which is presumably due to non-specific binding. The shift due to exposure to buffer solution with no DNA is 0.001°, which is within the noise level of the prism coupler rotation stage. It is anticipated that adjustments to the functionalization protocol will lead to larger waveguide resonance shifts upon exposure to complimentary DNA and reduced non-specific binding signals.

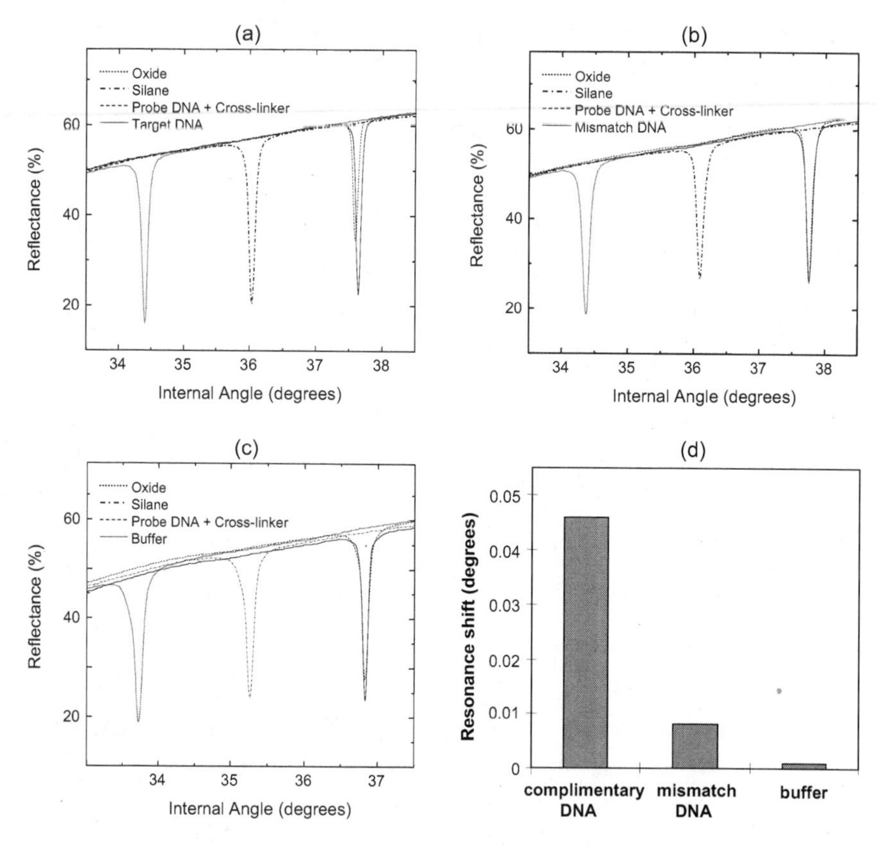

Figure 5. Experimental measurement of porous silicon waveguide resonance after several functionalization steps and exposure to either (a) complimentary DNA , (b) non-complimentary DNA, or (c) buffer solution. The small shift due to DNA hybridization is easily within the resolution of the prism coupler rotation stage of 0.002°. (d) Histogram summarizing the selectivity of the porous silicon waveguide sensor to 24 base pair DNA. DNA hybridization (complimentary DNA shift) can be clearly distinguished from the low level of non-specific binding (mismatch DNA shift) and the noise floor of the measurement system (buffer shift).

Potential for Sensitive, Label-free Detection of Toxins

The porous silicon waveguide biosensor has several advantages for the detection of toxins and other small molecules in real-world environments that contain non-purified target analytes. First, porous silicon has been demonstrated for selective detection of another small molecule, namely DNA. Second, the large surface area of porous silicon is well-suited to serve as both an active sensing medium and a material for preconcentration of target molecules. Third, the nanoscale pores naturally filter out material that is larger than the pore

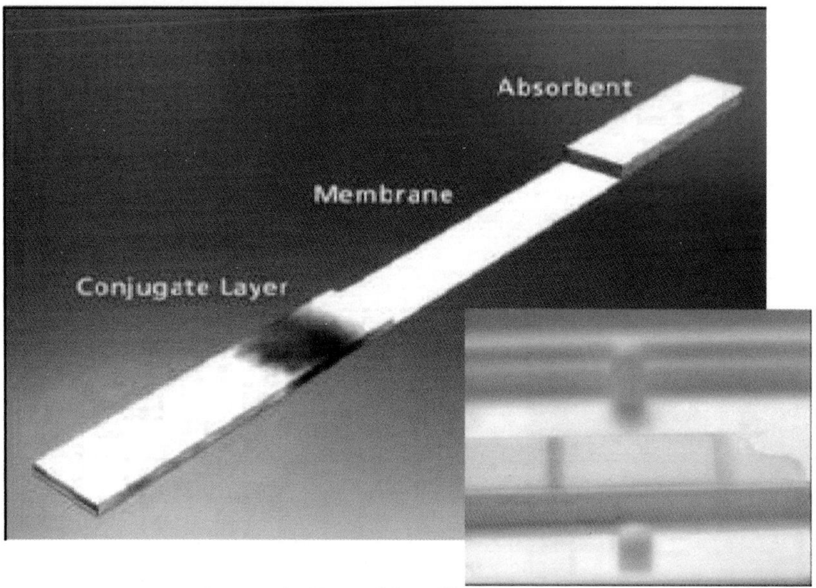

Figure 1. Typical hand held assay test (4).

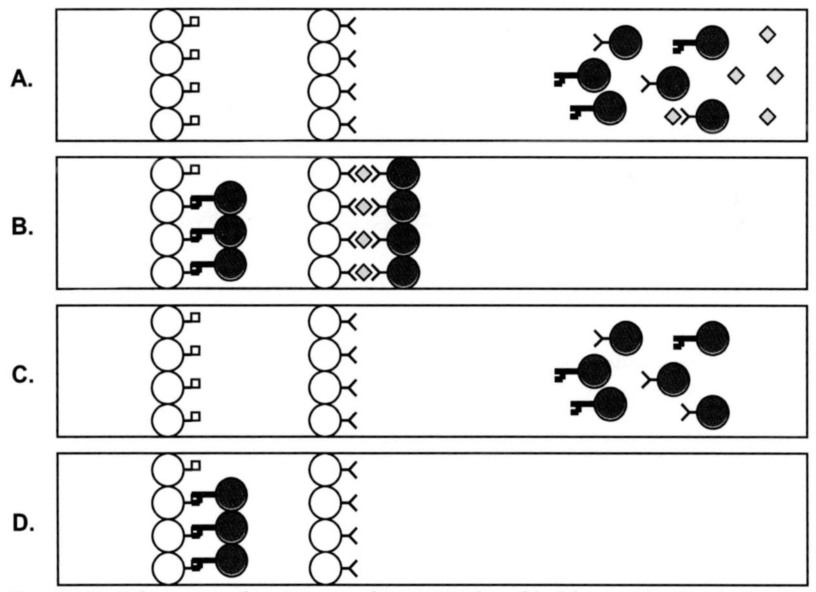

Figure 2. Schematic of positive and negative hand held assay tests. A. Target is present in the applied sample. B. The presence of target results in the formation of two purple lines on the test. C. No target is present in the applied sample. D. The lack of target results in the formation of one purple line on the test (4).

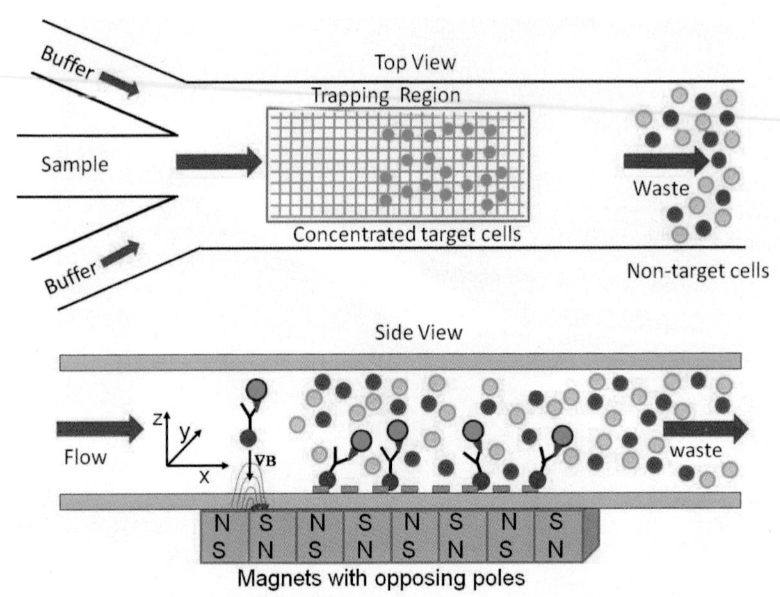

Figure 6. Diagram of a selection using a CTMACS device.

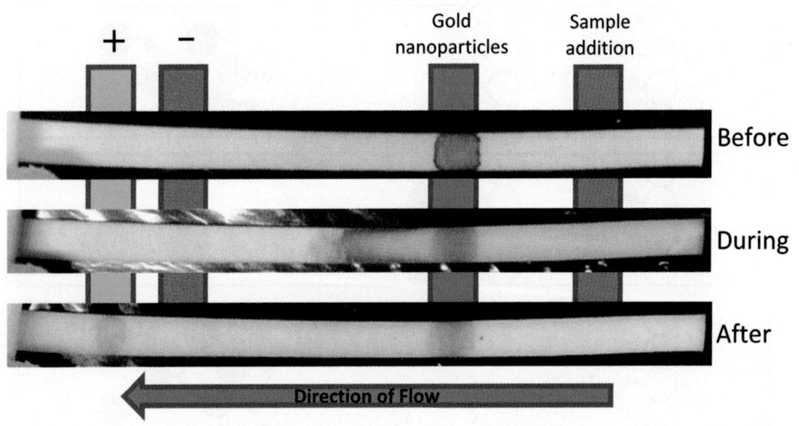

Figure 11. Cell based hand held assay for the detection of T7 antibodies.

diameter, such as cells and large proteins. As suggested in the inset to Figure 4, any debris non-specifically bound on the surface of the porous silicon waveguide sensor will not significantly affect the sensor response, in strong contrast to evanescent wave-based sensors such as surface plasmon resonance sensors. Finally, aptamers, single-stranded nucleic acid binding species, have been shown to be effective probes for the capture of harmful toxins (*29, 30*), and the functionalization protocol for attachment of aptamers to porous silicon is expected to directly follow from the functionalization protocol for DNA probes.

Summary

Porous silicon has several advantages as a biosensing material, including its large surface area that allows for the immobilization of many more probe molecules than is feasible for planar surfaces, its ability to concentrate target molecules within its volume, and the capacity to filter out contaminant material that is larger than the pore diameter. Porous silicon waveguides were fabricated to achieve a structure for which the majority of the electric field is concentrated in regions in which biomolecules are captured. The porous silicon waveguide sensor was demonstrated for the selective detection of DNA. The waveguide resonance shift resulting from DNA hybridization was six times larger than the small shift due to non-specific binding of non-complimentary DNA. Future modifications to the functionalization protocol are expected to improve the porous silicon waveguide sensor selectivity and extend its detection capabilities to other small molecules, including toxins.

Acknowledgements

This work was supported in part by the National Science Foundation (ECCS-0722143).

References

1. Homola, J.; Yee, S. S.; Gauglitz, G. *Sens. Act. B* **1999**, *54*, 3-15.
2. Abel, A. P.; Weller, M. G.; Duveneck, G. L.; Ehrat, M.; Widmer, H. M. *Anal. Chem.* **1996**, *68*, 2905-2912.
3. Rowe, C. A.; Tender, L. M.; Feldstein, M. J.; Golden, J. P.; Scruggs, S. B.; MacCraith, B. D.; Cras, J. J.; Ligler, F. S. *Anal. Chem.* **1999**, *71*, 3846-3852.
4. Uhlir, A. *Bell Syst. Tech. J.* **1956**, *35*, 333-347.
5. Watanabe, Y.; Arita, Y.; Yokoyama, T.; Igarashi, Y. *J. Electrochem. Soc.* **1975**, *122*, 1351-1355.
6. Cullis, A. G.; Canham, L. T.; Calcott, P. D. J. *J. Appl. Phys.* **1997**, *82*, 909-965.

7. *Properties of Porous Silicon*; Canham, L. T., Ed.; EMIS Datareviews Series 18; INSPEC: London, UK, 1997; pp 1-424.
8. Lehmann, V. *Electrochemistry of Silicon: Instrumentation, Science, Materials and Applications*; Wiley-VCH: Weinheim, Germany, 2002; pp 1-277.
9. Theiss, W. *Surf. Sci. Rep.* **1997**, *29*, 91-192.
10. Lugo, J. E.; delRio, J. A.; TaguenaMartinez, J. *J. Appl. Phys.* **1997**, *81*, 1923-1928.
11. Saarinan, J. J. (Lappeenranta University of Technology); Weiss, S. M. (Vanderbilt University); Fauchet, P. M. (University of Rochester); Sipe, J. E. (University of Toronto), *unpublished.*
12. Vincent, G. *Appl. Phys. Lett.* **1994**, *64*, 2367-2369.
13. Berger, M. G.; Arens-Fischer, R.; Thonissen, M.; Kruger, M.; Billat, S.; Luth, H.; Hilbrich, S.; Theiss, W. Grosse, P. *Thin Solid Films* **1997**, *297*, 237-240.
14. Pavesi, L *Rivista Del Nuovo Cimento* **1997**, *20*, 1-76.
15. Loni, A.; Canham, L. T.; Berger, M. G.; Arens-Fischer, R.; Munder, H.; Luth, H.; Arrand, H. F.; Benson, T. M. *Thin Solid Films* **1996**, *276*, 143-146.
16. Stewart, M. P.; Buriak, J. M. *Adv. Mater.* **2000**, *12*, 859-869.
17. Snow, P. A.; Squire, E. K.; St. J. Russel, P.; Canaham, L. T. *J. Appl. Phys.* **1999**, *86*, 1781-1784.
18. Cunin, F.; Schmedake, Th.A.; Link, J.R.; Li, Y.Y.; Koh, J.; Bhatia, S.N.; Sailor, M. J. *Nature Mater.* **2002**, *1*, 39-41.
19. Chan, S.; Horner, S. R.; Fauchet, P. M.; Miller, B. L. *J. Am. Chem. Soc.* **2001**, *123*, 11797-11798.
20. Mulloni, V.; Pavesi, L. *Appl. Phys. Lett.* **2000**, *76*, 2523-2524.
21. De Stefano, L.; Moretti, L.; Lamberti, A.; Longo, O.; Rocchia, M.; Rossi, A. M.; Arcari, P.; Rendina, I. *IEEE Trans. Nanotechnol.* **2004**, *3*, 49-54.
22. Ouyang, H.; Christophersen, M.; Viard, R.; Miller, B. L.; Fauchet, P. M. *Adv. Funct. Mater.* **2005**, *15*, 1851-1859.
22. Saarinen, J. J.; Sipe, J. E.; Weiss, S. M.; Fauchet, P. M. *Opt. Express* **2005**, *13*, 3754-3764.
23. Lukosz, W. *Biosens. Bioelectron.* **1991**, *6*, 215-225.
24. Taylor, H.F.; Yariv, A. *Proc. of IEEE* **1974**, *62*, 1044-1060.
25. Rong, G.; Weiss, S. M. (Vanderbilt University), *unpublished.*
26. Densmore, A.; Xu, D. X.; Waldron, P.; Janz, S.; Cheben, P.; Lapointe, J.; Delage, A.; Lamontagne, B.; Schmid, J. H.; Post, E. *IEEE Photon. Technol. Lett.* **2006**, *18*, 2520-2522.
27. Rong, G.; Weiss, S. M. *Proc. of SPIE* **2007**, *6477*, 647717.
28. Hermanson, G. *Bioconjugate Techniques*; Academic Press: San Diego, CA, 1996; p 470.
29. Ellington, A. D.; Szostak, J. W. *Nature* **1990**, *346*, 818-822.
30. Vivekananda, J.; Kiel, J. L. *Laboratory Investigation* **2006**, *86*, 610-618.

Chapter 15

A Microfluidic Sensor Array for Ricin Detection

Z. Hugh Fan[1,2], Qian Mei[1], Shouguang Jin[3]

[1]Department of Mechanical and Aerospace Engineering
[2]Department of Biomedical Engineering
[3]Department of Molecular Genetics and Microbiology
University of Florida, PO Box 116250, Gainesville, FL 32611, USA

We have developed a method for detecting toxins that inhibit protein synthesis. Biological synthesis of a protein starts from a gene to a messenger RNA and to a protein. This process can be implemented in a cell-free medium in a microfluidic array device. The device is comprised of reaction chambers for protein expression and feeding chambers as nutrient reservoirs. The device also incorporates dialysis membranes and microfluidic channels to connect chambers for supplying nutrients continuously and removing the reaction byproducts. To detect a toxin, a group of proteins are simultaneously synthesized in the array. The production yields of these proteins are inhibited differentially by the toxin. The toxin can thus be identified based on the unique response pattern (or signature) of the array. Three proteins have been synthesized in the device for the feasibility demonstration. The limit of detection for ricin, a bioterrorism agent, is determined at 10 pM.

Introduction

Detection and identification of toxins are important for medical diagnostics, food/water safety testing, and biological warfare defense. Methods to detect them include immunoassay, sensors, and mass spectrometry. Nucleic acid-based genetic analysis is not applicable to toxins because most toxins are proteins. One example of such toxins is ricin, which is listed as a Category B bioterrorism agent according to the Centers for Disease Control and Prevention (*1*) and was used as an agent in the letter sent to US congress in Feb. 2004. To detect them, immunoassay and immunosensors are often used due to their simplicity and rapid analysis. However, immunological assay requires an antibody that is specific to the agent of interest. It can not be used for detecting unknown or new agents because antibody is not available. With the increasing ability to modify and engineer potential warfare agents, the ability to detect agents that have not been identified previously becomes more important.

To address this challenge, we chose to develop a detection method based on the mechanism of toxin actions. One of the mechanisms by which toxins cause toxic effects is to inhibit protein synthesis in cells (*2, 3*). For example, ricin acts on the 28S ribosomal subunit and prevents the binding of elongation factor-2, a critical component in the process of protein translation. A number of other biological toxins exert their toxic effects through inhibition of protein synthesis, including Shiga toxin, diptheria toxin and exotoxin A (*3*). Therefore, protein expression could be used as a sensing mechanism for toxin detection.

While protein expression is commonly implemented using prokaryotic *E. coli* cells, it has also been realized in a cell-free medium employing a process called *in vitro* transcription and translation (IVT) (*4-7*). In IVT systems, a DNA template consisting of a coding sequence is transcribed into messenger RNA using RNA polymerases and an appropriate promoter; either eukaryotic or prokaryotic lysate is then exploited for providing ribosomes and additional components necessary for protein translation. The transcription and translation steps are coupled together and take place in the same reaction mixtures. Due to the absence of cellular control mechanisms, IVT overcomes the limitations (e.g., cytotoxicity) experienced by cell-based recombinant protein production (*4, 5*). IVT has been demonstrated for various applications (*8, 9*) and realized in microfluidic devices (*10-14*).

In this work, we develop a microfluidic sensor array for detecting toxins that inhibit protein synthesis. The sensor array consists of an array of microfluidic units; each unit is for expression of one protein and thus functions as one sensor. The sensor array will express a group of proteins in various expression systems; these proteins and expression systems are chosen judiciously so that the production yield of each protein is inhibited or affected differently by a toxin. The toxin can thus be profiled and identified based on the response pattern (or signature) of the array. New agents can also be identified by comparing the response pattern with signatures of known agents in a pre-collected database. The function of the device is demonstrated by synthesizing three proteins and studying differential inhibitory effects of two toxin simulants on protein expression. Finally, the sensor array is demonstrated for ricin detection.

Experimental

Reagents and Materials

The RTS 100 *E. coli* lysate, RTS 500 *E. coli* HY kit, RTS 100 wheat germ CECF kit, two expression vectors containing the genes encoding green fluorescent protein (GFP) and chloramphenicol acetyl-transferase (CAT), and anti-6xHis are obtained form Roche Diagnostics GmbH (Mannheim, Germany). TNT Quick coupled transcription/tranlation systems, T7 luciferase DNA vector, luciferase assay reagent, and nuclease-free water are from Promega Corporation (Madison, WI). Acrylamide-bisacrylamide (electrophoretic grade, 5% C), tetramethylethylenediamine, sodium dodecyl sulphate (SDS), ammonium persulfate, tris(hydroxymethyl)aminomethane (Tris), glycine, sodium chloride, glycerol, bromophenol blue, Tween-20, tetracycline, and cycloheximide are purchased from Fisher Scientific (Atlanta, GA). Polyvinylidene difluoride (PVDF) membranes (0.2 μm), and filter papers are from Bio-Rad Laboratories (Hercule, CA). Biotinylated secondary antibody and streptavidin-alkaline phosphatase are from Amersham Biosciences (Piscataway, NJ). The phosphatase staining solution (Bromo-chloro-indoryl phosphate/Nitro Blue Tetrazolium, BCIP/NBT) is obtained from KPL (Gaithersburg, MD). Ricin (A chain) is from Sigma-Aldrich (St. Louis, MO). Acrylic sheets are from Lucite-ES (Lucite International, Inc., Cordova, TN). The dialysis membrane with the molecular weight cutoff of 8 KDa is obtained from Spectrum Labs (Rancho Dominguez, CA), while a biocompatible epoxy (353ND-T) is bought from Epoxy Technologies (Billerica, MA).

Device Design and Fabrication

The design of the devices is shown in **Figure 1**. Figure 1a shows a picture of array devices forming 3 × 4 wells. The device is fabricated as previously described (*15*) and the well volume is about 13 μL, which is about 25 times smaller than the wells in conventional 96-well microplates. Figure 1b shows a design consisting of two parts. The top part consists of reaction chambers while the bottom part contains feeding chambers. A dialysis membrane is glued to the bottom of each reaction chamber using epoxy. As illustrated in Figure 1c, the membrane and microfluidic channel provide a path to connect two chambers (*14*). For both devices, the pitch (the distance between the well centers) is 9 mm, following the microplate standards defined by the Society for Biomolecular Screening (SBS) and accepted by the American National Standards Institute. To eliminate possible inhibition of enzymatic reactions, all devices are rinsed with nuclease-free water and then sterilized by exposure to UV light for 30 minutes.

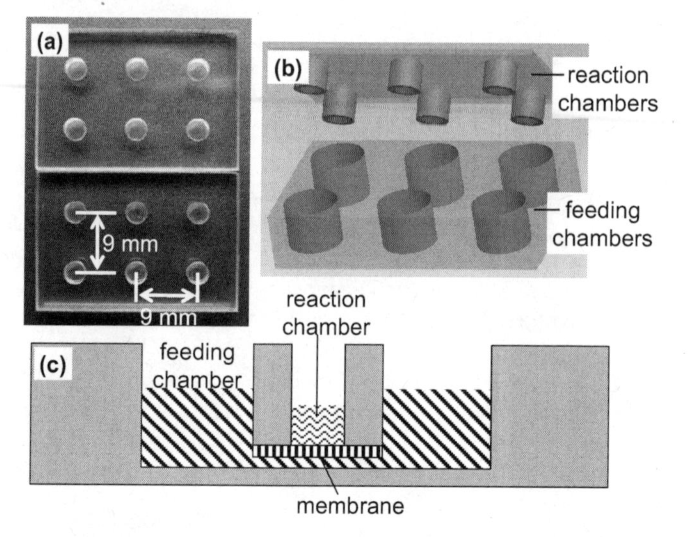

Figure 1. (a) Picture of two devices forming a 3 × 4 array. The diameter and depth of each well are 2.7 mm and 2.3 mm, respectively. (b) Design of a device consisting of reaction and feeding chambers. (c) Cross-sectional, expanded view of a reaction chamber nested in a feeding chamber, showing the dialysis membrane and microfluidic connection. (Reproduced from reference 14 . Copyright 2006 American Chemical Society)

Protein Synthesis

For demonstrating the toxin detection concept in the device in Figure 1a, GFP and CAT are expressed using a RTS 100 kit in *E. coli* lysate. Luciferase is expressed using rabbit reticulocyte lysate. For other experiments in the device in Figure 1b, GFP and CAT are expressed using a RTS 500 kit. The reaction and feeding solutions are prepared as follows. The reaction solution contains 0.525 mL *E. coli* lysate, 0.225 mL reaction mix, 0.27 mL amino acid without methionine, and 30 μL methionine. The feeding solution consists of 8.1 mL feeding mix, 2.65 mL amino acid without methionine, and 0.3 mL methionine. The reaction mix and feeding mix consists of proprietary composition, supplied in the kit by the manufacturer. To run the protein synthesis in the device, the reaction chamber is filled with 8 μL of the reaction solution containing GFP vector or CAT vector while 80 μL of the feeding solution is added to each feeding chamber. A biocompatible PCR tape (Corning, NY) is used to seal the reaction chambers to prevent evaporation. For comparison, 8 μL of reaction solution is dispensed into a microcentrifuge tube and incubated for the same time period as for device, though no feeding solution is used. Both the device and the tube are placed on a shaker with a speed of 100 rpm and incubated at room temperature for a period of time. At the end of GFP expression, the tubes and the device are stored at 4 °C overnight for appropriate folding. This step is not required for CAT. Western blotting is used to quantify the synthesized GFP and CAT. Gel images are scanned by a flatbed scanning and quantified using

ImageJ from the National Institute of Health (http://rsb.info.nih.gov/ij). The production of luciferase is carried out using a RTS 100 wheat germ expression kit. The amount of luciferase synthesized is determined by a Sirius luminometer from Berthold (Pforzheim, Germany). The expression product of 2 μL is added to a tube containing 40 μL of luciferase assay reagent and mixed evenly. The luminometer is programmed to have a two-second delay, followed by a five-second measurement of luciferase activity.

Toxin Detection

For the toxin inhibition assay in Figure 1a, 1.5 μL of tetracycline or cycloheximide are added into 6.5 μL of protein expression mixture. For each set of experiments, a positive control (without inhibitor) and a negative control (without the expression vector) are included.

To demonstrate ricin detection, a series of concentrations of ricin solutions are prepared from a stock solution of 35 μM. Then, 2 μL of each ricin sample is added into 6 μL of the reaction solution in the reaction chamber in Figure 1b. The volume of the feeding solution remained at 80 μL. For the positive controls in the same device, 2 μL of water is added. The negative controls contain no luciferase vector, providing with the background signal.

Results and Discussion

Toxin Detection Array

As discussed in the *Introduction*, detection of unknown or engineered agents is important due to the universal access of the recombinant technology. A multiplexed sensor array is a unique approach to obtain the fingerprint of a new agent. The concept of the sensor array for detecting toxins using IVT is demonstrated as follows. In the array of IVT wells in Figure 1a, each well is used to express one protein and thus functions as a sensor for detecting the inhibitory effects of a toxin. The top row of wells is for the positive controls to express each of 3 proteins, the second row for the negative controls, and the third and fourth rows are for the sample, allowing one repeat to enhance the precision. Use of the positive and negative controls and comparison of the signal from the sample wells with those in the control wells will reduce false positives and negatives.

Three proteins expressed in the array device are green fluorescent protein (GFP), chloramphenicol acetyl-transferase (CAT), and luciferase. Their production yields are inhibited differentially by two toxin simulants, tetracycline (TC) and cycloheximide (CH). **Figure 2a** shows the response pattern of the IVT array when TC is used whereas the response pattern of the same IVT array for CH is illustrated in **Figure 2b**. The signals for the positive control are from the first row of 3 wells in a device, in which GFP, CAT, and luciferase are synthesized in their respective expression mixture. These wells are free of toxins. The signals for the negative control are from the second row of 3 wells, in which the expression vector is not added. The signals for the samples are from the remaining two rows of 3 wells, in which either 17 ng of CH or 25 ng of TC is added into the protein expression system. The significant difference in the response patterns between CH and TC clearly indicates that it is feasible to use the IVT array to detect and identify a toxin.

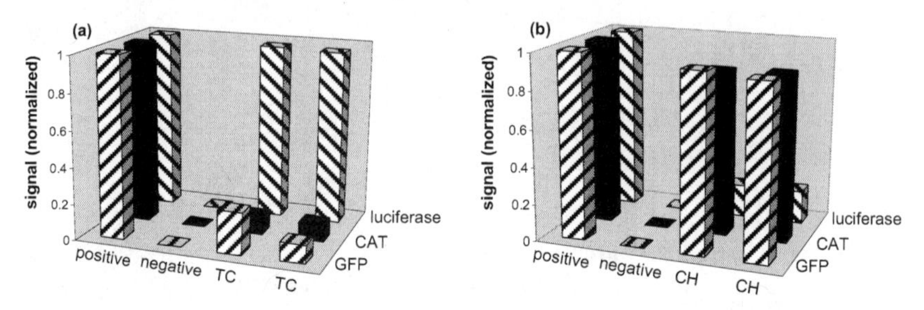

Figure 2. Response pattern of the array for two toxin simulants, tetracycline (TC, a) and cycloheximide (CH, b). (Reproduced from reference 15 . Copyright 2005 American Chemical Society)

Device with Fluid Manipulation

The nested well device in Figure 1b possesses flow control whereas simple wells in Figure 1a do not. Each nested well unit is for expression of one protein. Both gene transcription and protein translation take place in the reaction chamber, while the feeding chamber functions as a nutrient reservoir. The feeding chambers contain amino acids, adenosine triphosphate (ATP), guanosine triphosphate (GTP) and buffer. ATP is critical to activate amino acid substrates and GTP is the energy source required for the function of ribosomes. The reaction chamber contains cell-free expression system with other reagents as in the feeding chamber.

As illustrated by Spirin et. al (*5, 16*) and Roche's RTS (*17*), continuous supply of nutrients and selective removal of small molecule byproducts are critical to achieving high protein expression yield. We obtain these desirable features by incorporating a dialysis membrane and a fluidic connection as shown in Figure 1c. The dialysis membrane has the molecular weight cutoff of 8 KDa, allowing continuous feeding of small-molecule nutrients (e.g., ATP and GTP). It also lets small molecular byproducts (e.g., hydrolysis products of triphosphates) diffuse away while retaining the synthesized proteins in the reaction chamber. A picture of two parts of the sensor array is shown in **Figure 3**.

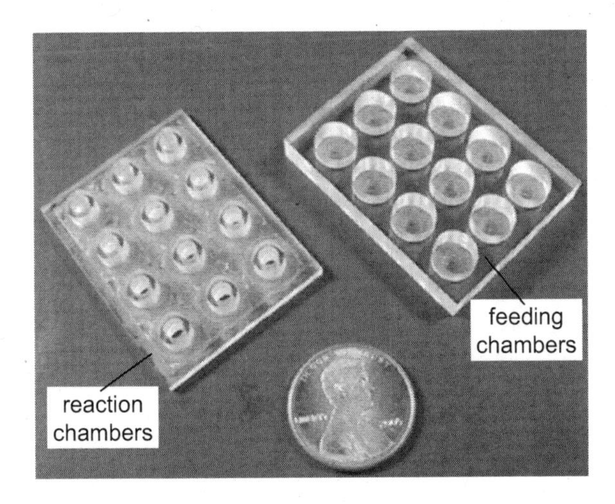

Figure 3. Picture of the top part (the reaction chambers) and the bottom part (the feeding chambers).

The fluidic connection between the reaction and feeding chambers is shown in the cross-sectional view in Figure 1c. The height of the connection ranges from hundreds microns to a few millimeters. One of the mechanisms for supplying nutrients from the feeding chambers to the reaction chambers and for removing byproducts from the reaction chambers to the feeding chambers is diffusion, a net flow resulting from the difference in concentration of solutes between two solutions separated by the membrane. In addition, the flow to supply fresh solution from the feeding chamber to the reaction chamber is augmented by a hydrostatic pressure, which is caused by the difference in the solution level between the reaction and feeding chambers. When the feeding chamber has slightly higher solution level than the reaction chamber, the pressure difference resulting from the height difference will drive solution from the feeding chamber into the reaction chamber.

Incorporation of the dialysis membrane and fluidic connection enables continuous supply of nutrients and selective removal of small molecule byproducts. The nutrient-feeding solution will avoid fast depletion of the energy source, and the accumulation of small molecular byproducts is reduced. This device design leads to higher protein expression yield as discussed in the next section.

Protein Synthesis in Nested Well Array

To demonstrate the function of the sensor array with fluid manipulation, GFP is first expressed using commercial *E. coli* system with a GFP coding sequence. **Figure 4a** shows the protein expression yield as a function of reaction time; the signal intensity is obtained from the western blot images. The result indicates that GFP is continuously synthesized in the device for up to 20 hrs. To show the effects of the flow manipulation, we did the same reactions in a microcentrifuge tube and found that protein expression ceased after 4 hrs.

This result clearly illustrates the importance of incorporation of the dialysis membrane and fluidic connection in the device. Continuous supply of nutrients and selective removal of small molecule byproducts enable protein synthesis to continue for up to 20 hrs in the device, which is 5 times longer than the expression time in the tube. The longer expression time results in higher expression yield. The amount of GFP synthesized in the device is about 13 times larger than in the tube.

Similar expression has been achieved for other two proteins, CAT and luciferase. CAT is expressed in the same *E. coli* expression system; success of the protein expression is confirmed by Western blotting. The signal intensity as a function of the reaction time is shown in **Figure 4b**. The result also suggests that longer expression time can be achieved in the device than in the tube. The production yield increased more than 22 fold in the device. This confirms that we have achieved the desired fluid manipulation, both continuous feeding of the nutrient solution and removal of the byproducts, in the device. Synthesis of luciferase is carried out using wheat germ expression system. Detection of the expression product is achieved by monitoring the intensity of luminescence after mixing the product with luciferin and ATP. A similar result (data not shown) is obtained.

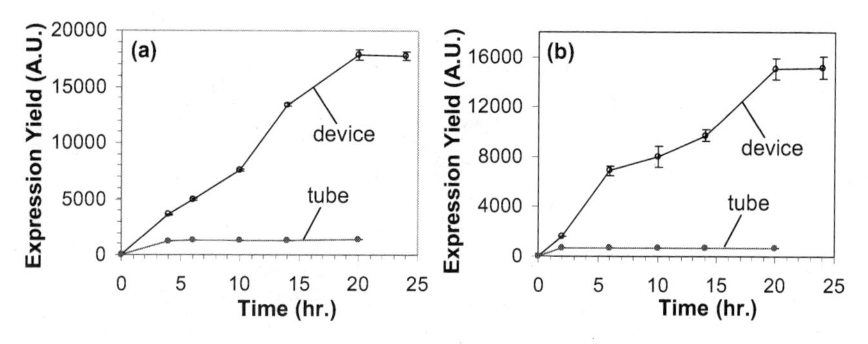

Figure 4. Comparison of protein synthesis between a device and a microcentrifuge tube. (a) The production yield of GFP as a function of the expression time. (b) The production yield of CAT as a function of the expression time.

Ricin Detection

To detect ricin, we studied its inhibitory effects on luciferase expression by adding a series of concentrations of ricin into the IVT reactions in the array device. To achieve a lower detection limit, 4-hr protein expression was used, though ricin detection can be achieved in as short as 5 minutes. Ricin sample size was 2 μL. As shown in **Figure 5a,** the expression yield of luciferase, indicated by luminescence, decreases with the concentration of ricin (solid circles). However, the expression yield remained the same when the ricin is heat denatured and its toxicity is deactivated (open circles). The error bar of each data point indicates the standard deviation that is obtained from three repeat experiments. The calibration curve is obtained by plotting the detection signal (i.e., the difference between the samples and the controls) as a function of the ricin concentration (**Figure 5b**). A linear relationship exists from 0.035 to 0.69 nM. For a larger concentration range, a non-linear behavior of inhibition reactions is observed as expected (*13*). The detection limit is calculated to be 0.01 nM (0.3 ng/mL) by using the criterion that the signal-to-noise ratio is three times the standard deviation of the blank.

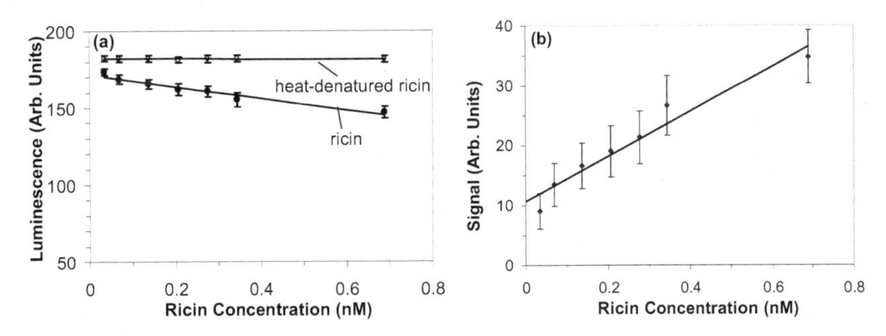

Figure 5 (a) The inhibitory effects of ricin on the production yield of luciferase. The expression yield of luciferase is plotted as a function of the concentration of ricin (solid circles) and heat-denatured ricin (open circles). (b) *Calibration curve for ricin detection. (Reproduced from reference 14 . Copyright 2006 American Chemical Society)*

Conclusion

A novel concept for toxin detection is presented based on toxin's inhibition of biological protein synthesis. We demonstrated the feasibility of the concept by confirming differential inhibitory effects of two toxin simulants on the expression yields of three proteins. The unique response pattern is obtained from the 3 × 4 IVT sensor array for each toxin simulant.

In addition, an array of nested wells with microfluidic connection has been developed for protein expression. The array device consists of reaction chambers nested in corresponding feeding chambers. A dialysis membrane and a microfluidic channel are used to connect reaction and feeding chambers. We demonstrate higher expression yield of GFP, CAT, and luciferase in the device than in a conventional microcentrifuge tube. The microfluidic sensor array is also demonstrated for detecting ricin from 0.035 to 0.69 nM.with a detection limit at 10 pM.

Acknowledgment

This work is supported in part by the Research Opportunity Incentive Seed Fund from the University of Florida and a grant from Defense Advanced Research Projects Agency (DARPA) via Micro/Nano Fluidics Fundamentals Focus Center at the University of California at Irvine. We would like to thank our colleagues Carl Fredrickson, Wei Lian, and Andrew Simon for their contribution cited in this chapter.

References

(1) CDC *(The Centers for Disease Control and Prevention)*, www.bt.cdc.gov/agent/agentlist-category.
(2) Teter, K.; Holmes, R. K., *Infect. Immun.,* **2002,** 70, 6172-9.
(3) Salyers, A. A.; Whitt, D. D., *Bacterial pathogenesis: a molecular approach.* 2nd ed.; ASM Press: Washington, D.C., 2002.
(4) Kigawa, T.; Yabuki, T.; Yoshida, Y.; Tsutsui, M.; Ito, Y.; Shibata, T.; Yokoyama, S., *Febs Letters* **1999,** 442, 15-19.
(5) Spirin, A. S., *Trends Biotechnol.,* **2004,** 22, 538-45.
(6) Katzen, F.; Chang, G.; Kudlicki, W., *Trends Biotechnol.,* **2005,** 23, 150-6.
(7) Shimizu, Y.; Inoue, A.; Tomari, Y.; Suzuki, T.; Yokogawa, T.; Nishikawa, K.; Ueda, T., *Nat, Biotechnol.,* **2001,** 19, 751-5.
(8) He, M.; Taussig, M. J., *Nucleic Acids Res.,* **2001,** 29, E73-3.
(9) Tao, S. C.; Zhu, H., *Nat. Biotechnol.,* **2006,** 24, 1253-4.
(10) Angenendt, P.; Nyarsik, L.; Szaflarski, W.; Glokler, J.; Nierhaus, K. H.; Lehrach, H.; Cahill, D. J.; Lueking, A., *Anal. Chem.* **2004,** 76, 1844-1849.
(11) Tabuchi, M.; Hino, M.; Shinohara, Y.; Baba, Y., *Proteomics* **2002,** 2, 430-5.
(12) Nojima, T.; Fujii, T.; Hosokawa, K.; Yotsumoto, A.; Shoji, S.; Endo, I., *Bioprocess Eng.* **2000,** 22, 13-17.
(13) Harley, S. M.; Beevers, H., *Proc. Natl. Acad. Sci. USA,* **1982,** 79, 5935-5938.
(14) Mei, Q.; Fredrickson, C. K.; Lian, W.; Jin, S.; Fan, Z. H., *Anal. Chem.* **2006,** 78, 7659-64.
(15) Mei, Q.; Fredrickson, C. K.; Jin, S.; Fan, Z. H., *Anal. Chem.* **2005,** 77, 5494-500.
(16) Spirin, A. S.; Baranov, V. I.; Ryabova, L. A.; Ovodov, S. Y.; Alakhov, Y. B., *Science* **1988,** 242, 1162-4.
(17) Betton, J. M., *Cur.r Protein Pept. Sci.* **2003,** 4, 73-80.

Chapter 16

Enhanced Raman Scattering of Nitroexplosives on Metal Oxides and Ag/TiO_2 Nanoparticles

Samuel P. Hernández-Rivera[1], Julio G. Briano[2], Edwin de la Cruz-Montoya[1], Gabriel A. Pérez-Acosta[2] and Jacqueline I. Jeréz-Rozo[1]

Chemical Imaging Center / Center for Chemical Sensors Development
[1]Department of Chemistry, [2]Department of Chemical Engineering
University of Puerto Rico-Mayagüez
PO Box 9019, Mayagüez, PR 00681

A central goal in the emerging field of nanotechnology is to fabricate nanostructures with very large surface area. This property is essential when designing sensors of very small footprint. Surface Enhanced Raman Scattering is normally obtained from nanoactive surfaces or metallic colloidal sols of group II-B elements, in particular silver and gold. In this study another type of nanosurface has been explored seeking for SERS activity. Oxides of elements of the fourth transition period were tested for enhanced detection of nitroexplosives. Crystalline nanoparticles of TiO_2, SnO_2 and Sc_2O_3 were used in the experiments. Polymorphism seems to play an important role in the Raman signal enhancement when using metal oxides: high rutile percent mixtures with anatase gave higher Raman scattered signal enhancement. TiO_2 coated silver nanoparticles were also successfully prepared and tested for SERS activity in high explosives detection. Understanding the nano-structure and the electronic properties of these systems will help in developing sensor applications for the detection of nitroaromatic explosives at the sub-picogram level.

Semiconductor particles at the nano scale are currently under investigation as examples of non-molecular materials that show effects of quantum and dielectric confinement. The chemical modification or surface capping with organic functional molecules on the nanoparticles surfaces produces unusual physical and chemistry properties, such as photoreactivity and phototoelectric conversion with high efficiency; a prominent enhanced Raman Effect and large nonlinear optical properties *(1)*. Core-shell nanostructures are also of interest because of their potential applications to different fields, such as microelectronics, optoelectronics, catalysis, and optical devices *(2)*. Among such nanocomposite structures, the TiO_2/Ag has attracted attention not only because TiO_2 is a useful material of many applications including photoelectrochemical activity, solar energy conversion, and photocatalysis *(3)* but also because silver nanomaterials display some unique activities in chemical and biological sensing. Such effects are based on Surface Enhanced Raman Scattering (SERS), Localized Surface Plasmon Resonance (LSPR), and Metal-Enhanced Fluorescence (MEF). Moreover, TiO_2/Ag nanostructures may lead to enhanced optical and catalytic properties because of the fabrication of a silver core and TiO_2 sheath as well as new chemical activities owing to the electron transfer between photoexited TiO_2 and silver *(4)*.

Surface Enhanced Raman Scattering has generally been observed on Ag, Cu, Au and several transition metals, with certain restricted use. Until recently, SERS has been reported on conducting polymers *(5)*, semiconductors *(6)* and single molecules *(7-8)*. SERS makes Raman Spectroscopy to be more sensitive and competitive when compared to more robust techniques such as Infrared Spectroscopy and Electronic Fluorescence. Sensitivity is a term often used to refer either to the lowest level of chemical concentration that can be detected or to the smallest increment of concentration that can be detected in the sensing environment. However in HE detection, environmental monitoring and toxic and chemical agents detection is necessary only to know when the concentration of a single or small number of chemicals exceeds certain alarm levels *(8-10)*. The focus of this work is to find a variety of SERS substrates using films and nanoparticles of TiO_2 aimed at high explosives (HE) detection.

Different mechanisms have been proposed for the enhancement in the Raman scattered intensity, but only two mechanisms are fundamentally sound: an electromagnetic enhancement and a chemical bond effect *(11-12)*. The electromagnetic field-surface-molecule interaction arises when the electric field of the exciting light is modified by discontinuities at the metallic surface. The **E** field changes dramatically at the local discontinuities or surface roughness giving rise to a high local field at the molecule-surface binding site. This effect is considered electromagnetic enhancement mechanism.

The molecular orbital molecule-surface interaction generates new electronic states that can be resonant with light laser at lower frequencies than the states of either the molecule or the surface. This is the so called chemical enhancement, the other SERS important mechanism *(12)*. The results presented here demonstrate that the signals of the Raman spectrum of the target nitroaromatic HE: TNT and DNT are enhanced strongly by contact with the metal oxide particles. Since Raman signal enhancement is clear in this experiment, other mechanisms could be responsible for the enhancement factors found.

Enhanced Raman Scattering Substrates

Several experiments were carried out to observe the enhancement of the Raman signatures of TNT and DNT with metal oxides. In these experiments TiO_2 was used in the rutile and anatase phases and included depositing finely divided particles to increase the surface area of the substrate. Anatase was converted to rutile by heating in a furnace at a temperature slightly higher than the phase transition temperature for several minutes. Upon examination of the Raman spectra of both sources of TiO_2 it was confirmed that the conversion was only partial.

In other experiments, twenty (20) mg of KBr and one (1) mg of TNT were mixed and ground with a mortar and pestle. Samples were transferred to a glass slide and Raman spectra were collected at various spots. The procedure was repeated for each of the oxides as substrates. To improve sample preparation method, KBr/TNT and oxide/TNT samples were mixed, ground and pressed formed into pellets of the same type used for solid samples in IR spectroscopy absorption/transmittance measurements.

Synthesis of Ag/TiO_2

The preparation of Ag/TiO_2 nanoparticles can be considered as a combination of two processes that occur sequentially in a single reaction mixture: the formation of the silver core and the coating with TiO_2. The starting reaction mixture was prepared from two solutions. First, Ag sols were prepared using the method of Lee and Meisel *(13)*. All glassware was rigorously cleaned using *aqua regia* (strong acid mixture of HCl/HNO_3; 4:1) which was handled with extreme care as copious amounts of gas were liberated. After the *aqua regia* wash, the glassware was rinsed with triple distilled water before washing with detergent. Finally, the glassware was rinsed again with triple distilled water.

One hundred (100) mL of triple distilled water was transferred to the flask. The water was stirred vigorously and heated to approximately 90°C while purging with dry high purity N_2 and 0.024 g of $AgNO_3$ was added. Almost at the point of boiling, a solution of 1% trisodium citrate, in the amount required for each experiment, was added. The mixture was held at boiling for 1 hr and then allowed to cool.

The second solution contained equimolar amounts of isopropoxide and isopropanol in water at a concentration approximated 5.75 mM for each component. Freshly prepared 20 mL of the first solution and 5 mL of the second one were mixed in a round bottom flask and stirred while heating at the reflux temperature. After 90 minutes the colloids attained a green-black coloration.

Raman Spectroscopy Measurements

A Renishaw Raman Microspectrometer RM2000 system was employed for the vibrational spectroscopy measurements. The system was equipped with a

Leica microscope. The excitation sources were a Spectra Physics EXCELSIOR diode pumped 532 nm green laser with a variable output power (200 mW max., Spectra-Physics/Newport Cor. Mountain View, CA) and a Coherent INNOVA 308 argon ion laser (Ar^+) operating at 514.5 and 488 nm (Coherent Laser Group, Santa Clara, CA). Leicá objectives with 10x magnifications were used. The other excitation sources were a Coherent VERDI diode pumped 532 nm green laser with an variable output power up to 6 W and a two portable Raman spectrometers R-3000 QE High Power 785 nm (300 mW max output power) and 532 nm diode laser (35 mW max power) from Raman Systems, Inc. (Austin, TX). The laser beams were focused onto capillary tubes through objectives with 10x magnifications or with fiber optic cables. The spectra were obtained in the 100-3000 cm^{-1} range using 20 scans and 10 s of integration time.

Characterization

The prepared nanoparticles were characterized by UV-VIS absorption. Colloids were also evaluated for ERS by observing the enhancement produced on a 10^{-4} M adenine solution. Figure 1 shows the UV-VIS absorption spectra of colloids of: (a) Ag nanoparticles; (b) Titania gel; and (c) Titania-coated Ag nanoparticle suspension. The concentration of Ag nanoparticles in the suspension decreased after coverage with TiO_2 as evidenced by UV-VIS absorption measurements. The spectrum of the suspension of Ag nanoparticles had a surface plasmon of 420 nm arising from typical silver nanoparticles with yellow color colloidal suspension, which is characteristic of Ag colloidal dispersion *(8)*. Similar surface plasmon absorptions due to Ag nanoparticles were observed for colloidal suspensions of Titania coated Ag nanoparticles at longer wavelengths (438-444 nm) than for the suspension of Ag nanoparticles. This absorption shift is caused by refractive index of Titania and suggests coverage of Ag nanoparticles with Titania.

Scanning Electron Microscope (SEM, JEOL,6460LV) coupled with Energy-Dispersed Analysis X-Ray photoemission (EDAX) was employed to observe the microstructure of the samples. X-ray microanalyses provided an indication of which elements present in the sample and were used to assess the crystallinity (or lack of it) of the titanium dioxide nanoparticles and of the Ag/TiO_2 colloidal suspensions deposited on test surfaces. The electron microscope parameters used were: 3300x of magnification; beam spot size of 500 μm. The secondary electron image was taken with on energy dispersed X-ray spectrometer operating at 10 kV. Figure 2 shows peaks for O, Ag, and Ti and evidences the effectiveness in coating Ag^o nanoparticles with TiO_2 nanoparticles *(14)*.

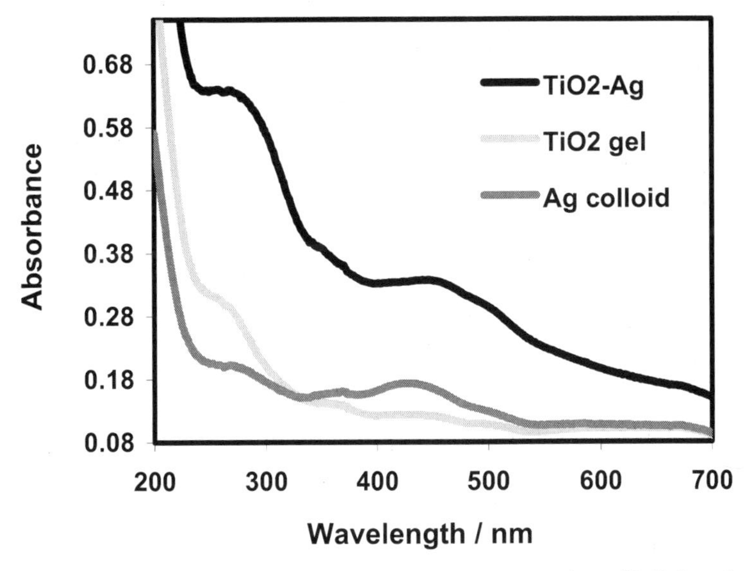

Figure 1. UV-VIS spectra of: (a) Ag colloidal nanoparticles; (b) TiO_2 gel; (c) TiO_2 coated Ag nanoparticles suspension.

ERS Spectroscopy of Nitroexplosives on Metallic Oxides

Metal oxides mediated Enhanced Raman Scattering in our research group came as a result of continued pursuit of improving Raman signatures of explosives on test surfaces: stainless steel, silicon, glass, compact disks and soils. Figure 3 shows white light micrographs obtained with a high magnification objective of a mineralogical microscope *(15)*. On the left, Figure 3 (a) shows DNT crystals deposited on gold coated glass microscope slide by direct solution transfer after solvent evaporation. The spontaneous or normal Raman Scattering (RS) spectrum is illustrated in Figure 4 (a). Figure 3 (b) shows the appearance of the sample on the left after spraying commercial TiO_2 (Anatase) nanoparticles on top of the DNT crystals and shaking off the excess of oxide. The large enhancement found in the oxide mediated ERS spectrum of this mixture is shown in Figure 4 (b). To collect the Raman spectrum of this sample the power of a solid state diode laser operating at 785 nm had to be turned down to its minimum level and 3 s of integration saturated the detector at low Raman Shifts (< 700 cm^{-1}), which corresponds to the TiO_2 vibrational signatures region. No *a priori* selective band enhancements can be identified, but rather all the vibrational bands of the nitroaromatic explosive seem equally enhanced.

Figures 5-7 contain examples of use of oxides for Enhanced Raman Spectroscopy using unconventional substrates. Scandium (II) oxide and tin (IV) oxides were used as "proof of concept" experiment. Oxides used are semi-conductors and in these cases signal enhancement was significant. The ERS spectra obtained using oxides to enhance Raman signals are quite different than ordinary SERS or Surface Enhanced Resonance Scattering (SERRS).

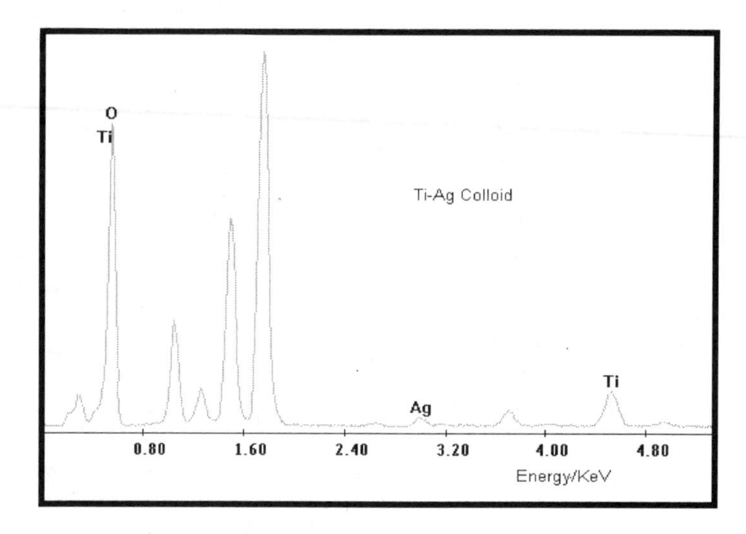

Figure 2. EDAX spectra for Ag /TiO_2 colloids obtained with JEOL 6460LV SEM. (Reproduced with permission from reference 14.)

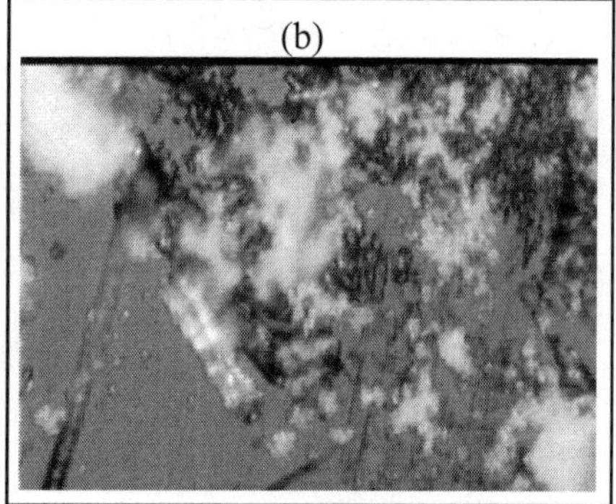

Figure 3. White light micrographs of analyte deposited on stainless steel plate: (a) DNT crystals on stainless steel plate; (b) image of DNT after adding TiO_2. (Reproduced with permission from reference 15.)

All vibrations present in the neat, crystalline explosive seem to be enhanced by the same factor (non-selective enhancement). Moreover, the spectra resemble solid phase normal Raman (NR) spectra, with fully resolved, non-broadened peaks. The reproducibility in band intensities is slightly better than in SERS spectra (~ 20-30%), but is prone to sample variation due to the lack of homogeneity in the solid-solid mixtures. KBr-analyte pellets were used as reference, since no evidence of ERS was found for this substrate. Experiments using cobalt (II) oxide did not result in observation of enhanced Raman signals under the experimental protocol followed (Figure 7).

Results with oxides tested are compatible with results of Liao et al. for titanium dioxide single crystals *(16)*. It is postulated that the enhancement depends on the physical form of the oxide: type of polymorph present; bulk, cluster or nanoparticle present and oxidation state of the metal cation *(15)*. An energy transfer mechanism has been suggested for the ERS phenomenon, but it is very likely that other mechanisms such as transfer of oscillator strength, also contribute to the Raman signal enhancement observed in these systems *(16-17)*.

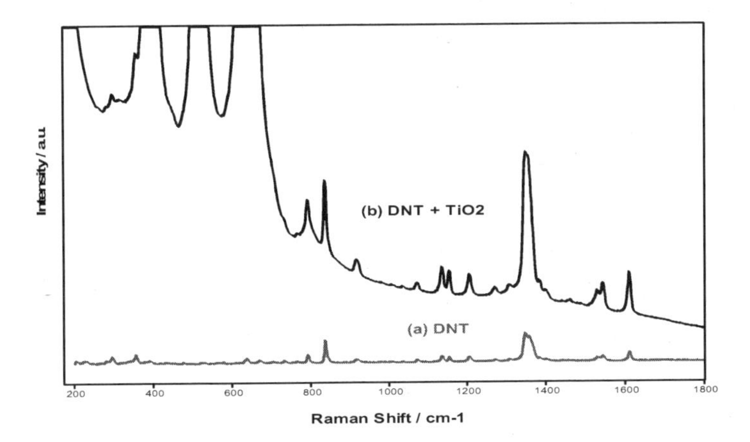

Figure 4. Raman spectra of (a) neat 2,4-DNT; (b) DNT + TiO_2.

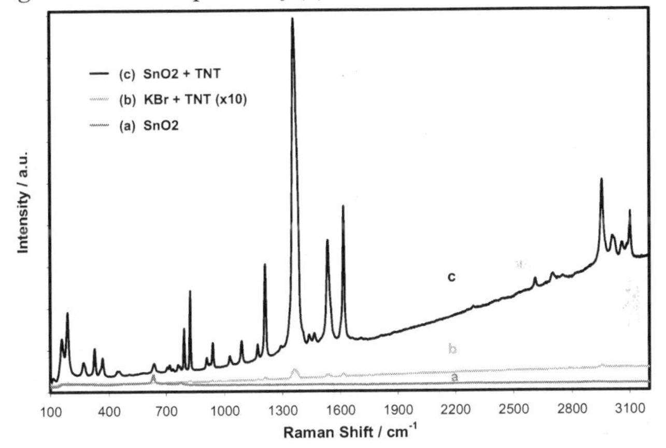

Figure 5. Raman spectra of: (a) SnO; (b) TNT + KBr (intensity x10); (c) SnO + TNT (20:1 mix).

SERS spectra of nitroaromatic HE were obtained with Ag/TiO_2 colloidal suspensions. A series of 10^{-3} M TNT solutions at different pH values were studied for SERS activity. Figure 8 shows typical SERS spectra obtained with 488 nm laser excitation at pH=6.5 (trace a) and pH=10.5 (trace b). The largest enhancement of the NO_2 asymmetric stretching mode (ca. 1365 cm^{-1}) was observed for pH = 10.5. Spectral characteristics such as narrowing of bands, large number of enhanced vibrations and presence of luminescence background (as observed in trace e) demonstrate SERS of TNT in the Ag/TiO_2 system. The aromatic ring breathing mode was observed near 1018 cm^{-1}. This band was not prominent in spectra acquired with 532 nm excitation. Other bands observed were 1555-1587 cm^{-1}: NO_2 asymmetric stretch and 1300-1357 cm^{-1}: NO_2 symmetric stretch. The band pattern was highly prominent at pH 10.5 and decrease to nearly noticeable at slightly acidic media (pH 6.5) as illustrated in Figure 8. These vibrations were observed at very low concentrations in experiments with 532 nm excitation. In addition, the band 1213 cm^{-1} is shifted to higher wavenumbers and NO_2 symmetric stretch band at 1365 cm^{-1} shifts to lower wavenumbers: 1330-1357 cm^{-1}. This can be attributed to very aggressive chemistry in the presence of alkali that generates nitro containing stable products which bind strongly to the SERS substrates.

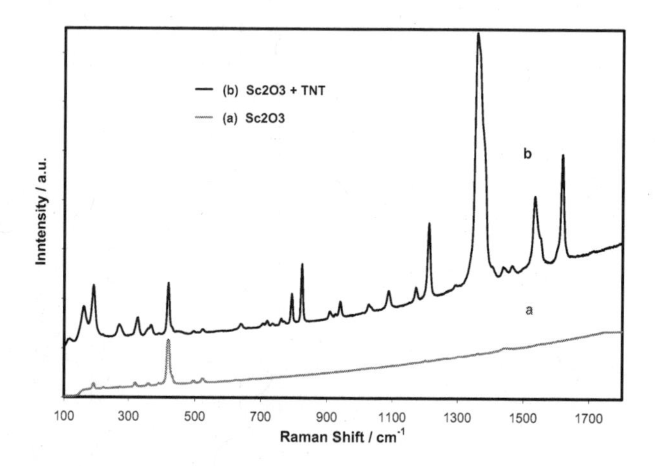

Figure 6. Raman spectra of: (a) Neat Sc_2O_3 ; (b) Sc_2O_3 + TNT: 20:1 mix.

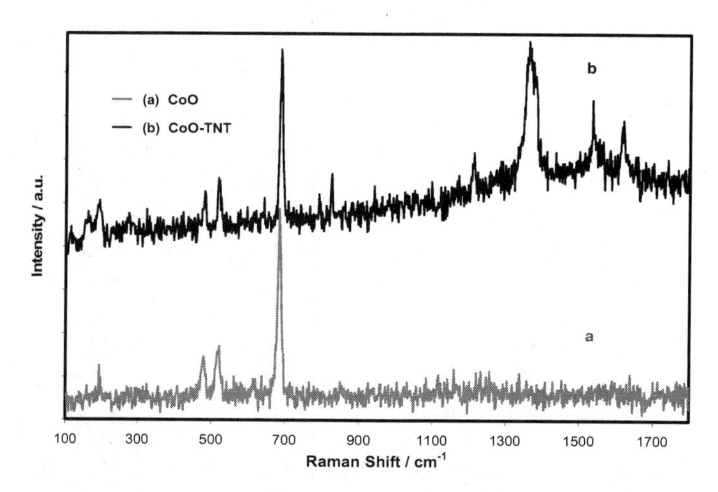

Figure 7. (a) CoO; (b) CoO + TNT (20:1 mix). CoO does not participate in Raman signal enhancement under current experimental conditions.

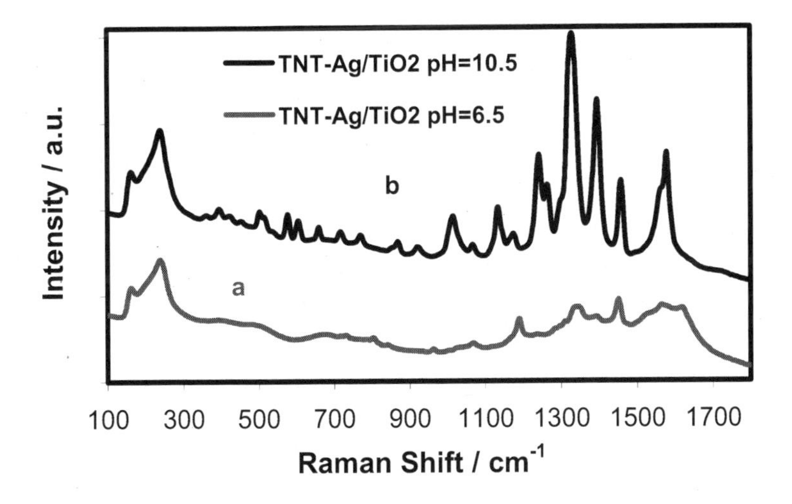

Figure 8. SERS spectra recorded with 488 nm laser of *TNT with Ag –Ti O $_2$ colloid at different pH: (a) 6.5; (b) 10.5.*

Figure 9 shows that nitro bands were easily detected at very dilute water solutions of TNT when excited with a blue monochromatic source:488 nm, Ar^+ laser. The pH of these solutions was 10.5. Results show a decrease of intensity of NO_2 stretching mode (1300-1370 cm^{-1}) until a concentration of $1x10^{-12}$ M was reached *(19-21)*. However, spectra are well spectroscopically defined for aqueous TNT solutions of concentrations ranging from $1x10^{-4}$ to $1x10^{-10}$ M.

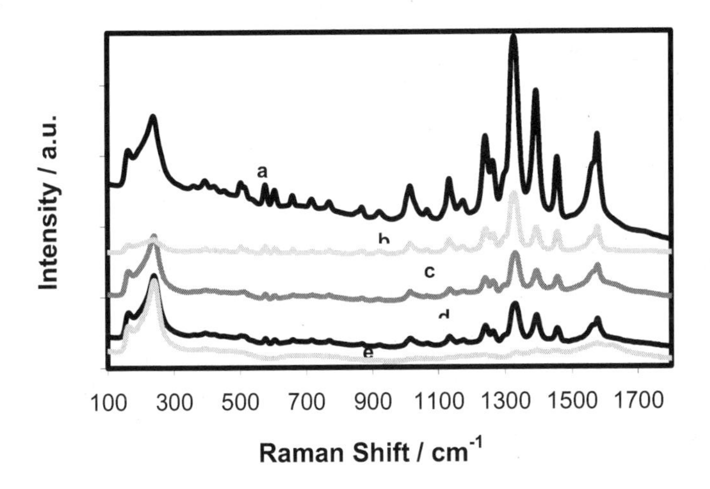

Figure 9. SERS spectra of TNT solutions on Ag/TiO$_2$ sols (488 nm; pH=10.5) for decreasing concentrations: (a) 1x10^{-4}M; (b) 1x10^{-6} M; (c) 1x10^{-8} M; (d) 1x10^{-10}M; (e) 1x10^{-12}M.

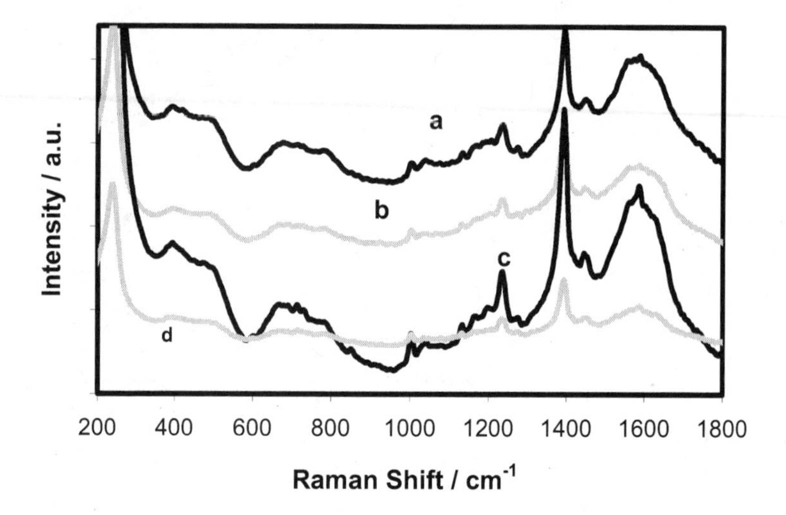

Figure 10. DNT SERS spectra *at 488 nm laser with TiO_2/Ag colloid at pH 12.3: a. 10^{-3} M, b. 10^{-4} M; c. 10^{-6} M; d. 10^{-8} M.*

Figure 10 shows the SERS spectra DNT with Ag/TiO_2 colloids at pH = 12.3 for solutions at different concentrations. At this pH vibrational enhancement assisted by Ag/TiO_2 showed the highest increase, in comparison with the other pH, of the NO_2 asymmetric stretch mode at 1583 cm^{-1}, symmetric NO_2 stretching vibration at 1348 cm^{-1} and for the 744 cm^{-1} band which can be assigned to out-of-plane C-H and C-N bend *(19)*. Attempts directed to obtain meaningful SERS spectra at lower DNT concentrations failed since it was not possible to observe characteristic spectral signatures of this nitroaromatic HE.

Conclusions

The results presented demonstrate that vibrational bands of the Raman spectra of the target nitroaromatic high explosives used: TNT and DNT are strongly enhanced on direct contact with TiO_2, Sc_2O_3 and SnO. Cobalt (II) oxide, CoO did not show ERS from the HE tested under for the same procedure used for the other oxides. Colloidal suspensions of Ag^o coated TiO_2 nanoparticles in which TiO_2 was assumed to cover the silver nanoparticles were successfully prepared and tested for SERS activity on TNT and DNT aqueous solutions. The oxide doped metal assembly of Ag/TiO_2 nanoparticles suspensions exhibited shifts to longer wavelength with respect to normal colloidal silver sols, which may be indicative of the semiconductor covering al least partially the core metal spheres or spheroids. The SERS spectra of TNT and DNT were enhanced strongly by colloidal suspensions at the different wavelength of laser 488, 785 and 532 nm. This was observed particularly for the SERS spectra of TNT solutions from 10^{-4} to 10^{-10} M at pH 10.5 excited with 488 nm laser. The aromatic ring breathing mode was observed near 1018 cm^{-1}, the asymmetric stretching NO_2 was detected in the Raman Shift range of 1555-1587cm^{-1} and the symmetric NO_2 at 1300-1357 cm^{-1}. On the other hand, DNT solutions with pH = 12.3 showed an increase of the NO_2 asymmetric stretching mode the 1583 cm^{-1}, symmetric NO_2 stretching vibration at 1348 cm^{-1} and 744 cm^{-1} vibrational mode which can be assigned to out-of plane C-H and C-N bend recorded with 532 nm laser. Further studies are needed to investigate the sensitivity and selectivity of metallic oxides enhanced Raman signal from nitroaromatic and nitroaliphatic compounds.

Acknowledgments

This work was supported in part by the U.S. Department of Defense, University Research Initiative Multidisciplinary University Research Initiative (URI)-MURI Program, under grant number **DAAD19-02-1-0257**. The authors also acknowledge contributions from Aaron LaPointe of Night Vision and Electronic Sensors Directorate, Department of Defense and from the Department of Chemistry of the University of Puerto Rico–Mayagüez.

Parts of the work on the last stage of the project was supported by the U.S. Department of Homeland Security under Award Number **2008-ST-061-ED0001**. The views and conclusions contained in this document are those of the authors and should not be interpreted as necessarily representing the official policies, either expressed or implied, of the U.S. Department of Homeland Security.

The authors would like to acknowledge The International Society for Optical Engineering (SPIE), for allowing the use of Figures 2 and 3 (adapted from references 14 and 15), previously published by the authors as part of the SPIE sponsored Defense and Security Symposia IV and VI, respectively.

References

1. Arenas, J. F.; Pelaez, D.; Lopez Ramirez, M. R ; Castro, J. L.; Otero, J.S. *Opt. Pur. and Apl.*, **2004,** *37* (2), 673.
2. Dabbousi, B.O.; Rodriguez-Viejo, J.; Mikule, F. V.; Heine, J. R.; Mattousi, H.; Ober, R.; Jensen, K. F.; Bawendi, M. G. *J. Phys. Chem. B*. **1997,** *101*, 9463.
3. Danek, M.; Jensen, K. F.; Murray, C. B.; Bawendi, M. G. *Chem Mater*. **1996,** *8*, 173.
4. Peng, X. G.; Schlamp, M. C.;. Kadavanich, A. V.; Alivisatos, A. P. *J. Am. Chemin. Soc.* **1997**, *119*, 7019.
5. Hadjiivanov, K.; Mihaylov, M.; Abadjieva N.; Klisurski, D. *J. Chem. Soc.* Faradya Trans **1994,** *98*, 3371.
6. Quim, M.; Mill, G. *J. Phys. Chem.* **1998**, *98*, 9840.
7. Du, J.; Zhang, J.; Liu, Z.; Han, B.; Jiang, T. *Lagmuir*. **2006**, *22*, 1307.
8. Silvia, J. M.; Janni, J. A.; Klein, J. D. *Anal Chem*, **72,** 5834-5840, **2000**.
9. Vogt, F.; Booksh, K.; *The analyst*, **2004**, *129*, 492.
10. Chen-Chi Wang, Ying. J.Y. *Chem. Mater*. **1999**, *11*, 3113.

11. Aroca, R. F.; Clavijo, R.E.; Halls, M. D.; Schlegel, H. B. *J. Phys. Chem. A.*, **2000**, *104*: 9500.
12. Campion, A.; Kambhampati, P. *Chemical Society Reviews*, **1998**, *27*, 241.
13. Lee, P.C.; Meisel, T.J. *J. Phys Chem.* **1982**, *86*, 3391.
14. De La Cruz-Montoya, E., Blanco, A., Balaguera-Gelves, M., Pacheco-Londono, L., Hernandez-Rivera, S. P., *Proc. SPIE*, **2005**, 5778, 359.
15. De La Cruz-Montoya, E., Pérez-Acosta, G., Luna Pineda, T., Hernández-Rivera, S. P., *Proc. SPIE*, **2007**, 6538: 653826.
16. De La Cruz-Montoya, E., Jeréz, J.I., Balaguera-Gelves, M., Luna-Pineda, T., Castro, M.E. and Hernández-Rivera, S.P., *Proc. SPIE*, **2006**, 6203: 62030X.
17. Jeréz Rozo, J. I., del Rocío Balaguera, M., Cabanzo, A., De La Cruz-Montoya, E., Hernández-Rivera, S. P., *Proc. SPIE*. **2006**, 6201, 62012G .
18. Liao, L. B.; Zhou, H. Y.; Xiao, X. M. *Chemical Physics*. **2005**, *316*, 164.
19. Sylvia, J.M.; Janni, J.A.; Klein, J.D. and Kevin M. Spencer, *Anal. Chem.* **2000**, 72: 5834.
20. Kneipp, K.; Wang, Y.; Kneipp, H.; Perelman, L.T.; Itzkan, I.; Dasari, R.R. and Feld, M.S. *Phys. Rev. Lett.*, **1997**, *78*:,1667..
21. Kneipp, K.; *Spectrochim, Acta part A*. **1995**, *51*, 2171.

Chapter 17

Enhanced Raman Scattering of TNT on Nanoparticles Substrates: Ag, Au and Bimetallic Au/Ag Colloidal Suspensions

Ana María Chamoun-Emanuelli, Oliva M. Primera-Pedrozo, Marcos A. Barreto-Caban, Jackeline I. Jerez-Rozo and Samuel P. Hernández-Rivera

Chemical Imaging Center / Center for Chemical Sensors Development Department of Chemistry, University of Puerto Rico-Mayagüez PO Box 9019, Mayagüez, PR 00681

Nanoparticles are of fundamental interest since they possess unique size-dependent properties which are quite different from the bulk and atomic state. Bimetallic nanoparticles are of particular interest because they combine the advantages of the individual monometallic counterparts. Ag, Au and Au/Ag bimetallic colloids were synthesized by chemical reduction methods and used for detecting TNT in solution with high sensitivity and molecular specificity. The nanoparticles were characterized using techniques such as UV-VIS spectroscopy, Scanning Electron Microscope, High Resolution Transmission Electron Microscopy and Raman Spectroscopy. Detection of TNT deposited on gold and silver colloids was achieved at wavelengths of 532 (green) and 785 (NIR) nm. The detection was performed as an indirect form via alkaline hydrolysis of TNT using a strong base. Results indicate that there is an increase in intensity of the vibrational signals due to SERS of TNT degradation products which was still detectable at 10^{-15} g by the presence of NO_2 out-of-plane bending modes at 820 and 850 cm^{-1} and NO_2 stretching mode at 1300-1370 cm^{-1}.

Modern security systems require detection and identification of trace explosive at very low detection limits, down to picogram and even femtogram levels. In this respect, Raman spectroscopy of explosive materials has been reported and specifically, surface-enhanced Raman scattering (SERS) has been shown to be a sensitive spectroscopy technique at these detection limits. In the Raman Effect incident light is inelastically scattered from a sample and shifted in frequency by the energy of its characteristic molecular vibrations. Since its discovery in 1927, the effect has attracted attention from a basic research point of view as well as a powerful spectroscopic technique with many practical applications *(1)*. Since its postulation by Smekal in 1923 *(2)* and experimental verification by Raman and Krishnan *(3)* in 1928, the development of the phenomenon of inelastic light scattering by matter as a technique useful in spectrochemical analysis was hampered by numerous technological as well as fundamental conditions inherent to the technique. Being an inelastic event, Raman scattering suffers from the weakness of having very low cross sections: about one photon in 10 million undergoes a Raman event. This fact had the effect that since its beginning practitioners were in constant seek of increasing the measured Raman signals: better sources of excitation, more efficient detectors, and more practical spectrometers. In 1974 Fleischmann, Hendra and Quillan *(4)* while working with silver electrodes exposed to electrochemical cycles observed a “giant Raman effect” in a phenomenon now called Surface Enhanced Raman Scattering (SERS). The increase in Raman intensity is typically observed when the sample is absorbed on roughened metals. Noble metallic colloidal suspensions or sols exhibit the effect in which the scattering cross-sections are dramatically enhanced for molecules adsorbed on metallic surfaces with certain degree of non uniformity *(5-6)*. Silver is known to be an ideal substrate for SERS studies. However; the stability of Ag colloids is low. The SERS effect is less pronounced for gold though the stability of the colloid is excellent *(7)*. Therefore, the preparation of Au/Ag bimetallic colloids allows for the combination of the SERS activities of both metals permitting the use of a broader range of the electromagnetic spectrum in the visible region *(8)*. Despite their simple preparation, these systems are rather difficult to control and characterize because of the enormous range of particle sizes and shapes that sometimes manifest in the lack of reproducibility of SERS experiments. The methods of preparation of colloidal suspensions involve the reduction of Au and Ag metal ions in aqueous solutions by reducing agents such as trisodium citrate, sodium borohydride, or hydroxylamine. Sodium borohydride and sodium citrate are widely used as reducing agents since they have been proven to produce effective sols, but the methods of preparation based on citrate are more common because it provides relatively high monodispersity in colloid size. The citrate preparation method involves the reduction of Au and Ag metal ions in aqueous solutions by trisodium citrate, which acts as a capping agent as well as a reducing agent. The nanoparticles were characterized using techniques such as UV-VIS spectroscopy, Scanning Electron-Microscopy (SEM) and High Resolution Transmission Electron Microscopy (HR-TEM). The later was combined with X-Ray point analysis and wide area imaging to obtain chemical and physical information on the bimetallic colloidal nanoparticles.

Metallic Colloidal Suspensions Synthesis

Methods have been developed for the preparation of silver colloids by chemical synthesis. All glassware was meticulously cleaned using *aqua regia* (strong acid mixture) of $HCl:HNO_3$ 3:1 which was handled with extreme care as copious amounts of gas were liberated during cleaning. Glassware was then rinsed with water before washing with detergent. Finally, the glassware was rinsed well with ultra high pure water (18.1 MΩ). Silver colloids preparation generally follows a modified Lee and Meisel method *(9)*. Briefly, 100 mL of a 10^{-3} M aqueous solution of $AgNO_3$ were heated to boiling and then 25 mL of a 1% trisodium citrate ($C_6H_5O_7Na_3$) solution was added. The mixture was kept boiling for 1 hour and then allowed to cool to room temperature and stored at 4°C. The resultant colloidal mixture typically had a dark grey color. For the Ag-coated Au colloids, different aliquots of a 1 mM $AgNO_3$ solution were added dropwise to 20 mL of a gold sol. After adding an appropriate aliquot of trisodium citrate solution, the mixture was boiled and stirred for an hour. Silver colloids were also prepared by the method of Leopold and Lendl by dissolving milligram amounts of silver nitrate ($AgNO_3$) in 100 mL water and adding hydroxylamine hydrochloride ($NH_2OH{\cdot}HCl$) dissolved in 0.05M NaOH *(10)*. The mixture was rapidly added to $AgNO_3$ solution and in a few seconds a grey-brown solution was obtained.

For the gold colloids, milligram amounts of $HAuCl_4$ were weighted, dissolved in 200 mL water and kept in amber bottles. The solution was brought to boiling while stirring vigorously and a 1 % (v/w) trisodium citrate solution was added. Boiling was maintained for 30 min, according to Rivas et al. *(11)*. Mixed silver/gold colloids were also prepared by Rivas et al. (12). Specifically, for the Ag-coated Au colloids, different aliquots of 1 mM $AgNO_3$ solution were added dropwise to 20 mL of a gold sol. After adding an appropriate aliquot of trisodium citrate solution, the mixture was boiled while stirring for an hour.

Physical and Chemical Characterization of Metallic Sols

Proper characterization of SERS active surfaces (solid, sols or semi-sols) is important in determining the usefulness of the metallic surface in the sought Raman application. This is even more critical when dealing with colloidal metallic suspensions since there are many more variables that need to be controlled when preparing a SERS active sol. preparing SERS active sols. The first characterization done after colloids preparation was to determine the wavelength location of the plasmon band by UV-VIS spectrophotometry. Obtaining λ_{max} is a key element to determine optimum laser excitation lines for SERS excitation. The magnitude of the enhancement obtained is proportional to the plasmon band absorbance at λ_{max}.

UV-VIS Spectrophotometry

The absorption spectra of metallic nanoparticles are characterized by a strong broad absorption band that is absent in the bulk spectra. Classically, this giant dipole band, referred to as the plasmon absorption band, and is ascribed to a collective oscillation of the conduction electrons in response to optical excitation *(13)*. The absorption maximum of the colloidal suspensions provides information on the average particle size, whereas its full-width at half-maximum (FWHM) gives an estimation of particle size distribution. In many cases aggregation occurs during preparation or storage. All UV-VIS spectra were obtained by diluting 1 μL of the colloidal suspensions in 3 mL water. Neat silver sols are highly absorbing suspensions and will cause detector saturation of most spectrophotometers. A Varian CARY 100 was used for the absorbance measurements. Scanning range was from 200 nm to 800 nm. Typical UV-VIS spectra obtained are shown in Figure 1. Ag colloids prepared using four different concentrations of citrate show differences in their spectral absorption characteristics, which may be attributed to morphological differences.

The Ag colloids prepared with hydroxylamine had absorption maxima centered about 450 nm and those of prepared with citrate typically ranged from 410 to 445 nm, depending on the preparation. In general the spectra showed small differences in average particle sizes and were comparable with results obtained in previous studies *(9-10, 14)*. The FWHM of the Ag/hydroxylamine sol was wider than those of Ag-citrate. Ag/hydroxylamine colloids showed absorption at longer wavelength regions, indicating presence of larger particle sizes, which may be due to the partial aggregation of particles. The colloidal suspensions of silver particles showed a bright yellow-greenish color due to the intense bands about the excitation of the surface plasmon resonance. When the metal is in the form of colloidal particles, the condition for excitation of a plasmon is shifted when compared to the optical properties of the bulk metal. The optical absorption of Au and Ag colloids has been a subject of study for years. These colloids show absorption maxima at about 420 and 520 nm for Ag and Au, respectively, because of the different plasmon excitation resonance of each metal *(12)*. Figure 2 shows the UV-VIS absorption spectra and actual photographs of several of the colloidal suspensions prepared.

Specially prepared, citrate-reduced silver sols were synthesized by preparing seeds (very small particle size nanoparticles) of Ag and Au and then further reducing more metal on them. Ag/Ag sols prepared this way had absorption maxima between 450 and 470 nm (Fig. 1). The red-shift is due to increase in average particle size of the nanoparticles ensemble. On the other hand, the UV-VIS absorption spectra of Ag-coated Au colloids showed only an absorption band at approximately 520 nm, which shows similar shape and location to the plasmon resonance of Au particles. The appearance of only one absorption band resembling Au sols indicates that homogeneous mixed colloidal particles of both metals are formed without significant formation of independent particles. The narrowing of the absorption band indicates that there is a change from a highly polydispersed system (Ag) to a relatively monodisperse system (Au). There is a blue shift in the position of the surface plasmon, which is consistent with smaller Au particles or that the surface is coated with Ag, whose plasmon resonance lies in the blue region of the electromagnetic spectrum.

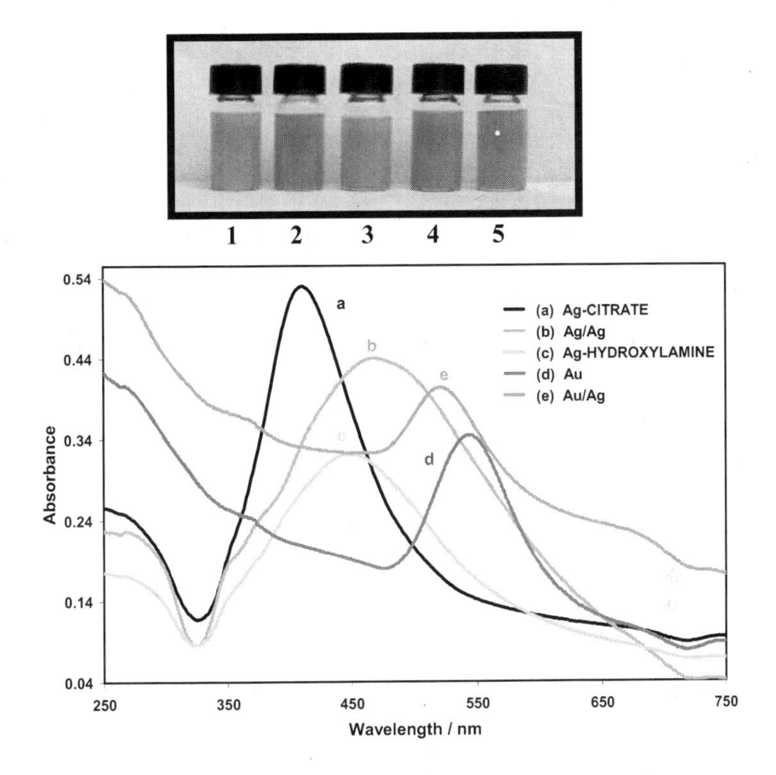

Figure 1. Top: Photographs of sols: (1-4) citrate (cit) reduced Ag sols at different Ag concentrations; (5) hydroxylamine (NH_2OH) reduced Ag sol. Bottom: (a) Au-cit, λ_{MAX} = 411 nm; (b) Ag/Ag, λ_{MAX} = 465 nm; (c) Ag-NH_2OH, λ_{MAX} = 449 nm; (d) Au, λ_{MAX} = 520 nm; and (e) Au/Ag, λ_{MAX} = 543 nm.

Morphology of Colloids: SEM, TEM and X-Ray Micro Probe Fluorescence

The shapes and sizes of these particles can be visualized using electron scanning electron microscopy (SEM). However, during specimen the sample can be vulnerable to agglomeration and represents the sizes and shapes that result upon drying of solvent on the test surfaces. Colloidal preparations yield particles that are approximately spherical as depicted in Figures 2 and 3. The average size and size distribution of nanoparticles in the silver colloidal are also dependent on the increase in temperature. To maintain steady growth of nanoparticles, as the size distribution sharpens, is very important to maintain a constant temperature. Colloids prepared by chemical methods tend to form clusters because of their complicated double layer structure. Sols with much higher morphological heterogeneity are obtained, since they are made of nanoparticles with a wide spectrum of sizes and shapes.

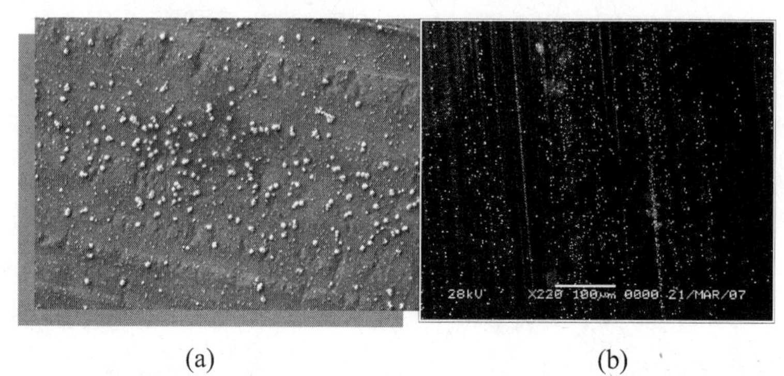

(a) (b)

Figure 2. SEM images of Ag colloids deposited on stainless steel surface: (a) reduced with citrate; (b) reduced with hydroxylamine.

The shapes and sizes of these particles are better characterized using TEM. Colloidal preparations yield particles that are approximately spherical as seen in Figures 2 and 3. According to TEM images, the particle size of gold nanoparticles is 54 ± 5 nm; for silver the average diameter is 60 ± 11 nm and 30-35 nm for bimetallic Au/Ag. The average size and size distribution of the nanoparticles in the silver colloids are dependent on temperature increase. Silver, as demonstrated in previous results is highly polydispersed. The temperature necessary to maintain steady growth of nanoparticles, as the size distribution sharpens, is very important to achieve a large electromagnetic enhancement on the metallic colloidal surface. Colloids in aqueous media, prepared by chemical methods, tend to form clusters because of their complicated double layer structure. These clusters have much higher morphological heterogeneity, since they are composed by nanoparticles with a wide spectrum of sizes and shapes. This correlates with the broader plasmon resonance extinction band. Bimetallic Au/Ag nanoparticles prepared were very similar in shape and size to the Au seed nanoparticles (about 35 nm in size and nearly spherical in shape) although they were significantly smaller than the typical gold particles used for SERS (~ 50-60 nm).

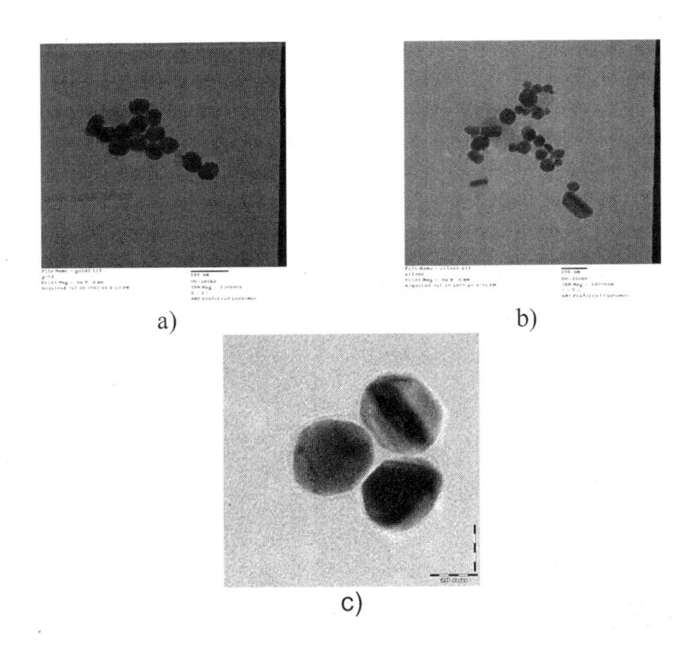

Figure 3. TEM images of citrate reduced sols.
a) Au nanoparticles; b) Ag nanoparticles c) Au/Ag nanoparticles.

Energy Dispersive X-Ray (EDX) analysis of several nanoparticles indicated that the Au/Ag nanoparticles were bimetallic random alloy nanoparticles. A typical EDX run is shown in Figure 5. Atomic Au and Ag emission lines have been marked. The elemental lines intensity ratio Au-L/Ag-K was 10.7 (not calibrated for the instrumental response). The particles are much richer in the gold component. Moreover, silver coverage of gold seed particles seemed to be incomplete and diffuse. This was demonstrated when an X-Ray line map was performed using the AuM and AgL lines. These results are shown in Figure 4. The gold and silver X-Ray emissions detected and mapped correspond to the nanoparticles shown in the white line drawn. Gold emissions map two particles of roughly of 20-30 nm (core substrate). Silver emissions correspond to a very thin atomic “shell”.

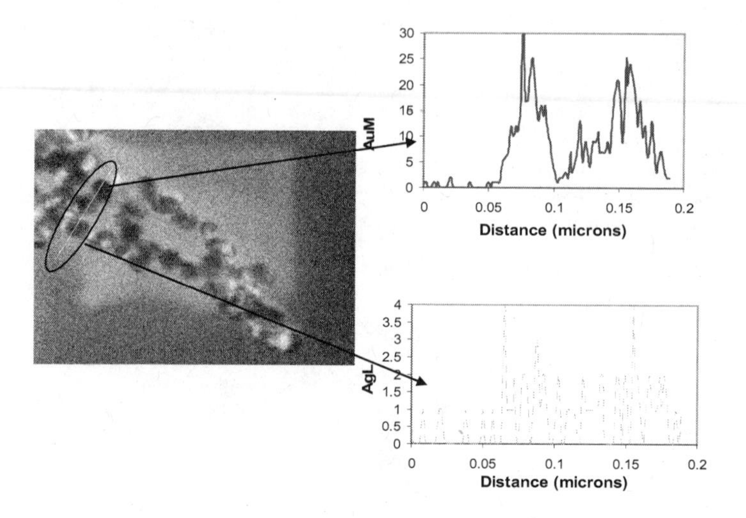

Figure 4. X-Ray fluorescence line scan mapping performed on Au/Ag nanoparticles using the Ag-L and Au-M atomic emission lines.

Tests for SERS Activity

Three main chemical systems (compounds) are normally used to test for SERS activity in metallic colloidal suspensions: pyridine, adenine and 1,2-Bis (4-pyridyl)-ethane (BPE). These analytes were prepared in aqueous solutions in the 10^{-3} to 10^{-5} M concentration range, as well as the other analytes studied. The final concentration was 10 times lower. Figure 5 shows the intensity enhancement that is typically obtained.

Since Raman spectroscopic systems used are microspectrometers the interrogation volume in a capillary tube as sample container when using a 10x objective is of the order of 400 pL ($4.0x10^{-10}$ L) and for a 10^{-5} M solution represents 4 fmol ($4.0x10^{-15}$ mol) or 2400 million particles (molecules). Since typical target explosive have molar masses in the range of 200-250 $g{\cdot}mol^{-1}$, the upper limit of the mass contained in the interrogation volume for a 10^{-5} M solution is about1 pg ($1x10^{-12}$ g).

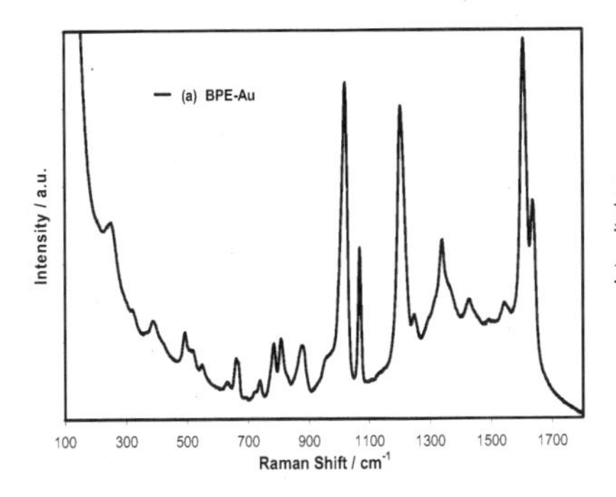

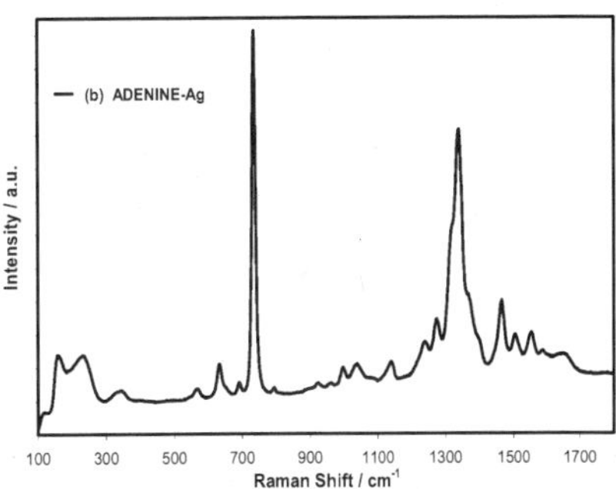

Figure 5. Compounds used to test the SERS effectiveness of colloidal suspensions: (a) BPE-Au colloids; (b) adenine-Ag colloids.

SERS of 2,4,6-Trinitrotoluene (TNT)

Nanoparticles are very suitable as SERS substrates as was demonstrated in previous section. However, for 2,4,6-TNT a strong base was needed to increase the intensity. The method for the preparation of colloidal solutions was optimized to have the maximum SERS effect on TNT vibrational signatures. The concentration of the reducing agent, the addition rate of the reducing agent, and the stirring rate were also evaluated in the previous context.

A strong relation was found between the pH of the TNT solution and the SERS signal. TNT explosive shows enhancement of the NO_2 out-of-plane bend in the 820-850 cm^{-1} region and the NO_2 stretching mode in the 1300-1370-cm^{-1} region for gold colloids. The nitro stretching modes are observed at 1356 cm^{-1} for TNT, the aromatic ring breathing mode close to 1000 cm^{-1} (15). The key spectral regions for TNT are the nitro stretching region centered about 1350 cm^{-1} and the out-of-plane bending modes about 820 cm^{-1}. Figure 6 illustrates the effect of adjusting the pH of the sample solution. In the original hypothesis it was thought that by changing the ionic strength of the adsorbing species, TNT molecules would be stimulated to bind stronger to the metallic surfaces. However, a higher enhancement of Raman signals than postulated were found at very high pH values (pH > 12).

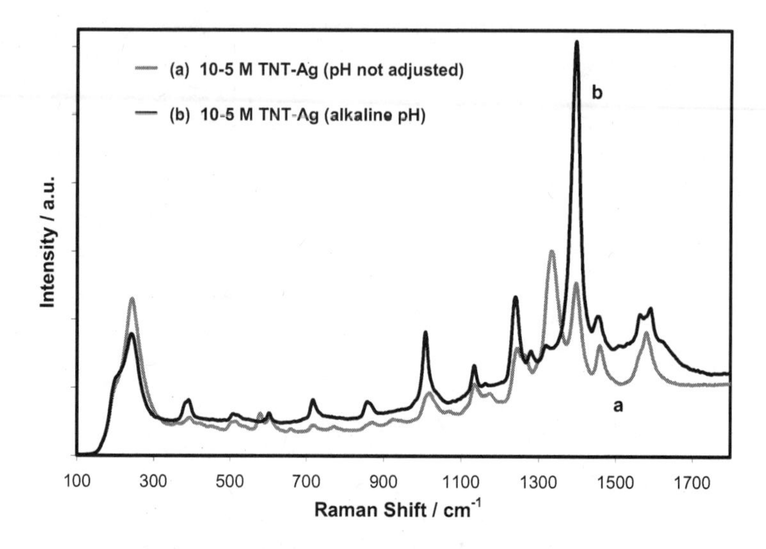

Figure 6. SERS spectra of TNT: (a) pH of sample not adjusted; (b) pH of sample adjusted to high values (pH > 12).

As a means to further increase the enhancement of TNT detection in diluted solutions, the effect of pH on the SERS spectra of TNT solutions was also investigated. The study involved the measurement of SERS spectra of aqueous TNT solutions (10 μg/mL; $5x10^{-5}$ M) prepared at different pH values. To optimize the adsorption process of the analyte, the pH values of the solution were changed from 2 to 13. The optimum pH values were in the alkaline end: larger than 13, since more and better defined bands was obtained (Figure 7) at very high pH. It is important to emphasize that when the pH was increased the color of the solutions changed due to the formation of chromophore groups which can in turn arose from a chemical reaction that led to the formation of degradation products of TNT, including bicyclical adducts as determined by HPLC-MS. Our findings are in accordance to previous studies reported by Thorn *et al.* that have confirmed the propensity of TNT to alkaline hydrolysis *(16)*. This susceptibility of TNT to basic media is being used by our group as an indirect mean for the detection of TNT in aqueous solution in combination with our SERS active substrates *(17)*.

For these highly basic solutions the symmetric nitro stretch typically found in the normal Raman spectra of TNT ca. 1358 cm^{-1}, was shifted to 1392 cm^{-1}. This suggests that the modification retains the basic nitroaromatic structure of the molecule, maintaining not only the nitro group but also the aromatic ring as evidenced by the presence of the totally symmetric ring breathing (1006 cm^{-1}) mode. Also, the adsorption process was probably more efficient and intimate at basic pH, allowing a higher enhancement due to the electromagnetic nature, which is mostly related to the roughness on the surface.

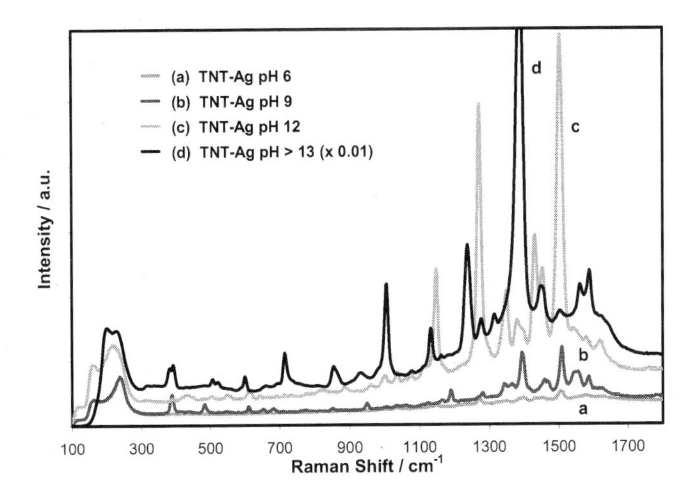

Figure 7. Effect of pH on TNT SERS signal observed. Maximum enhancements were found for pH values of 13 and higher.

The pH dependence of SERS signals from colloids revealed that the increase in pH value results in the enhancement of the Raman spectrum, a decrease in bandwidths of signals, disappearance of luminescence background ramp at high Raman Shift values, a change in the location of bands and a huge enhancement in the intensity of the bands. This is because of the progressive transition from a highly covered surface SERS spectrum to a less covered one. The differences in the relative intensities of the surface-enhanced bands illustrate selective enhancement resulting from the molecular adsorption geometry on the metal surface. The appearance of the strong band at about 1006 cm^{-1} (ring breathing mode) in the SERS spectrum is particularly interesting. This vibration is very weak in the "normal" Raman spectrum of TNT *(18)*. The relatively strong enhancement of this mode in the surface-enhancement spectrum suggests that TNT-modified molecules on colloidal silver are oriented perpendicular to the particle surface (Figure 6). The bands at 1359 and 1204 cm^{-1} (normal Raman) are slightly shifted to higher wavenumbers, and an additional band appears at 1239 cm^{-1}. Also, when the pH was changed, an enhancement of the out-of-plane cm^{-1}) takes place, confirming that the adsorption of the modified-TNT or degraded TNT occurs parallel to the roughened surface of the colloid. Figure 8 shows the SERS spectra of TNT adsorbed on several of the metal sols used in this work.

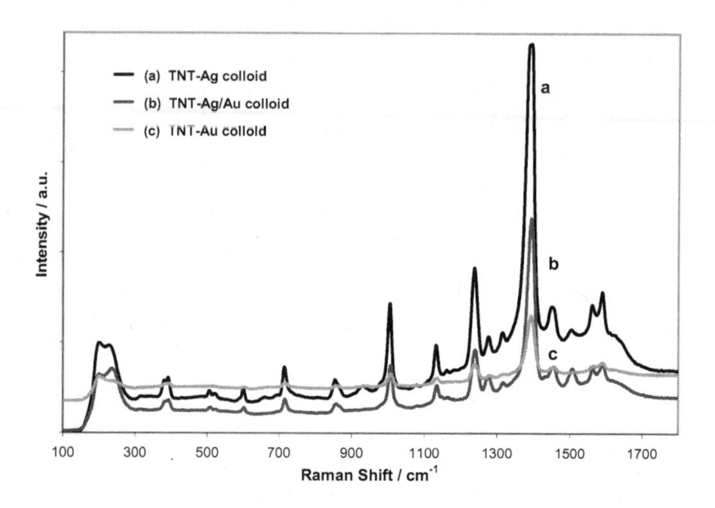

Figure 8. SERS spectra of TNT adsorbed on metallic sols. (a) Ag colloids; (b) Au colloids; Ag coated Au nanoparticles. SERS activity of TNT: Ag-sols > Au/Ag sols > Au sols.

Figure 9 shows a normal or spontaneous Raman spectrum of a saturated solution of TNT in methanol (~$10x^{-5}$ M) and of neat, freshly recrystallized TNT crystals. No methanol signals are observed after subtraction of the solvent. Neat TNT and its methanolic solution can be characterized by 4 principal bands: 1358 cm^{-1} (symmetric nitro stretching), 1551 cm^{-1} (asymmetric nitro stretching), 1212 cm^{-1} (C-H ring bend and in-plane rocking) and 1621 cm^{-1} (2,6-NO_2, asymmetric ring stretch). Since the symmetric nitro stretching is the most intense band it was used for surface enhancement factor (SEF) calculations. SEF were calculated according to Kneipp by using equation 1 *(18)*. Methanol does not exhibit an enhanced Raman Effect under normal conditions and for that reason it can be used as an internal standard for estimation of SEF. Table 1 shows the SEF for some of the colloids prepared. Our results confirm that silver citrate colloids are better for the detection of this nitroexplosive. SEF variations (standard deviations) were confirmed by using adenine as standard for SERS activity and were consistently lower for Ag metallic suspensions. Similar results have been reported for other analytes using silver and gold nanoparticles. Silver nanoparticles usually show higher SERS effect than gold metallic suspensions *(19-20)*.

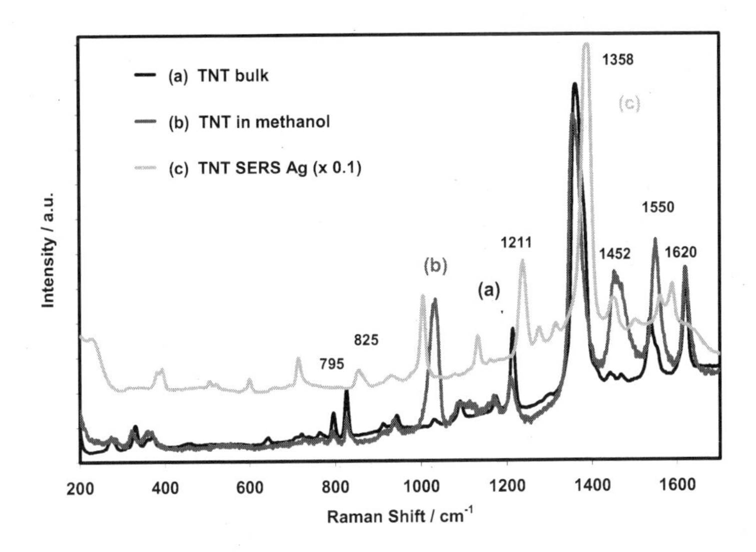

Figure 9. Comparison of TNT spectra excited at 514.5 nm: (a) normal or spontaneous Raman spectrum of TNT; (b) subtracted TNT spectrum in methanol solution; (c) SERS TNT spectrum on Ag nanoparticles. Peak labels correspond to SERS spectrum.

$$SEF = \frac{I_{NO_2,TNT}}{I_{Methanol}} x \frac{C_{Methanol}}{C_{TNT}} \qquad (1)$$

For the Au/Ag colloidal system the Raman intensity enhancement via SERS effect was lower than for colloidal silver but higher than gold colloids. Although previous research suggested that Au/Ag colloids are better than the individual metallic colloidal suspensions (i.e., Ag and Au) *(21)*, the highest SERS intensities are observable when the excitation wavelength is in resonance with the substrate plasmons and the analyte electronic absorption bands *(22)*. When sodium chloride was added to the colloids, the aggregation of the nanoparticles was induced and the plasmon absorption band shifted. In our results, the plasmon band of the gold/silver intermetallic alloys was closer to the neat gold colloids. For this reason the best SEF should have been obtained using NIR excitation (785 nm) based on the location of λ_{max} of the plasmon of the nanoparticles.

Table 1. Surface Enhancement Factors (SEF) for 2,4,6-TNT.

Colloid	SEF	STANDARD DEVIATION
Au colloid	5.10E+10	1.02E+10
Au/Ag colloid	1.44E+11	-----
Ag colloid	2.87E+11	3.82E+09

In these experiments, green line (532 nm) was used to excite SERS spectra. Since the colloids showed optical properties that were closer to those of neat Au nanoparticles *(23)*, they probably did not exhibit the strongest SERS effect at wavelength lower than red (632 nm) to NIR (785 nm) regions. Thus, the effectiveness of bimetallic colloids depends on the excitation wavelength used and a study of the influence of the excitation line on the SERS effect for TNT degradation products detection is in order.

Conclusions

In the present work wet chemistry methods of preparing silver nanoparticles reduced with different chemical agents including sodium citrate and hydroxylamine hydrochloride were discussed. Methodologies used for characterization of the nanoparticles were UV-VIS spectrophotometry, SEM and TEM images and X-Ray Microprobe Fluorescence. Finally, the metallic sols were tested for SERS activity using validated test compounds that exhibit strong Raman enhancements in the presence of the nanoparticles.

The benefits of forming bimetallic colloidal alloys by doping existing metal suspensions (seeds) has been clearly established since plasmon excitation these systems is much broader in wavelength range and allows for coupling with exciting single line lasers more conveniently. When working with neat gold colloid at 532 nm (green line, solid state diode laser) TNT is not Raman enhanced. However, when this sol is doped with silver and the solution treated with a strong base several bands of TNT degradation products are surface enhanced at the excitation wavelength (532 nm). Au-Ag alloy sols have been successfully used as substrates for SERS studies, due to the fact that they combine the advantages of both Au and Ag colloids and thus serve as interesting systems for further research. In the bimetallic particles prepared, the suspensions posses the physical stability of the typical neat gold metallic suspensions and the desired optical properties of silver, although a slight decrease of the SEF was observed compared to SEFs of neat silver colloidal suspensions.

Raman enhanced signals of TNT-modified compounds (degradation products) were detectable on metallic colloids due to the presence of the NO_2 out-of-plane bending modes at 820 and 850 cm^{-1} and the NO_2 stretching mode in the 1300-1370 cm^{-1} region. The aromatic ring breathing mode near 1000 cm^{-1} was also enhanced. These TNT related compounds were detected on silver sols with extremely high surface enhancement factors (10^{10}-10^{11}) on aqueous solution of 2,4,6-trinitrotoluene at very high pH. These very large enhancements cannot be attributed solely to typical SERS events and probably have contributions from resonance Raman events of colored TNT degradation products. A low limit of detection, lower than 10^{-16} g, of TNT sample in the laser path giving rise to enhanced Raman signals could be easily and reproducibly achieved.

Acknowledgments

This work was supported by the U.S. Department of Defense, University Research Initiative Multidisciplinary University Research Initiative (URI)-MURI Program, under grant number **DAAD19-02-1-0257.** The authors also acknowledge contributions from Aaron LaPointe of Night Vision and Electronic Sensors Directorate, Department of Defense. High Resolution TEM images and X-Ray fluorescence line scan study were measured at the Materials Characterization Center of the Rio Piedras Campus of the University of Puerto Rico by Mr. Oscar Resto, whom we recognize for his dedication and commitment to the scientific community in Puerto Rico.

Significant part of the work on the last part of the project was supported by the U.S. Department of Homeland Security under Award Number **2008-ST-061-ED0001**. The views and conclusions contained in this document are those of the authors and should not be interpreted as necessarily representing the official policies, either expressed or implied, of the U.S. Department of Homeland Security.

References

1. Kneipp, K.; Kneipp, H.; Itzkan, I.; Dasari. R. R.; Feld, M. S. *J. Phys.* **2002**, *14*, 597.
2. Smekal. A. *Naturwissenschaften*. **1923**, *43*, 873.
3. Raman C. V. Krishnan, K. S.; *Nature*. **1928**, *121*, 501.
4. Fleischmann, H.; Hendra, P. J.; MacQuillan, A. J. *Chem. Phys. Lett.* **1974**, *26*, 163.
5. Jeanmaire D. L.; Van Duyne, R. P. J. *Electroanal. Chem*. **1977**, *84*, 1.
6. Albrecht M. G.; Creighton, J. A. *J. Am. Chem. Soc.* **1977**, *99*, 5215.
7. Linyou, C.; Lianming, T.; Peng, D.; Tao, Z.; Zhongfan, L. *Chem. Mater.* **2004**, *16,* 3239.
8. Aroca, R. F.; Alvarez-Puebla, R. A.; Pieczonka, N.; Sanchez-Cortez S.; Garcia-Ramos, J. V. *Adv. Coll. Inter. Sci.* **2005**, *116*, 45.
9. Lee, P. C. ; Meisel, D. *J. Phys. Chem.* **1982**, *86*, 3391.
10. Leopold, N.; Lendl, B. *J. Phys. Chem. B*. **2003**, *107*, 5723.
11. Rivas, L.; Sanchez-Cortes, S.; Garcia-Ramos, J. V.; Morcillo, G. *Langmui*r. **2001, 17**, 574.
12. Rivas, L.; Sanchez-Cortes, S.; Garcia-Ramos, J. V.; Morcillo, G. *Langmuir*. **2000**, *16*, 9722.
13. Fujihara, S.; Maeda, T.; Ohgi, H.; Hosono, E.; Imai, H.; Kim, S.H. *Langmuir*. **2004**, *20*, 6476.

14. Camares, M. V.; Garcia, J. V.; Domingo C.; Sanchez, S. *J. Raman Spectrosc.* **2004**, *35*, 921.
15. Hostetler, M. J.; Wingate, J. E.; Zhong, C. J.; Harris, J. E.; Vachet, R. W.; Clark, M. R.; Londono, J. D.; Green, J. J.; Stokes, S. J.; Wignall, G. D.; Glish, G.L.; Porter, M. D.; Evans, N. D. Murray, R.W. *Langmuir.* **1998**, *14*, 17.
16. Thorn, K. A.; Thorne, P. G.; and Cox, L. G. *Environ. Sci. Technol.* **2004**, *38*, pp. 2224–2231.
17. Jerez-Rozo, J. I.; Primera-Pedrozo, O. M.; Barreto-Caban, M. A.; Hernandez-Rivera, S. P., *IEEE J. Sensors*, **2008**, **8**(6): 974-982.
18. Kneipp, K.; Wang, Y.; Dasari, R. R.; Feld, M. S.; Gilbert, B. D.; Janni, J.; Steinfeld, J. I. *Spectrochimica Acta* Part A. **1995**, *51*, 2171.
19. Garrell, R. L.; *Anal. Chem.* **1989**, *61*, 401.
20. Otto, A.; Mrozek, I.; Crabhorn, H.; Akemann, W. *J. Phys.: Condens. Matter*, **1992**, *4,* 1143.
21. Rivas, L.; Sanchez-Cortes, S.; García-Ramos, J.V.; and Morcillo, G. *Langmuir*, **2000,** *16*, No. 25, 9727.
22. Chourpa, I.; Lei, F. H.; Dubois, P.; Manfait, M.; and Sockalingum, G. D. *Chem. Soc. Rev.*, **2008**, *37*, 993–1000.
23. Primera-Pedrozo, O. M.; Jerez-Rozo, J. I.; De La Cruz-Motoya, E.; Luna-Pineda, T.; Pacheco-Londoño, L. C.; and Hernández-Rivera, S. P. *IEEE Sensors J.*, **2008**, *8 (6)*, 963-973.

Protection from Chemical and Biological Agents

Chapter 18

Catalytic Removal of Ethylene Oxide from Contaminated Airstreams by Alkali-Treated H-ZSM-5

Gregory W. Peterson[1],*, Christopher J. Karwacki[1], Joseph A. Rossin[2], and William B. Feaver[2]

[1]Edgewood Chemical Biological Center, AMSSB-ECB-RT-PF; Bldg E3549; 5183 Blackhawk Rd, APG, MD 21010-5423
[2]Guild Associates, 5750 Shier Rings Rd, Dublin, OH 43016-1234

Zeolite H-ZSM-5 was treated with sodium hydroxide in order to develop mesoporosity and increase ethylene oxide breakthrough time. An increase in mesoporosity, coupled with optimizing the acidity while maintaining the hydrophobicity of the zeolite resulted in the largest increase in the ethylene oxide breakthrough time.

Introduction

The filtration of toxic chemicals from contaminated air is of extreme importance in industrial environments to provide the necessary personal protection for workers to operate safely. Ethylene oxide (EtO), classified as a toxic industrial chemical due to its high worldwide production and toxicity, is one chemical that poses a unique challenge in workplace environments due to the inefficiency of removal by activated carbon, especially under conditions of high relative humidity (RH). EtO is known to undergo both acid and base catalyzed hydrolysis reactions (*1,2*). Although activated carbon impregnated with acidic or basic functional groups will facilitate hydrolysis reactions involving EtO, these materials are relatively ineffective under conditions of high RH (*3*).

A variety of acidic and basic filtration materials, including impregnated activated carbons, silicas, aluminas and zeolites (e.g. ZSM-5, Y and β) were evaluated for their ability to remove EtO under conditions of high RH (*4*). Of these materials, only H-ZSM-5, a high silica, acidic zeolite, demonstrated the ability to remove EtO over a wide range of RH's, from less than 15% RH to greater than 90% RH. The high silica content of ZSM-5 results in an adsorbent that is hydrophobic, adsorbing only a small amount of water at high relative humidity conditions (*5*), thereby minimizing the hydration of acid sites, which is key to facilitating the catalytic hydrolysis of EtO. EtO is removed using H-ZSM-5 via acid catalyzed hydrolysis reaction leading to the formation of ethylene glycol (*4,5*), which can undergo further oligomerization reactions:

$$C_2H_4O + H_2O \xrightarrow{H^+} HOC_2H_4OH \quad (1)$$

$$HOC_2H_4OH + C_2H_4O \xrightarrow{H^+} HOC_2H_4OC_2H_4OH \quad (2)$$

Karwacki and Rossin (*6*) have previously demonstrated the removal of ethylene oxide from streams of dry and humid air using H-ZSM-5. The reaction is facilitated by Brønsted acid-sites associated with H-ZSM-5. Once formed, ethylene glycol, and oligomers thereof, remain within the pores of the zeolite, physically blocking the channels and effectively poisoning the reaction. Although Karwacki and Rossin report that H-ZSM-5 has a high removal capacity for EtO, poisoning occurs as a result of product accumulation within the pores of the zeolite. Poisoning as a result of product accumulation inhibits intrazeolitic mass transfer by reducing channel diameter (*7-10*). One technique for reducing intrazeolitic mass transfer resistance is to alter the microporous, crystalline structure of ZSM-5 in favor of a more heterogeneous pore size distribution. The resulting increased mass transfer rates are expected to improve H-ZSM-5 removal capacity for EtO by diffusing reaction products away from acid sites.

Methods for modifying zeolite pore structures have been well documented in the literature, and two general methods exist; (1) developing larger pores during the zeolite growth process (*11-14*) and (2) post-synthesis modification of the zeolite (*12-15*). Zeolites that are treated during the growth process have activated carbon or carbon nanotubes included in the synthesis mixture. Following synthesis, the carbon structures are burned off, leaving voids in their place (*11-14*). The disadvantage to the synthesis procedure is that the crystal structure of the zeolite becomes irregular (i.e., unsymmetrical) (*12*), potentially decreasing the availability of Brønsted sites and reducing overall acidity.

Post-synthesis treatment of zeolites via steam, mineral acids or caustic solutions is an alternative method for altering the porosity of ZSM-5 (*12-15*). Steam and acid treatments have drawbacks in that aluminum is removed from the zeolite framework (*12*), therefore decreasing Brønsted acidity necessary for EtO removal.

An alternative post-treatment modification method involves treatment of the zeolite with a sodium hydroxide (alkaline) solution (*16-23*). Treatment of ZSM-5 with a sodium hydroxide solution extracts silica from the framework via hydrolysis, with minimal alumina extraction (*16-23*). Groen et al. (*12*) have previously reported mesopore development in the 2–50 nm range using this method. The development of mesopores, which is due to the enlargement of micropores (*24*), leads to significant increases surface area and pore volume. Several studies have shown that an alkaline post-treatment of ZSM-5 results in enhanced intrazeolitic mass transfer rates of adsorbates (*21-23,25*).

The objective of this study was to determine the effect of caustic treatments involving ZSM-5 on EtO filtration. Meeting this objective involved treating ZSM-5 crystals with alkali solutions of various concentrations, assessing the zeolite for changes in physical properties, then assessing the EtO removal capacity with breakthrough testing.

Experimental

Synthesis and Treatment of ZSM-5

Zeolite ZSM-5 was synthesized by adding 8,400 g of Ludox colloidal silica solution to a 5 gallon pail. To the colloidal silica solution was added 175 g of sodium hydroxide, 235 g of sodium aluminate, 925 g of tetrapropylammonium bromide and 5,250 g of DI water. The resulting synthesis gel was blended thoroughly using a high shear mixer. Upon completion of the mixing, the gel was added to a 5 liter Teflon lined autoclave and heated in a forced convection oven at 180°C for 72 hours. Upon completion of the crystallization process, the autoclave was removed from the oven and allowed to cool to room temperature. Once cool, product zeolite was removed from the autoclave, filtered from the mother liquor, and then washed repeatedly to neutrality using 4 gallons of DI water each time to remove the excess sodium. Following the final washing operation, the product zeolite was dried overnight at 110°C and then calcined at 650°C for 6 hours in order to remove the organic cation (tetrapropyl ammonium).

The sodium hydroxide treatment was performed as follows. To a Teflon jar was added 400 ml of DI water and the desired quantity of sodium hydroxide required to achieve the target treatment concentration. The sodium hydroxide solution in the Teflon jar was placed on a temperature-controlled stirring hot plate and heated to the desired temperature while stirring. Evaporation was minimized by loosely placing a lid over the jar. Once at temperature (typically 30 to 90 minutes following heating), 75 g of ZSM-5 crystals were added to the solution. The resulting slurry was stirred for the target period of time.

Upon completion of the alkaline treatment, product zeolite was filtered, washed until neutral, and then dried at 110°C for a minimum of 2 hours. Treated and untreated ZSM-5 crystals were ion-exchanged using an ammonium solution. The ion exchange was performed by dissolving 100 g of ammonium nitrate in 500 ml DI water in an 800 ml glass beaker. The solution was placed on a stirring hot plate and heated to approximately 90°C while stirring. Once the temperature exceeded 90°C, approximately 75 g of zeolite crystals were added to the solution. The beaker was covered using a watch glass to minimize evaporation. The slurry was mixed for 3 hours, with water added as needed to maintain the volume.

Upon completion of the ion exchange operation, the zeolite was separated from the solution via filtration, then washed once using 500 ml of DI water. Ion exchanged zeolite ZSM-5 (in the NH_4^+ form) was forwarded for preparation as particles. A small portion of the zeolite crystals were calcined at 550°C for 3 hours in order to decompose the ammonium complex, then forwarded for characterization.

ZSM-5 particles were prepared by blending ammonium exchanged ZSM-5 (NH_4^+-ZSM-5) with a clay binder in proportions of 100 parts zeolite, 25 parts binder. The blended dough was first dried at 110°C and then calcined at 525°C for 2 hours. The bound H-ZSM-5 was then crushed and sieved to 60x140 mesh particles.

ZSM-5 Characterization

The porosity of untreated and modified ZSM-5 was determined using an Autosorb 1C analyzer manufactured by Quantachrome. A 49-point N_2 BET adsorption isotherm was obtained using samples of zeolite crystals in the H-form. The surface area was determined using a multi-point Brunauer, Emmett and Teller (BET) method up to $p/p_o = 0.30$. This relative pressure was used because it accounts for only monolayer adsorption; higher relative pressures would include multilayer adsorption resulting in false readings. The total pore volume was determined at $p/p_o = 0.975$. The adsorption isotherm was used to determine the Barrett, Joyner and Halenda (BJH) pore size distribution of baseline and modified H-ZSM-5 samples. The adsorption isotherm, rather than the desorption isotherm, was employed in determining the pore-size distribution (*12*).

Water up-take was determined gravimetrically at 25°C using binder-free zeolite in the H-form. Prior to determining the water up-take, the zeolite was off-gassed in-situ at 350°C under vacuum ($< 10^{-4}$ torr) for 4 hours, then allowed to cool under dry N_2 at atmospheric pressure. The test cell was then placed in a temperature controlled bath at 25±0.2°C and exposed to flowing air (200 Nml/min) at varying levels of RH for a minimum of 10 hours. The water pick-up was determined by the weight change of the zeolite following exposure to humid air.

Ammonia temperature programmed desorption (NH_3-TPD) curves were recorded by placing 1.5 g of zeolite particles (60x140 mesh) into a tube furnace and heated to 125°C. Ammonia was supplied to the zeolite in dry air at a concentration of 1,000 mg/m^3 (1,450 ppm) and a flow rate of 2 L/min. The ammonia flow was terminated once the effluent concentration equaled the feed concentration, and clean air was subsequently purged through the zeolite until ammonia was no longer detected in the effluent. The temperature was then increased to 500°C over a 90 minute time period. Throughout the temperature ramp, the ammonia concentration was monitored in the effluent stream using an Amatek UV-Vis analyzer.

Proton Induced X-ray Emission (PIXE) analysis was performed by Lexington Analytical Services located in Lexington, KY.

Ethylene Oxide Breakthrough Testing

Breakthrough testing was conducted on zeolite packed beds in order to establish a correlation between the effect of alkaline treatment on physical properties and EtO breakthrough time. Sorbent beds were packed by storm-filling 13.2 cm^3 of filtration material into a 4.1 cm ID glass filtration tube. The filter bed was packed on a 200 mesh screen located near the bottom of the tube. The volume of filtration material was sufficient to produce a bed depth of 1.0 cm. Testing was initiated by passing either low (RH = 15%) or high (RH = 80%) humidity air through the filtration bed. For tests performed at 80% RH, the filtration material was prehumidified at 25°C, 80% RH overnight in an environmental chamber. At time zero, the EtO flow was introduced into the process stream at a rate sufficient to yield a feed concentration of 1,000 mg/m^3 at a flow rate of 5,200 ml/min (referenced to 1 atm pressure, 25°C). The flow rate employed in all testing corresponds to a superficial linear velocity of 6.6 cm/s. All tests were performed at 25±2°C. Breakthrough tests were conducted on small particles (i.e., 60x140 mesh) in order to minimize bulk diffusion mass transfer effects.

A Miran 1A infrared detector was used to continuously monitor the concentration of EtO in the feed stream, while a Hewlett Packard 5890 Series II gas chromatograph equipped with a flame ionization detector (GC/FID) was used to monitor the effluent concentration of EtO. The detection limit of EtO in the effluent stream was approximately 0.15 mg/m^3. The authors found that use of the Miran 1A infrared detector for detection of the low concentration breakthrough was not suitable due to a low level interference of small changes in water vapor concentration.

Results

Characterization of Alkali-Treated H-ZSM-5

The chemical compositions of treated and untreated samples of H-ZSM-5 were determined using PIXE analysis. The percent recovery was calculated based on the weight of zeolite before and after the alkaline treatment. The calculated SiO_2/Al_2O_3 ratio was determined based on the sample recovery assuming only SiO_2 was removed. Table 1 summarizes the findings.

Table 1. Effect of Alkaline Treatment on Silica Removal.

Sample	*SiO_2/ Al_2O_3 (PIXE)*	*Percent of Sample Recovered*	*Percent of Material Removed*	*Calculated SiO_2/Al_2O_3*	*Density (g/cm^3)*
H-ZSM-5 (U)	52	n/a	n/a	52	0.76
H-ZSM-5 (0.4N)	42	92%	8%	48	0.72
H-ZSM-5 (1.0N)	29	57%	43%	29	0.62
H-ZSM-5 (1.5N)	20	43%	57%	22	0.52

The similarity of the measured (PIXE) SiO_2/Al_2O_3 to the calculated SiO_2/Al_2O_3 indicates that sodium hydroxide treatment of H-ZSM-5 results in preferential removal of silica from the zeolite lattice. It is also apparent that higher sodium hydroxide concentrations lead to more aggressive silica removal, resulting in a decrease in the zeolite density. In fact, the linearity of the percent recovery with respect to sodium hydroxide concentration suggests a stoichiometric reaction:

$$2\,NaOH + SiO_2 \rightarrow Na_2SiO_3 + H_2O \qquad (3)$$

The loss of silica from the alkali treatment results in a change to the pore structure of the H-ZSM-5 lattice. Nitrogen adsorption isotherms, as shown in Figure 1, were recorded in order to assess the effect of alkaline desilication on the porosity of the resulting zeolite.

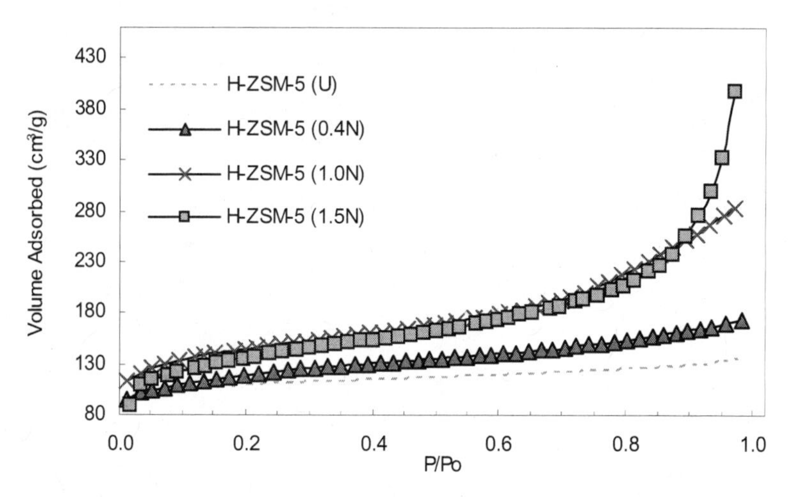

Figure 1. Nitrogen Adsorption Isotherms of H-ZSM-5 Samples.

Each of the treated samples shows increases in nitrogen uptake as compared to the untreated sample, indicating an increase in both surface area and pore volume. The BET surface area, pore size distribution and pore volume were calculated using N_2 BET adsorption data, rather than desorption data, in order to discount hysteresis and tensile strength effects (*26*). These data are reported in Table 2.

Table 2. Porosity of H-ZSM-5 Samples.

Sample	*Surface Area (m^2/g)*	*Pore Volume (cm^3/g)*	*Low-Range Mesopore Fraction*, 12-30 Å*
H-ZSM-5 (U)	368	0.209	28.7%
H-ZSM-5 (0.4N)	399	0.268	35.3%
H-ZSM-5 (1.0N)	476	0.438	17.4%
H-ZSM-5 (1.5N)	454	0.616	15.2%

**Represents the percentage of low-range mesopores (12-30 Å) volume to the total pore volume*

Each treated sample has a larger surface area as compared to the untreated zeolite, indicating that the microporosity has not been compromised by the treatment. An increase in the zeolite pore volume with increasing NaOH solution concentration is also noted. Coupled with density measurements, data suggest that the more aggressive treatments essentially dissolve the zeolite from the outside surface inwards; whereas the H-ZSM-5 (0.4N) zeolite has an increase in low-range mesopores (12-30 Å), totaling approximately 35% of the total pore volume, the more aggressive samples show almost a 50% decrease in low-range mesopore volume. These results are substantiated by the BJH pore size distribution data (Figure 2), which show that the H-ZSM-5 (0.4N) sample contains more mesopores in the 12-30 Å region than the untreated sample.

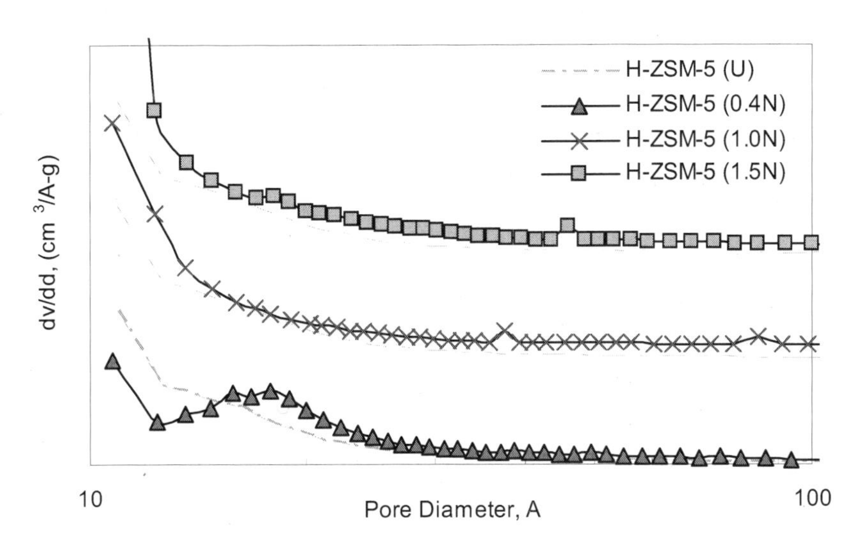

Figure 2. BJH Pore Distribution Differential Data of H-ZSM-5 Samples.

The more aggressively treated samples, H-ZSM-5 (1.0N) and H-ZSM-5 (1.5N), also show an increase in mesopore volume over the same (12-30 Å) region; however, there is also an increase in high-range mesopore (greater than 100 Å) volume and low-range macropore volume for these samples, further indicating excessive dissolution of the zeolite lattice.

NH_3-TPD curves were recorded in order to determine the effect of the alkaline treatment on the extent and nature of acid sites present on the treated samples. Figure 3 reports the NH_3-TPD profiles for the treated samples. The NH_3-TPD profile recorded the untreated sample is shown for comparison.

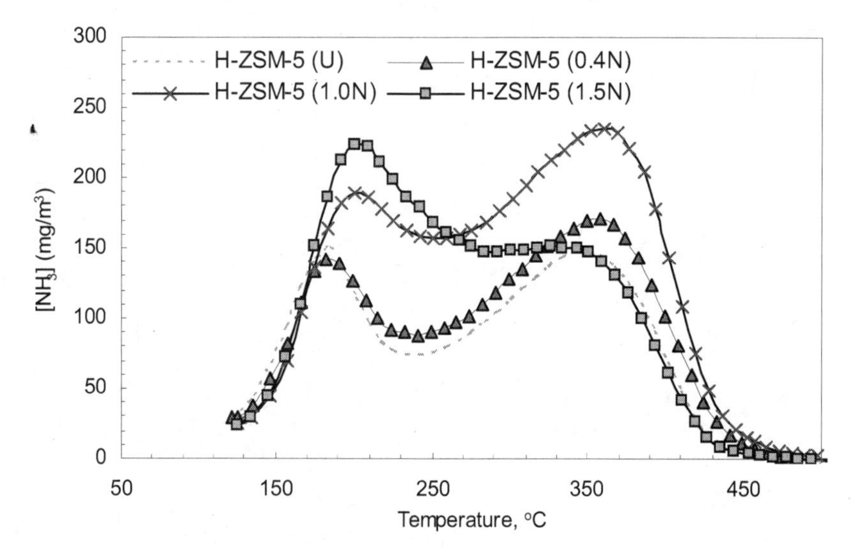

Figure 3. NH_3-TPD Curves of H-ZSM-5 Samples.

The NH_3-TPD curve for the untreated H-ZSM-5 has peaks of similar magnitude at approximately 180-200°C and 340-360°C. These peaks are attributed to adsorption of NH_3 onto weak acid sites, such as terminal hydroxyl groups, and stronger acid sites, such as Brønsted and/or Lewis acid sites, respectively (*24,25,27,28*). In general, the lower temperature peak increases in intensity with increasing alkaline treatment concentration. The development of larger voids via aggressive silica removal likely results in an increased number of terminal or nested hydroxyl groups, therefore increasing the number of weak acid sites. The H-ZSM-5 (0.4N) sample, however, does not show an increase in the number of weak sites, indicating that the treatment is not strong enough to extract substantial silica from the zeolite lattice. Instead, it is likely that the less aggressive treatment (0.4N NaOH) results in the removal of silica from defects in the zeolite, such as the edge of micropores. This is further demonstrated by the pore distribution, which shows that mesopores are developed at the expense of some micropores (Figure 3). Conversely, the 1.0N and 1.5N alkaline solutions may be strong enough to create large voids in the lattice near the edge of the zeolite crystal, yet leave most of the micropore channel structure intact. This conclusion is also supported by the pore distribution, which shows that the

H-ZSM-5 (1.0N) and H-ZSM-5 (1.5N) samples retain microporosity while adding mesopore and macropore volume.

The higher temperature peak shows an increase in aluminum site density for the H-ZSM-5 (0.4N) and H-ZSM-5 (1.0N) samples as compared to the untreated sample, also indicating an increase in the number of Brønsted acid sites. However, deconvolution of the NH_3-TPD profile for the H-ZSM-5 (1.5N) zeolite shows significantly less ammonia adsorption, indicating a higher SiO_2/Al_2O_3 and a reduced acidity. Therefore, even though the overall SiO_2/Al_2O_3 is lower than all other samples (Table 2), a majority of the alumina is not contributing to Brønsted acidity. Table 3 summarizes the percentage of alumina available as strong acid sites, or the measured SiO_2/Al_2O_3 divided by the SiO_2/Al_2O_3 calculated from the NH_3-TPD curves. Additionally, the number of acid sites (H^+) per unit volume of the H-ZSM-5 particles is reported.

Table 3. Effect of Alkaline Treatment on Zeolite Acidity.

Sample	*SiO_2/Al_2O_3 (PIXE)*	*SiO_2/Al_2O_3 (NH_3-TPD)*	*PIXE / NH_3-TPD*	*H^+ (mmols/cm^3)*
H-ZSM-5 (U)	52	66	79%	0.29
H-ZSM-5 (0.4N)	42	54	78%	0.36
H-ZSM-5 (1.0N)	29	38	76%	0.49
H-ZSM-5 (1.5N)	20	88	23%	0.15

The SiO_2/Al_2O_3 determined from chemical analysis is a measurement of the chemical composition of the zeolite, whereas the SiO_2/Al_2O_3 determined from the NH_3-TPD data is a measurement of the alumina available as acid sites. The ratio of these two values, therefore, is the overall percentage of alumina available as acid sites in the zeolite. It is apparent that as the alkaline concentration of the treatment solution increases, the percentage of acid sites available for reaction remains constant up to a concentration of 1.0N. Additionally, the overall number of acid sites per unit volume increases with increasing alkaline concentration up to treatment using 1.0N NaOH solution. This is a direct result of silica removal, which causes an increase in alumina content per unit volume. The most aggressively treated sample (1.5N NaOH), however, has the lowest number of acid sites per unit volume; the high alkaline concentration of the treatment solution likely led to the destruction of the zeolite crystal structure, likely extracting both silica and alumina from the crystals.

Water adsorption isotherms were recorded in order to determine the impact of the alkaline treatments (i.e. changes in porosity) on water adsorption. The adsorption of water within the pores of ZSM-5 is important to EtO filtration. The surface of ZSM-5 is hydrophobic. Water adsorption on H-ZSM-5 occurs at silanol groups and Brønsted acid centers. Excessive water adsorption has the potential to reduce EtO filtration by fully hydrating the Brønsted acid site, thereby minimizing the adsorption and subsequent hydrolysis of EtO. Water adsorption isotherms are presented in Figure 4.

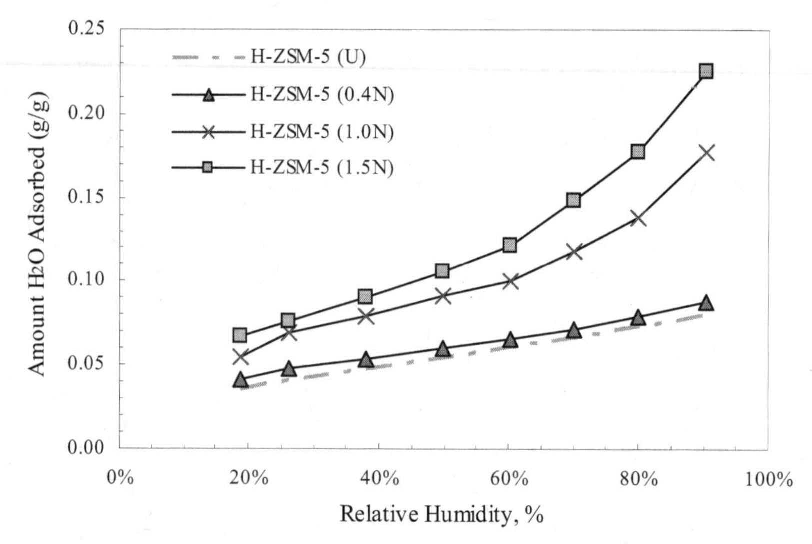

Figure 4. Water Adsorption Isotherms of H-ZSM-5 Samples.

The H-ZSM-5 (0.4N) sample has a similar water adsorption isotherm to the untreated sample, indicating that the treatment was not extensive enough to affect the hydrophobicity of the zeolite. Isotherms for the more aggressively-treated samples (1.0N and 1.5N solutions), however, deviate significantly from that of the untreated sample, adsorbing in excess of twice the water at high RH. Results presented in Figure 4 suggest that the hydrophobic nature of H-ZSM-5 treated using 1.0N and 1.5N NaOH solutions has been compromised.

Ethylene Oxide Breakthrough Testing

Alkaline-treated H-ZSM-5 samples were evaluated for their ability to remove EtO from streams of dry and humid air. Breakthrough curves recorded at 15 and 80% RH are shown in Figures 5 and 6, respectively, and illustrate that all samples show some ability to remove EtO. The breakthrough concentration of EtO is 1.8 mg/m^3. Data presented in Figures 5 and 6 illustrate the low concentration portion of the EtO breakthrough curves. These data, rather than full breakthrough curves are illustrated in order to evaluate the performance of the filtration material at breakthrough.

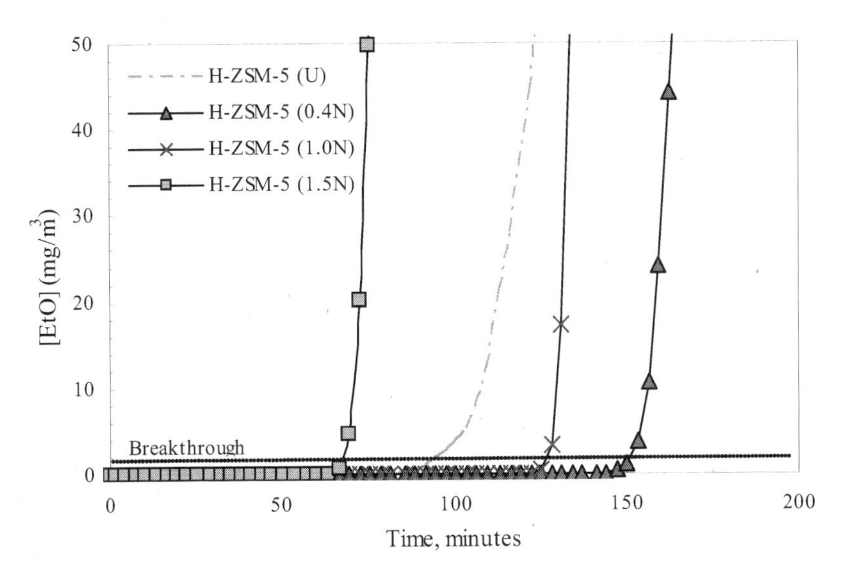

Figure 5. Ethylene Oxide Breakthrough Curves at 15% RH. 1,000 mg/m^3 Feed, 1 cm Bed Depth, 6.6 cm/s Velocity, 15% RH, As-received 60x140 Mesh Particles.

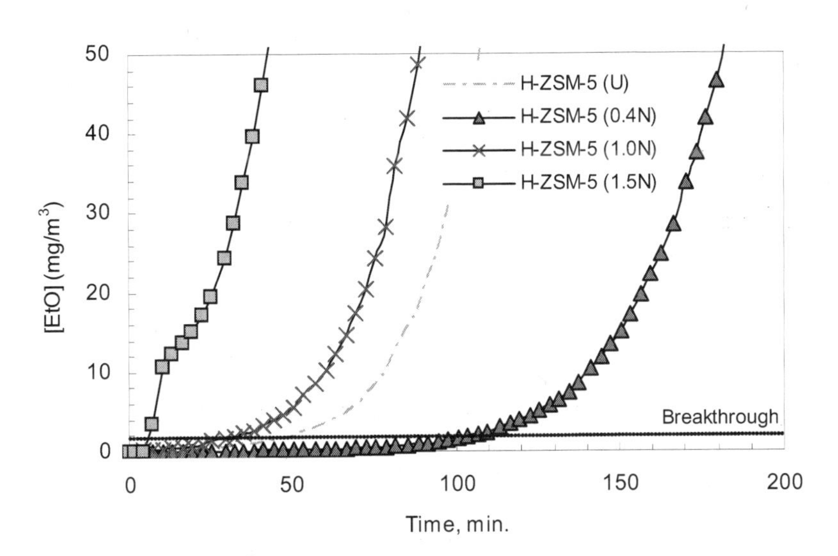

Figure 6. Ethylene Oxide Breakthrough Curves at 80% RH. 1,000 mg/m^3 Feed, 1 cm Bed Depth, 6.6 cm/s Velocity, 80% RH, Prehumidified 60x140 Mesh Particles.

Data recorded under both low (15%) and high (80%) relative humidity conditions show that the H-ZSM-5 (0.4N) sample yields improved EtO filtration performance relative to the untreated material. The H-ZSM-5 (1.0N) sample produces similar breakthrough curves to the untreated sample while the ethylene oxide filtration performance of the H-ZSM-5 (1.5N) sample is significantly reduced as compared to the untreated sample.

Discussion

The development of mesopores by alkaline treatment leads to an increase in EtO breakthrough time. Due to the nature of the EtO removal mechanism by H-ZSM-5, reaction products eventually poison the catalytic reaction by blocking channels and reducing intrazeolitic mass transfer rates (*7-10*). The zeolite treated with a 0.4N alkaline solution has the highest percentage of low-range mesopores and exhibits the greatest EtO breakthrough time. As the percentage of low-range mesopores (12-30 Å) decrease, EtO removal decreases due to byproduct poisoning. It is apparent that the 1.0N and 1.5N alkaline solutions remove substantial silica from the zeolite; however, as illustrated by the differential pore distribution curves, silica is removed to create high-range mesopores and macropores, neither of which contribute to a reduction in poisoning.

More aggressive alkaline treatments result in a lower SiO_2/Al_2O_3 ratio, thus generally increasing the acid site density of the resulting crystals. The H-ZSM-5 (0.4N) and H-ZSM-4 (1.0N) samples both improve EtO filtration at low RH conditions, yet only the H-ZSM-5 (0.4N) sample improves EtO filtration at high RH conditions. This was unexpected as the highest acid site density material, H-ZSM-5 (1.0N), should provide the best EtO filtration; however, moisture uptake results demonstrate that the H-ZSM-5 (1.0N) adsorbs substantially more moisture than the H-ZSM-5 (0.4N) sample, indicative of acid-site hydration. The H-ZSM-5 (1.5N) sample actually exhibits a lower acidity compared to the untreated material and, coupled with the reduction in hydrophobicity, subsequently displays the shortest EtO breakthrough time. It follows that, in addition to increasing the mesoporosity of the zeolite substrate, the acidity and hydrophobicity must be retained in order to optimize EtO filtration.

Conclusions

Treatment of ZSM-5 using a 0.4N NaOH solution resulted in an increased EtO breakthrough time relative to the untreated zeolite. The increased breakthrough time is attributed to the formation of low-range mesopores. The presence of low-range mesopores improves the EtO breakthrough time by enhancing intrazeolitic mass transfer. The formation of the low-range mesopores associated with the sample treated using the 0.4N solution do not significantly alter the acidic and hydrophobic properties relative to the untreated zeolite. Overly aggressive treatments (greater than 1.0N NaOH) result in less than optimal filtration performance. This is attributed to a decrease in the hydrophobic properties of the zeolite.

References

1. Buckles, C.; Chipman, P.; Cubillas, M.; Lakin, M.; Slezak, D.; Townsend, D.; Vogel, K.; Wagner, M. Ethylene Oxide User's Guide. Second Edition. August 1999 (available at www.ethyleneoxide.com).
2. Parker, R. E.; Isaacs, N.S. Mechanisms of Epoxide Reactions. *Chem. Rev.* **1959**, *59*, 737.
3. Pollara, F. J.; Liddle, L.W. U.S. 4,612,026, 1986.
4. Peterson, G.W.; Rossin, J.A. *Unpublished Data.*
5. Hill, S. G.; Seddon, D. *Zeolites* **1985**, *5,* 173.
6. Karwacki, C. J.; Rossin, J.A. U.S. Patent #6,837,917, 2005.
7. Choudhary, V. R.; Nayak, V. S.; Mamman, A. S. *Ind. Eng. Chem. Res.* **1992**, *31*, 624-628. *Ind. Eng. Chem. Res.* **1994**, *33*, 26-31.
8. Uguina, M. A.; Sotelo, J. L.; Serrano, D. P.; Valverde, J. L.
9. Choudhary, V. R.; Nayak, V. S.; Choudhary, T. V. *Ind. Eng. Chem. Res.* **1997**, *36*, 1812-1818.
10. Abdullah, A. Z.; Bakar, M. Z.; Bhatia, S. *Ind. Eng. Chem. Res.* **2003**, 42, 5737-5744.
11. Jacobsen, C. J. H.; Madsen, C.; Houzvicka, J.; Schmidt, I.; Carlsson, A. *J. Am. Chem. Soc.* **2000**, *122*, 7116-7117.
12. Groen, J. C.; Jansen, J. C.; Moulijn, J. A.; Perez-Ramirez, J. *J. Phys. Chem. B* **2004**, *108*, 13062-13065.
13. Tao, Y.; Kanoh, H.; Abrams, L.; Kaneko, K. *Chem. Rev.* **2006**, *106*, 896-910.
14. Fang, Y.; Hu, H. *J. Phys. Chem.* **2006**, *128*, 10636-10637.
15. Yamamoto, S.; Sugiyama, S; Matsuoka, O.; Kohmura, K.; Honda, T.; Banno, Y.; Nozoye, H. *J. Phys. Chem.* **1996**, *100*, 18474-18482.
16. Lietz, G.; Schnabel, K. H.; Peuker, Ch.; Gross, Th.; Storek, W.; Volter, J. *Journal of Catalysis* **1994**, 148, 562-568.
17. Čižmek, A.; Subotić, B.; Aiello, R.; Crea, F.; Nastro, A.; Tuoto, C. *Microporous Materials* **1995**, *4*, 159-168.
18. Le Van Mao, R.; Ramsaran, A.; Xiao, S.; Yao, J.; Semmer, V. *J. Mater. Chem.*, **1995**, *5(3)*, 533-535.
19. Čižmek, A.; Subotić, B.; Aiello, R.; Crea, F.; Nastro, A.; Tuoto, C. *Microporous Materials,* **1997**, *8*, 159-169.
20. Suzuki, T.; Okuhara, T. *Microporous and Mesoporous Materials* **2001**, *42*, 83-89.
21. Ogura, M.; Shinomiya, S.; Tateno, J.; Nara, Y.; Nomura, M.; Kikuchi, E.; Matsukata, M. *Applied Catalysis A: General* **2001**, *219*, 33-43.
22. Su, L.; Liu, L.; Zhuang, J.; Wang, H.; Li, Y.; Shhen, W.; Xu, Y.; Bao, X. *Catalysis Letters* Vol. 91, Nos. 3-4, December 2003.
23. Groen, J.C.; Bach, T.; Ziese, U.; Paulaime-van Donk, A.; de Jong, K.P.; Moulijn, J.A.; Perez-Ramirez, J. *J. Am. Chem. Soc.* **2005**, *127*, 10792 – 10793.
24. Jung, J.S.; Park, J.W.; Gon, S. *Applied Catalysis A: General* **2005** *288* 149-157.
25. Song, Y.; Zhu, X,; Song, Y.; Wang, Q.; Xu, L. *Applied Catalysis A: General* **2006**, *302*, 69-77.

26. Groen, J.C.; Peffer, L. A. A.; Perez-Ramirez, J. *Microporous and Mesoporous Materials* **2003**, *60*, 1-17.
27. Farneth, W. E.; Gorte, R. J. *Chem. Rev.* **1995**, *95*, 615-635.
28. Hunger, B.; Heuchel, M.; Clark, L.A.; Snurr, R.Q. *J. Phys. Chem. B* **2002**, *106*, 3882-3889.

Chapter 19

Applications of Nanocrystalline Zeolites to CWA Decontamination

V.H. Grassian, and S.C. Larsen

Department of Chemistry, University of Iowa, Iowa City, IA 52242

Nanocrystalline zeolites (crystal sizes of less than 100 nm) are porous aluminosilicate nanomaterials with increased surface areas relative to conventional micron-sized zeolites. Nanocrystalline zeolites are potentially useful for CWA decontamination applications because of their enhanced absorptive and catalytic properties. This chapter describes FTIR and solid state NMR studies of the thermal reactivity of two CWA simulants, 2-CEES (2 chloroethyl sulfide) and DMMP (dimethylmethylphosphonate) on nanocrystalline zeolites, such as NaZSM-5 (silicalite, purely siliceous form of ZSM-5) and NaY, with crystal sizes of approximately 25 nm.

Introduction

Decontamination of chemical warfare agents (CWA's) is important not only for battlefield applications, but for cleanup as well. Some common CWA's are VX-(*O*-ethyl *S*-(2iisopropylamino)ethyl methylphosphonotioate), which is a nerve gas, and HD (mustard gas), which is a blistering agent. For much of the CWA decontamination research, CWA simulants which have similar chemical structures and properties as CWA's but are much less toxic, are used. In the work described here, DMMP (dimethylmethylphosphonate) and 2-CEES (2 chloroethyl sulfide) will be used as CWA simulants for VX and HD, respectively. The chemical structures of VX and DMMP and HD and 2-CEES are shown in Figure 1.

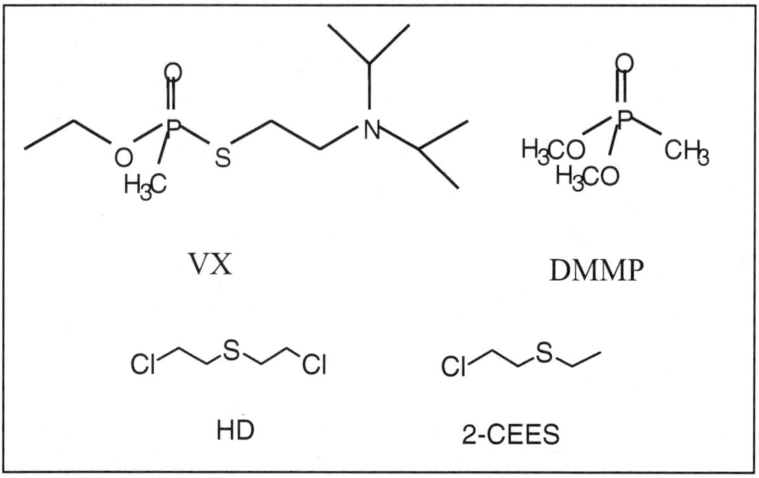

VX DMMP

HD 2-CEES

Figure 1. Two CWA's and their simulants

Nanocrystalline solid adsorbents and catalysts show great potential for decontamination. Recent work by Klabunde and coworkers has demonstrated that nanocrystalline metal oxides are promising destructive adsorbents for a variety of CWAs and CWA simulants (*1-8*). The unique properties of the nanocrystalline metal oxides as destructive adsorbents are attributed to the unusual crystal shapes, polar surfaces and high surface areas of these materials relative to conventional metal oxides. For example, nanocrystalline MgO exhibits high reactivity for the dehydrochlorination of 2-CEES while microcrystalline MgO is unreactive (*3*). The decontamination of CWAs and simulants on nanoporous materials, such as zeolites which are porous aluminosilicates, has also been investigated. Decontamination of VX and HD on NaY and AgY was investigated by Wagner and Bartram (*9*). They found that AgY was effective for decontamination of HD, but NaY was not and that VX could be successfully decontamintated by NaY and possibly AgY depending on the acceptability of certain toxic intermediates. Bellamy also found that HD could be decontaminated by 13X zeolites (*10*). These previous studies suggest that nanomaterials may have unique properties and promising reactivity for the decontamination of CWAs.

Nanocrystalline zeolites are zeolites with crystal sizes of 100 nm or less. Nanocrystalline zeolites are porous nanomaterials with internal surface porosity and a much greater external surface than conventional micron-sized zeolites. We have synthesized nanocrystalline zeolites, such as NaY, NaZSM-5 and silicalite (purely siliceous form of ZSM-5) with crystal sizes of approximately 25 nm (Table 1). The synthesis and characterization of these materials in our laboratory has been described previously (*11-14*). The nanocrystalline zeolites have external surface areas >100 m^2/g which accounts for about 30% of the total zeolite surface area compared to 2% of the total surface area for micron-sized zeolites as shown in Table 1. The enhanced external surface has the potential to be a new surface for adsorption and catalytic reactivitiy.

Nanocrystalline zeolites, NaY and NaZSM-5 and silicalite, synthesized by us, were evaluated for the decontamination of the CWA simulants, DMMP (*15*) and 2-CEES (*16,17*). The increased surface area and external surface reactivity of nanocrystalline zeolites provides potential advantages for CWA decontamination that are not found in conventional zeolite materials. The adsorption capacity and the thermal oxidation of these simulants on nanocrystalline zeolites was investigated using FTIR spectroscopy and solid state NMR spectroscopy and the reactivity on different materials was compared.

Table 1. External and Internal Surface Areas of Nanocrystalline Zeolites

Zeolite	*Crystal size (nm)*[a]	*Total (External) Surface Area(m^2/g)*[b]	*Adsorbed 2-CEES*[c] *mmol/g*
Silicalite	23	473 (138)	3.7 (1.3)
Silicalite	1000	343 (2)	--
NaZSM-5	15	556 (208)	4.5 (2.6)
NaY	22	584 (178)	--

[a] determined from SEM images, [b]BET surface area measured after (before) calcination to remove the organic template. [c]The total adsorption of 2-CEES with the adsorption of 2-CEES on the external surface given in parentheses (*16*).

Adsorption and thermal oxidation of a mustard gas simulant on nanocrystalline zeolites.

The 2-CEES molecule has a chemical structure similar to mustard gas (Figure 1) without one chlorine atom but it is significantly less toxic. Zawadski and Parsons have shown that 2-CEES and mustard gas have very similar crystal structures (*18*), therefore it is expected that 2-CEES will closely mimic the reactivity of mustard gas. FTIR spectroscopy and flow reactor measurements were used to investigate the adsorption, desorption and thermal oxidation of the mustard gas simulant, 2-CEES, on nanocrystalline zeolites. Thermal oxidation reactivity of 2-CEES on nanocrystalline NaZSM-5 (15 nm) is compared to that on nanocrystalline silicalite-1 (23 nm) and NaY(22 nm) (*16*).

The adsorption of 2-CEES on nanocrystalline NaZSM-5 and silicalite was monitored using a flow reactor apparatus and the thermal conductivity detector (TCD) of a gas chromatograph. During the initial adsorption period, the total amount of 2-CEES adsorbed on silicalite (ZSM-5) at room temperature was measured and quantified as listed in Table 1. 3.7 and 4.5 mmol/g of 2-CEES were adsorbed on silicalite (25 nm) and NaZSM-5 (15nm), respectively. The desorption during a room temperature helium purge was measured and was found to be 1.3 and 2.6 mmol/g for silicalite (23 nm) and NaZSM-5 (15 nm), respectively. The amount of 2-CEES desorbed during the room temperature helium purge was interpreted to be from 2-CEES desorbed from the external zeolite surface since approximately twice as much 2-CEES desorbed from NaZSM-5 (15 nm) compared to silicalite (23 nm) which is qualitatively similar to the ratio of the external surface areas. The 2-CEES adsrobed on the external surface is only weakly bound to the surface as suggested by its room temperature desorption. Overall, approximately 20% more 2-CEES was adsorbed on NaZSM-5 (15 nm) relative to silicalite (23 nm) and this was attributed to an increase in 2-CEES adsorption on the external surface of NaZSM-5. The external surface provides an additional adsorptive surface for 2-CEES that can also provide enhanced reactivity for 2-CEES decomposition.

The thermal oxidation of 2-CEES in nanocrystalline silicalite-1 and NaZSM-5 was investigated with FTIR spectroscopy to assess the reactivity of 2-CEES. Gas-phase products were monitored as a function of time as 2-CEES was oxidized on silicalite-1 (23 nm) and NaZSM-5 (15 nm) (*16*). FTIR spectra were collected every two minutes with 64 scans at 4 cm^{-1} resolution after saturating the sample with adsorbed 2-CEES and heating in the presence of 10 Torr of O_2 to T= 200 °C. Figure 2 shows representative spectra following the evolution of gas-phase products as a function of reaction time for 2-CEES oxidation on nanocrystalline NaZSM-5. The observed products detected include C_2H_4, HCl, CO_2, CO, SCO, CH_3CHO, CS_2, SO_2, and C_2H_5Cl.

In the presence of nanocrystalline NaZSM-5, all of the 2-CEES reacted within the first eight minutes. In the presence of nanocrystalline silicalite-1 (not shown), 2-CEES didn't completely disappear from the gas phase until longer times. CO_2 begins to form immediately in the thermal oxidation of 2-CEES on both silicalite-1 and NaZSM-5. HCl is formed and then disappears early in the 2-CEES/NaZSM-5 oxidation reaction but is not observed in the 2-CEES/silicalite-1 reaction.

The time course behavior of several of the major products is represented in the integrated absorbance versus time plots(not shown) (*16*). These results are presented in Table 2 in which the ratios of product formation rates and product amounts formed on nanocrystalline NaZSM-5 (15 nm) and silicalite-1 (23 nm) are listed. The reaction between 2-CEES and NaZSM-5 was found to be faster overall and produced higher concentrations of the final oxidation products than the reaction between 2-CEES and silicalite-1. The increased reactivity of NaZSM-5 to 2-CEES was attributed to the reactive aluminum sites in NaZSM-5 that are not present in silicalite-1.

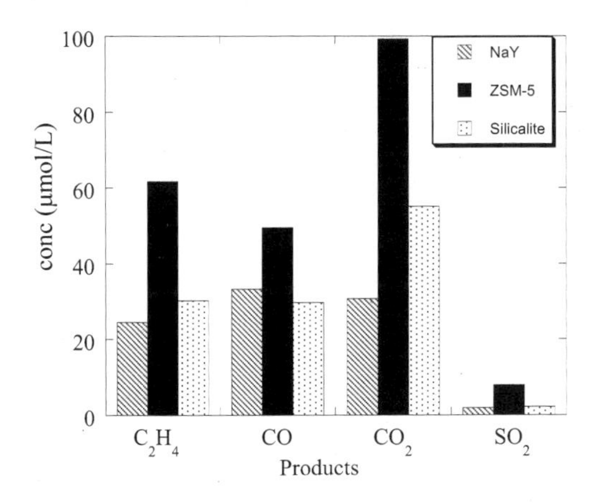

Figure 2. Gas-phase oxidation products of 2-CEES on nanocrystalline NaZSM-5 (15 nm particle size), p(O2)=10 torr and T=200°C. Reproduced with permission from reference (16). Copyright 2007 Elsevier.

Table 2. NaZSM-5: silicalite-1 ratio of initial rate and amount of product formation during the thermal oxidation of 2-CEES[a]

Gas Phase Product	*Relative Initial Rates of Product Formation*	*Relative Amounts of Product Formation*
Carbon monoxide, CO	1.8	2.0
Carbon dioxide, CO_2	1.7	1.8
Ethylene, C_2H_4	7.3	4.5
Acetaldehyde, CH_3CHO	4.1	1.3
Carbonyl sulfide, SCO	3.6	1.3
Sulfur dioxide, SO_2	7.5	3.1
Carbon monoxide, CO	1.8	2.0

[a] Reproduced with permission from reference (*16*). Copyright 2007 Elsevier.

Since aluminum was hypothesized to be responsible for the increased reactivity of nanocrystalline NaZSM-5 relative to silicalite, the 2-CEES reactivity on nanocrystalline NaY, which has a much lower Si/Al ratio than ZSM-5, was evaluated. The 2-CEES thermal oxidation activity of several different nanocrystalline zeolites was compared as shown in Figure 3. The amounts of selected products (CH_4, CO_2, CO and SO_2) formed on nanocrystalline NaY(22 nm), NaZSM-5(15 nm) and silicalite-1 (23 nm) is shown in Figure 3. The largest quantities of these products are formed on NaZSM-5 (15 nm), followed by silicalite (23 nm) and then NaY(22 nm). The higher aluminum content of NaY does not lead to increased reactivity suggesting that the shape selective properties of the zeolite may also play a role in the reactivity. NaZSM-5 has elliptical pores with ~5.6Å pore diameters that will provide a "better fit" for the adsorption of the linear 2-CEES relative to NaY which has super cages with 7.4Å pore windows.

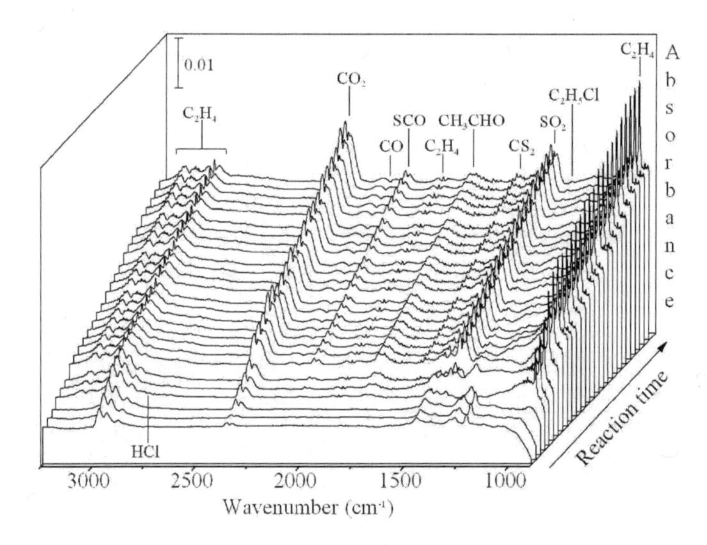

Figure 3. Comparison of gas phase products formed after reaction of 2-CEES with oxygen over NaZSM-5 (15 nm) = solid bar, silicalite (23 nm)=gray bar and NaY (22 nm)= striped bar

Adsorption and thermal oxidation of a nerve gas simulant, DMMP, on nanocrystalline NaY

Decontamination of chemical warfare agents, such as the nerve gas, VX, is required in a variety of situations including battlefields, laboratories, storage facilities, and destruction sites. Most research has dealt with battlefield decontamination due to the speed and ease of use of a decontaminant required in this situation. Because of the requirements of battlefield decontamination, reactive inorganic powders have been widely explored as possible catalysts for CWA decontamination. Studies have been performed examining the neutralization of VX on nanosized MgO (*3*), nanosized CaO (*2*), AgY (*9*), and nanosized Al_2O_3 (*1*) and other similar compounds on nanosized MgO (*19*).

The most common simulant of phosphorus-containing CWAs, such as VX, is dimethyl methyl phosphonate (DMMP). The adsorptive and reactive properties of VX and DMMP, on aluminosilicates, such as zeolites have also been examined. For use as decontamination materials, it has been shown by solid state NMR that VX hydrolyzes on NaY and AgY at room temperature through cleavage of the P–S bond to yield ethyl methylphosphonate (EMPA) (*9*). The reaction proceeds more quickly on AgY relative to NaY and the reaction continues on AgY as EMPA reacts further to form the desulfurized analogue of VX, 2-(diisopropylamino) ethyl methylphosphonate, also called QB. This is important since the QB has an LD_{50} that is 3 orders of magnitude less than the LD_{50} of VX. Recently, another zeolite application of chemical warfare agent detection was reported in which zeolite films immobilized on a quartz crystal microbalance were used as gas sensors for DMMP (*20*).

The use of nanocrystalline NaY for the thermal reaction of DMMP was evaluated thus taking advantage of the inherent reactivity of Y zeolites and the enhanced properties of the nanocrystalline form of Y zeolites. Nanocrystalline NaY with a crystal size of ~30 nm was evaluated for the adsorption and thermal reaction of the VX simulant, DMMP. FTIR and solid state ^{31}P NMR spectroscopy were used to identify the gas phase and adsorbed species formed during the adsorption and reaction of DMMP in nanocrystalline NaY. DMMP adsorbs molecularly on nanocrystalline NaY at room temperature.

The thermal reaction of DMMP and H_2O in nanocrystalline NaY was investigated using infrared spectroscopy. Gas phase products of the thermal reaction of DMMP, O_2 and H_2O in nanocrystalline NaY were monitored by FTIR spectroscopy as a function of time at 200°C. Gas phase FTIR spectra were acquired every minute for 5 h. Representative infrared spectra of the reaction of DMMP, O_2, and H_2O in nanocrystalline NaY at 200°C are shown in Figure 4. The major gas phase products observed are CH_3OH and CO_2. CH_3OCH_3 is another possible product, but a large degree of spectral overlap with methanol vibrational peaks makes quantification difficult. In this reaction, major DMMP peaks are observed at 1050 cm^{-1}, 1272 cm^{-1} and 1312 cm^{-1} throughout the reaction indicating that DMMP does not completely react on the surface of nanocrystalline NaY and remains in the gas phase before reacting to form products, such as CO_2 and CH_3OH.

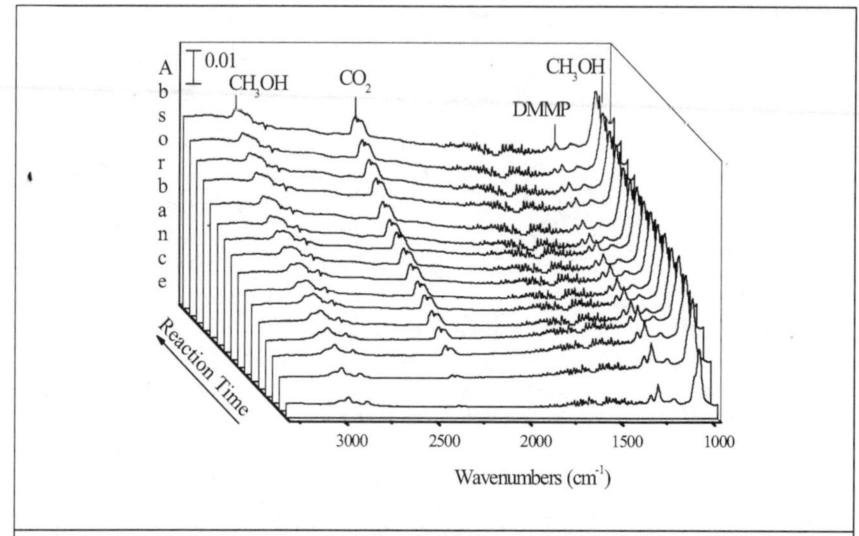

Figure 4. FTIR spectra of gas phase products of the thermal reaction of DMMP in the presence of water and oxygen on nanocrystalline NaY (30 nm) at 200°C. Representative spectra obtained every 20 minutes for four hours are shown here. Reproduced with permission from reference (15). Copyright 2006 American Chemical Society

The hydroxyl region of the FTIR spectrum reveals the external surface reactivity of nanocrystlaline NaY in the thermal oxidation of DMMP. The hydroxyl group region of the FTIR spectrum of nanocrystalline NaY before (solid line) and after thermal treatment (dashed line) with DMMP, O_2 and H_2O at 200°C for 5 h. is shown in Figure 5. As reported previously, several bands due to hydroxyl groups are observed in the FTIR spectrum of nanocrystalline NaY (*14,21-23*). In the OH stretching region of the FT-IR spectrum of nanocrystalline NaY zeolite, three absorptions are observed between 3765 and 3630 cm^{-1} (Fig. 5, top spectrum). The most intense and highest frequency band at 3744 cm^{-1} is assigned to terminal silanol groups that are on the external surface of the zeolite crystals. The absorption band at 3695 cm^{-1} is assigned to hydroxyl groups attached to Na^+. An absorption band at 3656 cm^{-1}, associated with hydroxyl groups attached to extra framework alumina (EFAL) species which have been shown to be located on the external NaY surface (*21-23*). The hydroxyl group region shows a complete loss of spectral intensity for all three absorption bands after reaction of DMMP on nanocrystalline NaY with O_2 and H_2O as shown in Figure 5, bottom spectrum. These results indicate that hydroxyl groups located on the external zeolite surface participate in these decomposition and reaction reactions of DMMP and thus represent active sites for the decontamination of DMMP.

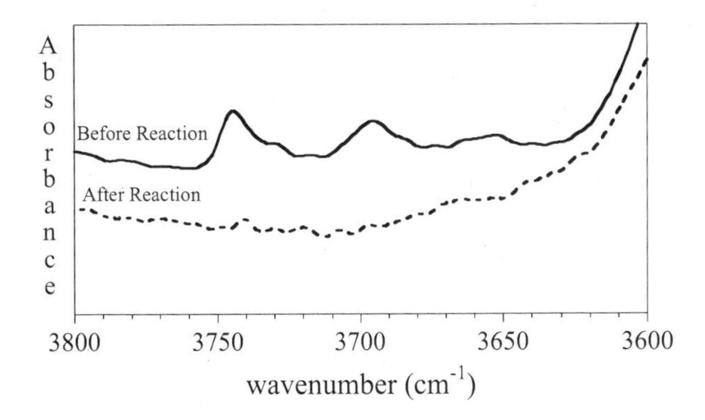

Figure 5. FTIR spectrum of the hydroxyl region of nanocrystalline NaY (30 nm) before (solid line) and after thermal treatment (dashed line) with DMMP, O_2, and H_2O at 200°C for 5 h. Reproduced with permission from reference (15). Copyright 2006 American Chemical Society

^{31}P MAS NMR experiments were conducted to investigate the phosphorus surface species (*9,24,25*) formed during the DMMP thermal reactions in nanocrystalline NaY since this information was difficult to obtain from the FTIR experiments due to spectral overlap and broadening. The ^{31}P MAS NMR spectra obtained after thermal reaction of DMMP, O_2 and H_2O all in nanocrystalline NaY at 200°C are shown in Figure 6. The NMR samples were contained in sealed sample tubes that were heated to 200°C ex-situ. All NMR spectra were recorded at room temperature under conditions of thermal equilibrium.

The ^{31}P chemical shift for neat DMMP (Fig. 6a) is 33 ppm. When DMMP is adsorbed in nanocrystalline NaY, the peak shifts slightly downfield to 34 ppm and spinning sidebands (marked with asterisks) appear in the NMR spectrum (Fig. 6b). The spinning sidebands indicate appreciable chemical shift anisotropy and a decrease in mobility suggesting that DMMP strongly adsorbs on the nanocrystalline NaY surface. When water is coadsorbed with DMMP, the peak shifts further downfield to 37 ppm and is narrower although spinning sidebands are still observed in the NMR spectrum (Fig. 6c).

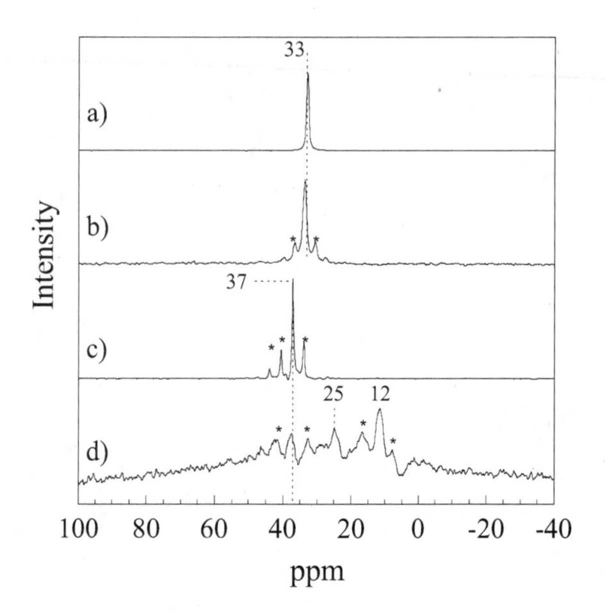

Figure 6. ^{31}P MAS NMR spectra of a) DMMP neat, b) DMMP adsorbed on nanocrystalline NaY (30 nm), c) DMMP adsorbed with H_2O on nanocrystalline NaY (30 nm), and d) after reaction of DMMP, O_2, H_2O and nanocrystalline NaY (30 nm) at 200°C for 5 h. Adapted from reference 15. Copyright 2006 American Chemical Society

DMMP was reacted with H_2O and O_2 at 200°C for 5 hours. The resulting ^{31}P MAS NMR spectrum is shown in Figure 6d and several different peaks are observed in the range 10 to 50 ppm. The peak at 37 ppm corresponds to unreacted DMMP adsorbed on nanocrystalline NaY in the presence of water. A peak at 25 ppm is observed in the ^{31}P MAS NMR spectrum which is close to the chemical shift of hydroxy methyl phosphonic acid ($OHCH_2PO(OH)_2$) or HMPA. Another peak appears at 12 ppm and this peak is close to the chemical shift for an authentic sample of dimethyl phosphite $(CH_3O)_2P(O)H$ or DMP. There is also a broad peak present in the ^{31}P NMR spectrum due to a strongly adsorbed immobile surface species. This peak is difficult to assign due to the width of the peak.

Thermal reaction of DMMP in the presence of water in nanocrystalline NaY at 200°C resulted in the formation of methanol, carbon dioxide and phosphorus decomposition products such as hydroxy methylphosphonic acid and dimethylphosphite. The external surface hydroxyl sites (silanol and EFAL) which are uniquely present in nanocrystalline NaY are important in the reaction and decomposition of DMMP. Future improvements in reactivity may be achieved by incorporating a reactive transition metal ion or oxide into the nanocrystalline NaY to provide additional reactive sites.

Outlook for Applications of Nanocrystalline Zeolites to CWA Decontamination

Nanocrystallline zeolites (Y, ZSM-5 and silicalite-1) with crystal sizes of 50 nm or less were synthesized, characterized and evaluated for CWA simulant adsorption and decontamination. The nanocrystalline zeolites have significantly increased external surface areas that account for up to 30% of the total zeolite surface area. The external surface sites have been characterized by FTIR spectroscopy. Specific signals in the hydroxyl region of the FTIR spectrum have been assigned to silanol groups and to hydroxyl groups near aluminum sites which are both located on the external zeolite surface.

The thermal decomposition of 2-CEES on nanocrystalline zeolites was probed by FTIR spectroscopy. Comparison of the reactivy of nanocrystalline NaZSM-5, silicalite and NaY indicated that NaZSM-5 was most effective for 2-CEES thermal oxidation and that external surface silanol sites were important to the zeolite reactivity. The adsorption and reaction of DMMP on nanocrystalline NaY was investigated using FTIR and ^{31}P solid state NMR spectroscopy. External surface silanol and EFAL sites were implicated in the thermal oxidation of DMMP on nanocrystalline NaY. Thus, the nanocrystalline zeolites can be envisioned as new bifunctional catalyst materials with active sites on the external surface playing an important role in the intrinsic reactivity of the material. Future studies will focus on optimizing the activity of nanocrystalline zeolites for CWA decontamination applications by tailoring the surface properties.

Acknowledgements

The authors would like to acknowledge Dr. Weiguo Song, Dr. Kevin Knagge, Matthew Johnson, and Shannon Stout for their contributions to this work. This material is based on work supported by the U.S. Army Research Laboratory and the U.S. Army Research Office under grant number W911NF-04-1-0160.

References

(1) Wagner, G. W.; Procell, L. R.; O'Connor, R. J.; Munavalli, S.; Carnes, C. L.; Kapoor, P. N.; Klabunde, K. J. *J. Am. Chem. Soc.* **2001**, *123*, 1636.
(2) Wagner, G. W.; Koper, O. B.; Lucas, E.; Decker, S.; Klabunde, K. J. *J. Phys. Chem. B* **2000**, *104*, 5118.
(3) Wagner, G. W.; Bartram, P. W.; Koper, O.; Klabunde, K. J. *J. Phys. Chem. B* **1999**, *103*, 3225.
(4) Li, Y. X.; Koper, O.; Atteya, M.; Klabunde, K. J. *Chem. Mater.* **1992**, *4*, 323.
(5) Li, Y. X.; Klabunde, K. J. *Langmuir* **1991**, *7*, 1388.

(6) Decker, S. P.; Klabunde, J. S.; Khaleel, A.; Klabunde, K. J. *Environ. Sci. Technol.* **2002**, *36*, 762.
(7) Lucas, E.; Decker, S.; Khaleel, A.; Seitz, A.; Fultz, S.; Ponce, A.; Li, W. F.; Carnes, C.; Klabunde, K. J. *Chem.-Eur. J.* **2001**, *7*, 2505.
(8) Lucas, E. M.; Klabunde, K. J. *Nanostructured Materials* **1999**, *12*, 179.
(9) Wagner, G. W.; Bartram, P. W. *Langmuir* **1999**, *15*, 8113.
(10) Bellamy, A. J. *J. Chem. Soc., Perkin Trans. 2* **1994**, 2325.
(11) Song, W.; Grassian, V. H.; Larsen, S. C. *Chem. Commun.* **2005**, 2951.
(12) Song, W.; Justice, R. E.; Jones, C. A.; Grassian, V. H.; Larsen, S. C. *Langmuir* **2004**, *20*, 8301.
(13) Song, W.; Justice, R. E.; Jones, C. A.; Grassian, V. H.; Larsen, S. C. *Langmuir* **2004**, *20*, 4696.
(14) Song, W. G.; Li, G. H.; Grassian, V. H.; Larsen, S. C. *Environ. Sci. Technol.* **2005**, *39*, 1214.
(15) Knagge, K.; Johnson, M.; Grassian, V. H.; Larsen, S. C. *Langmuir* **2006**, *22*, 11077.
(16) Stout, S. C.; Larsen, S. C.; Grassian, V. H. *Micropor. Mesopor. Mater.* **2007**, 100, 77-86.
(17) Stout, S. C. *MS Thesis, University of Iowa* **2005**.
(18) Zawadski, A.; Parsons, S. *Acta Cryst.* **2004**, *60*, 225.
(19) Rajagopalan, S.; Koper, O.; Decker, S.; Klabunde, K. J. *Chem.-Eur. J.* **2002**, *8*, 2602.
(20) Xie, H. F.; Yang, Q. D.; Sun, X. X.; Yu, T.; Zhou, J.; Huang, Y. P. *Sensor Mater.* **2005**, *17*, 21.
(21) Li, G. H.; Jones, C. A.; Grassian, V. H.; Larsen, S. C. *J. Catal.* **2005**, *234*, 401.
(22) Li, G. H.; Larsen, S. C.; Grassian, V. H. *Catal. Lett.* **2005**, *103*, 23.
(23) Li, G. H.; Larsen, S. C.; Grassian, V. H. *J. Molec. Catal.- A General* **2005**, *227*, 25.
(24) Wagner, G. W.; Bartram, P. W. *J. Mol. Catal. A-Chem.* **1995**, *99*, 175.
(25) Beaudry, W. T.; Wagner, G. W.; Ward, J. R. *J. Mol. Catal.* **1992**, *73*, 77.

Chapter 20

Effect of the Average Pore Size on the Adsorption Capacity and Off-Gassing Characteristics of Activated Carbon Fabrics for Decontamination of Surfaces

Brian MacIver[1], Ralph B. Spafford[2], James Minicucci[3], Ronald Willey[3], Adam Kulczyk[4], and Robert Kaiser[4]

[1] **Edgewood Chemical and Biological Center, Edgewood, MD21010**
[2] **FOCIS Division, SAIC, Birmingham, AL 35226**
[3]**Northeastern University, Boston, MA 02115**
[4]**Entropic Systems, Inc., Woburn, MA 01801**

Edgewood Chemical and Biological Center (ECBC) has been developing decontaminating absorbent/adsorbent wipes that are designed to remove the contaminant from a surface and entrap it into the structure of the wipe without leaving undesirable residues behind. Activated carbon fabric is the active component for these developmental wipes. The agent retention capacity and the agent off-gassing characteristics of the wipes are a function of the pore size distribution of the activated carbon fabric. Significant differences have been observed in the agent retention and off-gassing properties of microporous fabrics, where the majority of the pores are smaller than 2 nm, and of mesoporous fabrics that contain a significant population of pores that are larger than 2 nm. The adsorption capacity and off gassing properties of activated carbon fabrics that range in average pore size from less than 1 nm to 3 nm are presented for the chemical warfare agent HD, a vesicant, and for one of its simulants.

Introduction

In case of chemical attack, it is imperative that the war fighter be provided with an effective means of immediate decontamination. The US military currently utilizes kits that contain absorbent powders for this purpose. These powders have some inherent disadvantages and severe limitations, which include but are not limited to:

- the resulting contaminated powder that either remains on the treated areas or that spills onto the ground will still be a hazardous material,
- treating precision or sensitive equipment with such a powder is likely to harm this equipment and result in its malfunction.

Edgewood Chemical and Biological Center (ECBC) has been developing decontaminating absorbent/adsorbent wipes that are designed to remove the contaminant from a surface and entrap it into the structure of the wipe without leaving undesirable residues behind. Activated carbon fabric is the active component for these developmental wipes. The agent retention capacity and the agent off-gassing characteristics of the wipes are a function of the pore size distribution of the activated carbon fabric. Significant differences have been observed in the agent retention and off-gassing properties of microporous fabrics, where the majority of the pores are smaller than 2 nm, and of mesoporous fabrics that contain a significant population of pores that are larger than 2 nm.

The adsorption capacity of activated carbon fabrics that range in average pore size from less than 1 nm (micropores) to more than 2 nm (mesopores) was recently published for a vesicant simulant (chloroethyl ethylsulfide or CEES) (*1*). These results show a direct correlation between the average pore size of the fabric and its specific surface area with its adsoption capacity for CEES from CEES/methoxyperfluorobutane (sold commercially as Novec HFE-7100 by 3M Co.) solutions. The results of comparable measurements with a military grade vesicant, dichloroethyl sulfide (chemical warfare agent HD), are presented and compared to those obtained with CEES in this paper.

Adsorption Measurements

Experimental Materials and Procedure

The experimental materials and procedures are described in detail by Kaiser *et al.* (*1*), including the characterization of the activated carbon fabrics used such as BET aurface area and pore size distribution measurements by well established methods (*2, 3*), the composition of the carrier of the adsorption solution, and of the adsorption test procedure.. The key difference is the use of of a 70 ppm solution of chemical agent Bis (2-chloroethyl) sulfide (HD) in HFE-7100 as a challenge solution instead of a 70 ppm solution of CEES in HFE-7100. The HD was obtained from the US Army Edgewood Chemical and Biological Center's

Operation Directorate, distilled from munition grade material and used as received. The purity of the HD was measured to be 98.2% (area) by ^{13}C NMR.

Experimental Results

Table 1 presents a comparison of the CEES and HD adsorption capacities of four different activated carbon fabrics that were examined.

Table I. Comparison of CEES and HD Adsorption Capacities of Activated Carbon Fabrics

Supplier	Kothmex	Calgon	Calgon	Calgon
Material	1131 Felt	FM-100 Micro	FM-10 Meso	FM-100 Meso
Surface Area, m^2/g	1230	1360	995	655
Volume % Mesopores	13	18	60	84
Volume Mean Pore Diameter, AU	2	6	22	29
CEES Adsorption Capacity wt %				
Actual Data	1.87	2.48	2.66	3.24
Normalized to $1000m^2/g$	0.71	1.82	2.67	4.95
HD Adsoprtion Capacity, wt %				
Actual Data	2.12	3.67	5.34	5.54
Normalized to $1000m^2/g$	1.72	2.7	5.37	8.46
Ratio HD/CEES	2.44	1.48	1.89	1.71

The effect of volume average pore diameter of the fabric on its adsorption capacity for CEES and HD is presented in Figure 1. Figure 2 presents the same data corrected for differences in the specific surface areas of these fabrics.

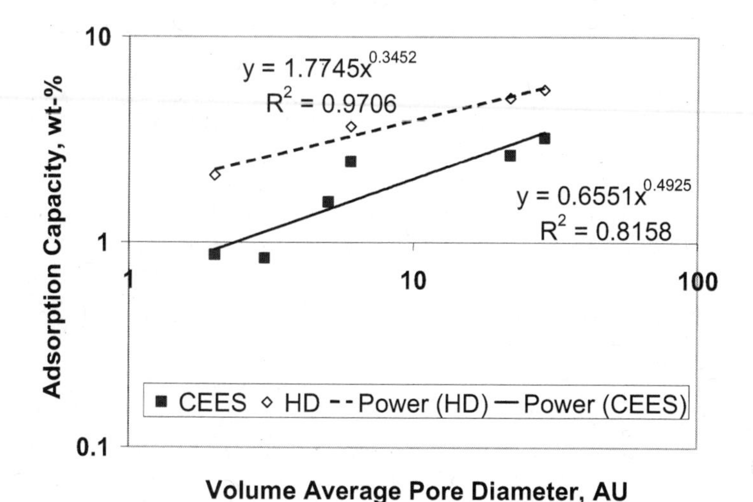

Figure 1. Activated Carbon Fabric Adsorption Capacity for HD and CEES vs. Volume Average Pore Diameter

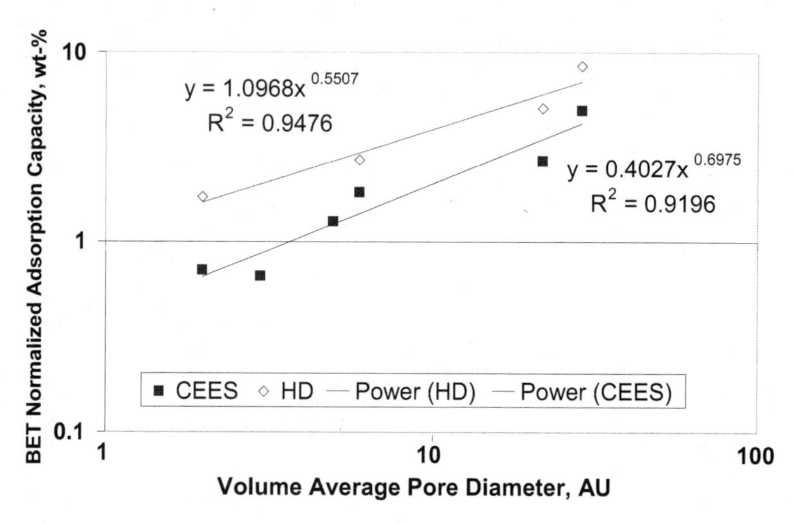

Figure 2. BET Normalized Adsorption Capacity of Activated Carbon Fabrics for CEES and Agent HD vs Volume Average Pore Diameter

Discussion of Results

Examination of the data indicates that the adsorption varies with the specific surface area and as a function of the volume average pore diameter, over the range of 2 Å to 29 Å ((0.2 nm to 2.9 nm). The adsorption capacity for HD increases from 1.72% to 8.46% (normalized to 1000 m^2/g), or about 4.9 foldover that range, while that of CEES increases from 0.71% to 4.95%, or about 6.7 fold. Given that CEES and HD have very similar molecular volumes (0.193 nm^3 for CEES, and 0.208 nm^3 for Agent HD, the higher adsorption capacity for HD is due in part to its higher molecular weight (159 Daltons vs 124.6 Daltons for CEES) and its somewhat lower solubility in HFE-7100 (3.4 vol-%) than CEES (12 vol-%). This similarity in the experimental results reinforces the argument for using CEES as a simulant for HD in liquid phase adsorption experiments, such as the ones described in this paper.

Measurement of Off-Gassing from Activated Carbon Fabrics

As compared to non-adsorbent fabrics that could also be used to remove hazardous liquids contaminants from solid surfaces, the adsorptive properties of the activated carbon fabrics mitigates off-gassing from used wipes. When the used wipes are repackaged in a sealable hermetic envelope, the adsorptive properties provide a redundant means of agent isolation. The used wipes can thus be safely handled until they are destroyed, for example, by incineration, or decontaminated by standard means, such as immersion in bleach solution.

Off-gassing tests were performed with CEES and with Agent HD to establish the effect of agent loading on swatches of activated carbon fabrics on the rate and extent of agent evaporation into an air stream flowing over a swatch.

Off-Gassing of CEES

Experimental Materials

Off-gassing tests were performed with 1.75 in (4.5 cm) diameter discs of the following Calgon Zorflex activated carbon fabrics: 50K knitted fabric, 100 micro woven fabric, and 100 meso woven fabric. The volume average pore size and specific surface area of each of these fabrics are as follows:

Fabric	50K	100 micro	100 meso
Vol-Avg. Pore Diameter, Å	6	6	29
Specific Surface Area, m^2/g	1100	1360	655
Total Pore volume, cc/g	0.72	0.75	0.81

CEES(98% 2-Chloroethyl ethyl sulfide (628-34-2)), was added to 1 mL of HFE-7100. This solution was added to a fabric disc already placed in an off-gassing cell. The CEES concentration was adjusted to obtain CEES fabric loadings of either 3.24 wt-%, 10 wt-% or 20 wt-%.

Off-gassing tests were also performed at CEES loadings of 3.24 wt-% and of 10 wt-% with the M 100 alumina powder from the M 295 decontamination kit. In these tests, the powder was sprinkled in a thin layer on the bottom of the off-gassing cell before adding 1 mL of the appropriate CEES/HFE-7100 solution. This powder has a specific surface area of 260 m^2/g and a total pore volume of 0.701 cc/g. It was found to be primarily mesoporous.

Off-Gassing Procedure

The off-gassing cell containing the contaminated coupon was sealed, and after a dwell time of 30 minutes, placed in the off-gassing apparatus shown in Figures 3 and 4. In this system, nitrogen gas was passed over the coupon at a constant flow rate of 500 mL/minute for 1 hour, at room temperature. The cell effluent gas passed (bubbled) through a liquid impinger containing 20 mL of GC grade 2-propanol (Aldrich 34863) to strip the volatilized CEES from the gas stream. The gas flow was interrupted periodically (5 min, 10 min, 15 min, 20 min, 30 min,. 45 min, and 60 min) to allow the liquid in the impinger to be replaced with fresh solvent. A 2mL sample of each scrubbing liquid was analyzed for CEES by gas chromatography with a flame ionization detector (GC/FID).

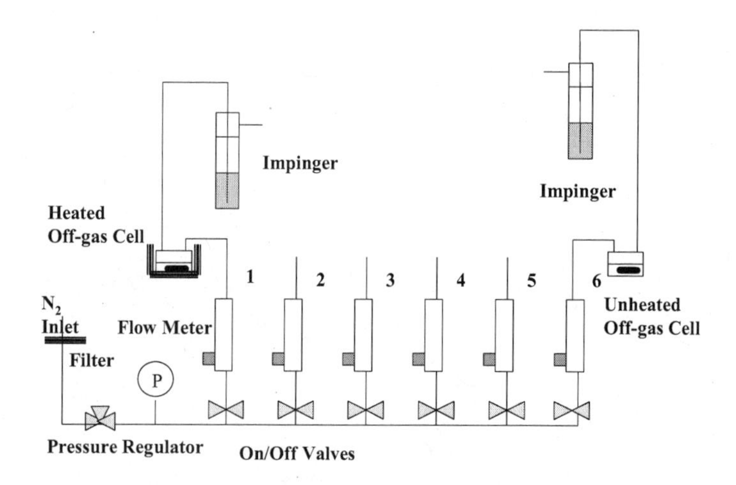

Figure 3. Flow Diagram of Coupon Off-Gassing Test System

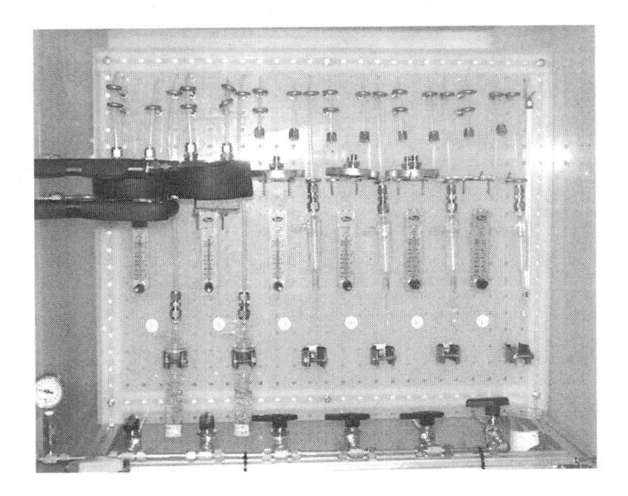

Figure 4. CEES Off-Gassing Test Stand

Off-Gassing Results

The amount of CEES off-gassed in one hour, as a percentage of the initial amount of CEES added to a sorbent, is presented in Figure 5, for the various sorbents and initial CEES loads examined. The amount of CEES off-gassed from the three activated carbon fabrics examined, at initial loads of 10 wt-% and 20 wt-%, are presented in Figure 6. For these fabrics, at an initial CEES loading of 3.24 wt-%, the amount of CEES off-gassed was not measurable.

Discussion of Results

In general, the amount of CEES off-gassed increased with the initial CEES loading on the sorbent, and, at any of the concentrations examined, there was significantly more off-gassing from the M 100 powder than from any of the activated carbon fabrics tested.

Off-gassing is expected mainly for the contaminant not adsorbed in the pores of the adsorbent. At low contaminant loadings, most, or all, of the contaminant molecules are expected to be in the pores. As the contaminant concentration increases, the contaminant capacity of the pores is reached, and excess contaminant disperses on the rest of the substrate which is non-binding. These excess molecules are free to evaporate.

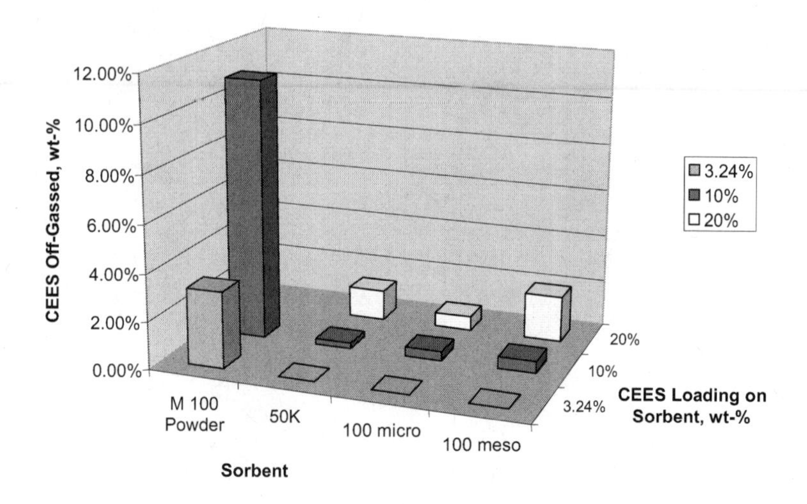

Figure 5. Effect of CEES Loading on its Off-Gassing from Sorbents of Interest

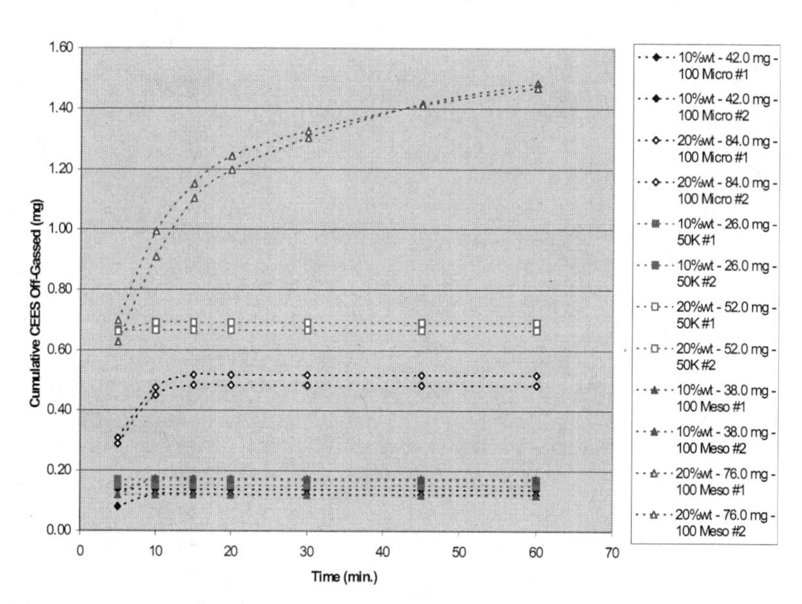

Figure 6. Off-Gassing of CEES from Selected Activated Carbon Fabrics at High (10 wt-% and 20 wt-%) CEES Loadings

For the three fabrics examined, it is presumed that, at an initial CEES loading of 3.24 wt-%, all the CEES molecules are absorbed in the pore structure. This initial loading is fairly close to the column breakthrough level observed for the liquid phase adsorption experiments described in the first part of this paper. At a 10 wt-% initial loading, the amount of off-gassed CEES is 1% or less for all three fabrics, at a 20 wt-% loading, it increases from about 1 % for the 100 micro fabric to slightly less than 4 % for the 100 meso fabric. These variations in off-gassing correlate well with the specific surface areas of the three fabric, with 100 meso having the lowest specific surface area, and 100 micro the highest.

The level of off-gassing observed from the M 100 powder is at least an order of magnitude higher than that observed from the activated carbon fabrics, over 2% evaporation at a 3.24 wt-% initial load, and over 10% at a 10 wt-% initial load. Because of the high levels of off-gassing obtained at a 10 wt-% initial load, this powder was not tested at a 20 wt-% initial load. This higher level of off-gassing occurs because the M 100 powder has a much lower specific surface area than the activated carbon fabrics.

Off-Gassing of Agent HD

Experimental

Confirming off-gassing tests were performed with Agent HD (sulfur mustard) on Zorflex 50-K fabric. In these tests, 1 mL of agent solution in HFE-7100 was added to a 5 cm diameter fabric disk. Agent concentration in the solutions was adjusted to obtain initial loadings of 1.1 wt-%, 2.0 wt-%, and 3.1 wt-%. This disk was allowed to dwell 30 minutes before being inserted into a vapor cup of the off-gassing system at ECBC. In this system, nitrogen gas was passed over the coupon at a constant flow rate of 300 mL/minute, at room temperature, for up to 10 hours. The cell effluent gas was analyzed by a Minicams.

Results

The results of the HD off-gassing tests are presented in Figure 7.

Discussion of Results

Examination of Figure 7 indicates that, except for the very first reading at the highest loading (3.1 wt-%), the measured off-gassing levels are below the threshold level of HD vapor concentration (0.0058 mg/m^3) specified by the Joint Program Interior Decontamination (JPID). For all concentrations examined, the off gassing level is less than a third of this value within one hour. Depending on the initial concentration, the off-gassing levels drop to zero within two to eight hours. As an aside, it should be noted that similar results have been obtained with nerve agents. These results indicate that while respiratory protection must be worn when an activated carbon fabric wipe is used to remove HD from a surface, the respiratory protection is no longer needed within a few minutes thereafter.

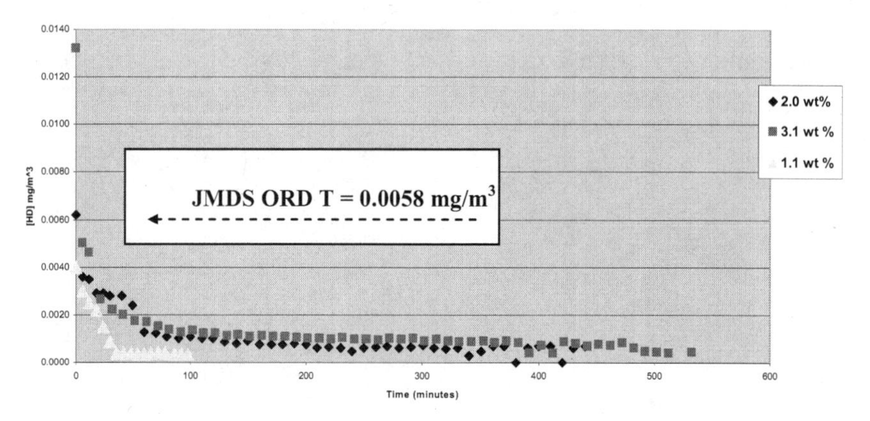

Figure 7. HD Off-Gassing from Zorflex 50K Activated Carbon Fabric

Conclusions

The adsorption capacity of activated carbon fabrics for CEES and HD dissolved in HFE-7100 increases with the specific surface area and as a function of the the volume average pore diameter, over the range of 0.2 nm to 2.9 nm. There is an excellent correlation between the HD adsorption values and the CEES adsorption values. Fabrics with a high mesopore concentration will exhibit significantly more liquid phase capacity for CEES and HD than microporous fabrics. Negligible off-gassing of CEES or HD is observed from activated carbon fabrics at low contaminant loadings when the contaminant is contained within the pores of the fabric.

For CEES loadings greater than the pore capacity for CEES, the level of CEES off-gassing from an activated carbon fabric, the level of CEES off-gassing from the fabrics varies as a function of the CEES load per unit surface area of the fabric, for both microporous and mesoporous fabrics. At comparable CEES loadings, there is significantly less CEES off-gassing from activated carbon fabrics than from the M 100 powder used in the M 295 decontamination kit. This is a result of the much lower specific surface area of this powder.

Acknowledgements

The authors acknowledge the support of the U.S. Army Edgewood Chemical and Biological Center for this effort.

References

1. Robert Kaiser, Adam Kulczyk, David Rich, Ronald J. Willey, James Minicucci and Brian MacIver, Ind. Eng. Chem. Research, **2007**, *46*, 6126-6132.
2. Coulter Omnisorp 100 360 Series Automated Gas Sorption Analyser, Reference Manual. 1991.
3. Barrett, E. P.; Joyner, L. G.; Halenda, P. P., *J. Am. Chem. Soc.* **1951,** *73*, 373-80

Chapter 21

Nanostructured Chem-Bio Non-Woven Filter

F. Tepper and L. Kaledin

Argonide Corporation, Sanford, Florida

An electrostatic filter containing powdered activated carbon (PAC) can retain high levels of microbes while the PAC within the structure has a high dynamic response as a chemical absorber. In water, a 2.5" diameter, 10" long pleated cartridge (without PAC) retained greater than 6 LRV (log retention value) of bacteria and >3 LRV of MS2 virus at flowrate of 1 gallon per minute (GPM). This study provides background data on the precursor filter, which uses a highly electropositive 2 nm alumina monohydrate fiber in a 2 μm pore size media. Sub-micron particles are retained primarily by electrostatic forces. Data compare the PAC version to the precursor. The PAC filter has a higher retention of virus, and at least an equivalent retention of bacteria and fine (~9 μm) test dust. Data are shown for adsorption of dissolved iodine and chlorine by the PAC media under short (~0.3 second) residence times, as compared to commercialized thin layer media containing granular activated carbon (GAC). The new PAC media has a dynamic adsorption rate at least two orders of magnitude greater than GAC media. The PAC media shows high promise as a chem-bio filter for purifying water. Preliminary data on air filter media are also presented.

Introduction

Filter media generally fall into two categories- non-woven and membranes. Membranes are very limited as dead end filters because of their high pressure drop and low dirt holding capacity. Cross-flow (tangential) filters mitigate those limitations but at the expense of complexity, requiring backflushing and chemical treatment in order to extend filter life. Moreover they generate a waste stream. Depth (non-woven) filters are inherently more reliable but have much poorer retention of fine particles such as microbes. Nano polymeric fibers are used to enhance the retention of fine and sub-micron particles by the mechanisms of impaction and interception. However such nano polymer fibers, are difficult to produce smaller than about 100 nanometers in large scale manufacture. And they tend to accumulate particles at the surface, resulting in early build up of a filter cake and poor dirt holding capacity.

Nano Alumina Monohydrate (NanoCeram®) Electropositive Filter

A novel water purification filter has been in development over the past seven years. The first version is a pleated depth filter cartridge that has high particle retention efficiency at moderate to high flowrates, while also having a high particle retention capacity. The filter media is also sold as non-woven media by Ahlstrom under the tradename Disruptor™. Argonide manufactures the filter cartridge from Disruptor media.

The filter's active component is a nano alumina monohydrate fiber, only 2 nm in diameter and with an external surface area of ~500 m^2/g. It has been identified as crystalline boehmite (AlOOH) using x-ray diffraction. The isoelectric point of raw fibers is approximately 11.1 (*1*). Particles with an electronegative charge are attracted and retained by the nano alumina. The nano alumina is bonded by a proprietary process to a microglass fiber that serves as a scaffold in a non-woven media.

Figure 1 shows a TEM of the nano alumina bonded to a microglass fiber (*2, 3*). Polymeric fibers, primarily polyester plus cellulose are added for formability and pleatability. The nano alumina fiber content is optimized at 35 weight percent, so that the nano alumina fibers completely occupy the available surface of the 0.6 µm microglass fiber scaffolding, thus increasing the density of exposed electropositive charges that would collect electronegative particles. Early in the development of the media, the electropotential for a 35 wt % sheet was in the range of +30 mV. As the manufacturing process became more refined, the electropotential increased to about +50 mV.

The nano alumina is end-attached and the fibers project out about 0.2 to 0.3 µm into the flow stream. This results in an open space, free for fluid to flow unimpeded through the 2 µm average pore size, allowing moderate to high flowrate at low pressure drop. Computations show that there is a local electropositive field that projects out up to about 1 µm beyond the nano alumina (*2*) that attracts nano size particles (e.g.-virus) that pass close by, increasing the capture cross section. The filter media's thickness is about 0.8 mm thick, resulting in approximately 400 pores that a particle must transit before exiting as filtrate. The combination of a dense attracting field and a tortuous path before exiting, increases the retentivity of a particle so that it is equivalent or better than an ultraporous membrane, yet the flowrate is equivalent to a 2 μm depth filter. And the large density of positive charges results in a high particle or dirt holding capacity.

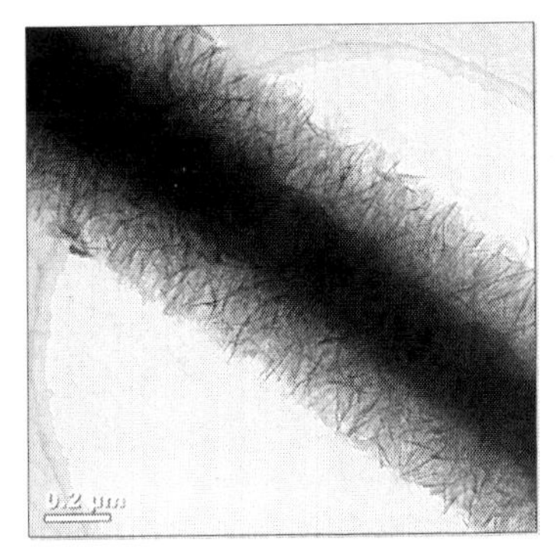

Photo courtesy of R. Ristau, IMS, Univ. of Conn.

Figure 1. TEM image of nano alumina monohydrate on microglass fibers

Figure 2 shows a filter media that had been immersed in a concentrated solution of T4 phage for four hours and was washed with ethanol. This phage has an ovoid head approximately 60 nm across and 90 nm long (*4*). A tail extends out about 200 nm. Note that the phage particles are crowded onto the media and appear to be tightly attached even after a 5 minute wash with ethanol.

Filter Capabilities

The filter has a high efficiency and high capacity for holding dirt. A2 fine test dust (available from PTI, Inc.) with average particle size (APS) of ~ 9 μm is often used to characterize the efficiency and dirt holding capacity of filters. A 2.5" diameter, 5" long cartridge was challenged by 250 NTU (nephelometric turbidity units) of A2 dust at a flowrate of 1.5 GPM, while a turbidometer (LaMotte 2020), with a sensitivity of 0.01 NTU was used to monitor turbidity. The test was terminated when the pressure drop exceeded 40 psi. During the test no turbidity could be detected in the effluent. The filter retained 17 mg/cm^2 of A2 fine dust.

Typically a 2.5" diameter, 10" long cartridge is capable of >6 LRV of moderate size and larger size bacteria (Raoultella terrigina, ~0.5μm, E coli, ~1.0 μm), and >5 LRV of a small bacteria (B. diminuta, ~0.3 μm). In clean water, the retention of virus (MS2 bacteriophage, 27.5 nm (*5*)) by a single layer is >3 LRV. MS2 virus was found to have a moderate to high retentivity at pH 9 and in concentrated (30 g/l NaCl) salt solution. The filter also has a high efficiency for adsorbing proteins, RNA/DNA, dyes and endotoxins.

The retention of monodisperse latex beads of various diameters (0.030, 0.2, 0.5, 1 and 4.3 μm) was measured using a turbidometer. In every case but with the 30 nm beads, there was no measurable leakage until the filter was clogged. In the case of the 30 nm beads we got a breakthrough curve (*2*). The filter is therefore classified as '0.2 μm absolute' (*6*) with its greater than 99.9% removal efficiency of 0.2 μm beads.

Courtesy E. Helmke, Alfred Wegener Institute, Germany

Figure 2. T4 phage crowded onto nano alumina/microglass fibers

Modeling the Process

Breakthrough curves were developed by challenging filter media with 30 nm latex beads and measuring the turbidometry of the effluent. The object was to develop a model that described the adsorption process (*2*). Breakthrough curves were obtained as a function of flowrate, bed thickness, and pH from 5 to 9. The model was then tested with MS2 virus. An excellent fit (see Figure 3 and Refs. (*2, 7*)) of experimental data points was obtained as compared to projected values derived from the model. The model is being used for scaling filters for larger area and thicker bed depths and could be useful for projecting the retention of other sub-micron and nano size particulate contaminants. Since bacteria are much larger than the 30 nm beads or MS2 virus, they are mostly mechanical entrained in the filter obviating the model's use for bacteria and for other coarser particulate contaminants.

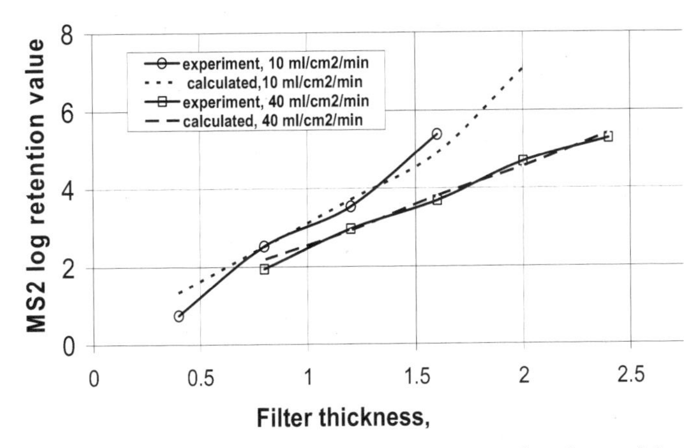

Figure 3. Retention of MS2 virus as compared to the model

Suspending Fine and Nanoparticles in a Non-Woven Web

We discovered that we could add fine and nano size particles while forming the non-woven paper-like sheet and that these particles were readily held by electrostatic forces. For instance, a handsheet was prepared containing 28 weight percent of fumed silica (APS~10 nm). When forming the handsheet (*8*), the fumed silica which is very difficult to filter from aqueous mixtures, was rapidly clarified to form a clear slurry. The resulting handsheet had a low pressure drop that is not characteristic of filtered colloidal silica. When examined under the electron microscope we noted that the nano silica particles would cling to the nano alumina (see Figure 4). Several layers of this nano silica would be deposited, suggesting that the electropositive field on the nano alumina caused a dipole on the silica particles. Handsheets were also formed containing 50 nm titanium dioxide (12%) and RNA (5%). The result was a nano engineered media that could retain and utilize the function of virtually any sub-micron or nanosize particle.

A practical application focused on integrating PAC (90% less than 625 mesh size (~20 μm)) into the non-woven structure. Activated carbon has a large amount of its surface area within micro-pores of about 0.2 to 2 nm (*9*) in diameter. At the same time, the dimensions of the pore have a significant impact upon diffusion rates of adsorbed molecules through the granule. Generally, diffusion rates are determined by the mean free path length of the molecules being adsorbed. The smaller the pores the longer is the mean free path length and the slower the diffusion rates (*10*). Therefore, the small pores in activated carbon deleteriously constrain the ingress of sorbates, particularly in the case in water filtration, where water molecules restrict the inward diffusion of sorbate, particularly if they are covalent compounds such as chloroform. By substantially reducing the particle's size one could reduce the length through the tortuous meso and micropore structure, thereby increasing the adsorption rate.

Figure 4. Nano silica enveloping the nano alumina structure

Granular activated carbon (GAC) is used in water purification, including drinking water, in many industrial applications, including the pharmaceutical industry and for manufacturing beverages. In air purification, GAC is used to control odors and gaseous and vapor contaminants in hospitals, laboratories, restaurants, animal facilities, libraries, airports, commercial buildings and respiratory equipment. It is the principal adsorbent for retaining nerve and mustard chemical warfare agents. A disadvantage of GAC beds is that these filters have large interstitial spaces to ensure that the filter bed has a low pressure drop. As a result, these filters do not have a high efficiency at very short residence times (thin beds). If the particle size of the carbon is substantially reduced then the filter would have too high a pressure drop to be usable. Also, filters having very small pore sizes are easily and rapidly clogged due to debris accumulation on upstream surfaces. This causes a rapid decline in the ability of the filters to pass fluid without having to apply a prohibitively high pressure gradient across the filter. Fine particles can erode from the surface migrate, causing channeling and clogging and would likely contain a higher level of adsorbed toxic chemical agent into the fluid. High efficiency particulate filters are often necessary downstream of GAC filters.

Fibrous structural media containing GAC are used extensively as filters. Compared to a granular bed such as GAC, a non-woven fibrous structure minimizes channeling, allows significant filter design variations, and can be manufactured by low cost paper manufacture methods.

PAC is generally recognized as having superior adsorption kinetics to GAC, while having a higher external surface area and approximately equivalent iodine numbers. However, combining PAC into a non-woven matrix is difficult because adhesives are required to attach it to the fiber matrix. This causes at least some of the particles to be blinded or deactivated.

Chemical impregnants and catalysts are used to remove or destroy contaminants that are not readily physiosorbed by the carbon. For example, ASC Whetlerite (*11*), an activated carbon impregnated with copper, chromium and silver salts was used in military gas masks for many decades. These impregnants removed or destroyed chemical warfare agents such as cyanogen chloride, hydrogen cyanide, and arsine. Whetlerite has been replaced by ASZM-TEDA particularly designed to avoid the use of Cr^{+6}. In other examples, activated carbon is impregnated with citric acid to increase the ability of the activated carbon to adsorb ammonia or with hydroxides, such as sodium hydroxide or other caustic compounds, to remove hydrogen sulfide. The process for incorporating PAC could be adapted to such specialty carbons.

The primary object of this study relates to the adsorption of dissolved halogens by thin beds of filter media at high flowrates, comparing developed PAC/nano alumina media with flexible nano alumina media currently being used in water filtration.

Experimental Results

Iodine Adsorption

Figure 5 shows the adsorption of 20 ppm iodine by a 25 mm diameter disc at a flowrate of 20 ml/min. The media contained 57 wt % PAC (Calgon Carbon Corp., grade WPH, median particle size 16.8 μm (*12*)) and 14% nano alumina fibers, along with microglass, cellulose and a polyester fiber, serving to bind the larger fibers together. Iodine aliquots taken from the effluent were measured using Genesis-10 UV spectrophotometer at 290 nm. At 0.5 ppm, iodine becomes unpalatable and also exceeds regulatory limits. A breakthrough curve is shown for the PAC media where approximately 0.85 liter of water was retained by the PAC disc to below the 0.5 ppm level (detection limit ~0.1 ppm iodine). Also seen are data for 25 mm discs sectioned from the media of three different commercially available activated carbon filter cartridges. Their basis weights are somewhat comparable to the PAC. These media were inspected microscopically and at least two of the three appear to contain suspended GAC.

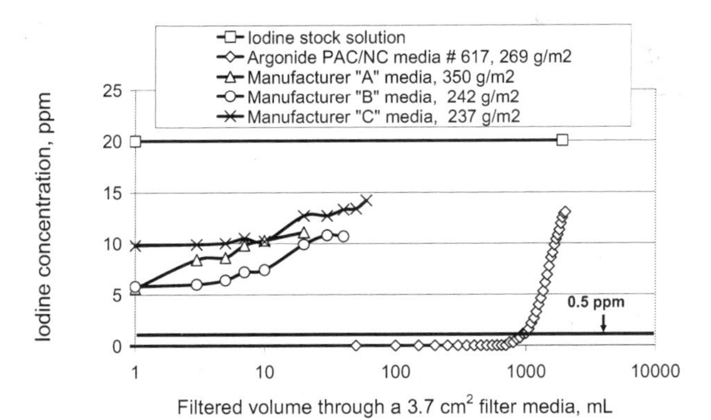

Figure 5. Adsorption of 20 ppm iodine by a 25 mm diameter disc at a flowrate of 20 ml/min

Not one of the other media could retain iodine below about 5 ppm iodine, even at the outset. The estimated volume retained to 0.5 ppm was less than 1 ml in all cases. Using the iodine number (>800 mg/g) provided by Calgon

Carbon Corp., we estimated that the utilization of the PAC media exceeded 50% of theoretical while the other media retained about 1% or less.

A stack of three layers of the respective media's were tested under the same conditions (see Figure 6). In this case the PAC media retained approximately 3,800 ml to the 0.5 ppm level while the other media's retained no more than about 8 ml. In a third series (not shown) we measured the retention of iodine from a 500 ppm iodine solution and found that a single layer of the PAC media could retain 50 ml to 0.5 ppm, while 3 and 6 layer stacks had volumetric capacities respectively of 280 ml and 900 ml.

Because there was interest in a filter with combined chemical and biological retention capabilities, the amount of PAC was reduced to 32%, allowing an increase in nano alumina from 14% to 25 %. Breakthrough curves in Figure 7 show a reduction in iodine removal efficiency at input concentration of 500 ppm. But by substituting a coconut based PAC with higher iodine number (Calgon Carbon Corp., grade 3164, iodine number ~1100) and with smaller median particle size of ~8 μm, the iodine removal efficiency for 32% media under dynamic conditions was almost equivalent to that of the 57% PAC media.

We also measured the retention of iodine from a 20 ppm iodine solution and found that a single layer of the 32%PAC media with median particle size of ~8 μm could retain 400 ml to 0.5 ppm limit, while the 57%PAC media with median particle size 16.8 μm had volumetric capacity of 800 ml.

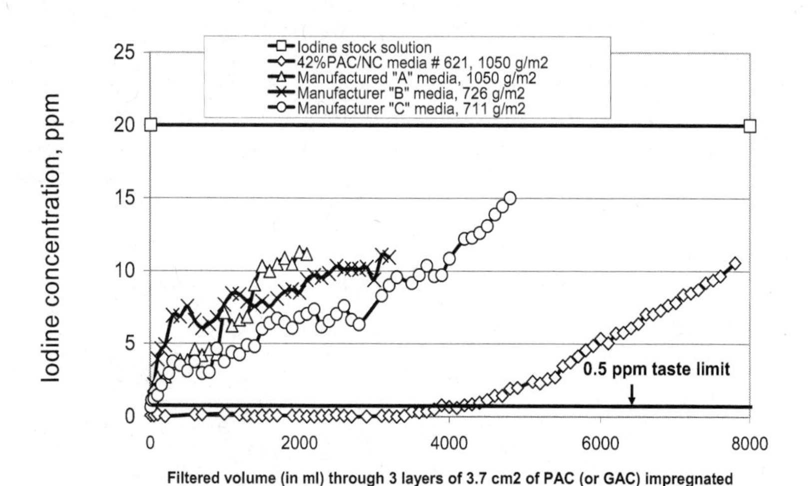

Figure 6. Iodine adsorption by 3 layers of media

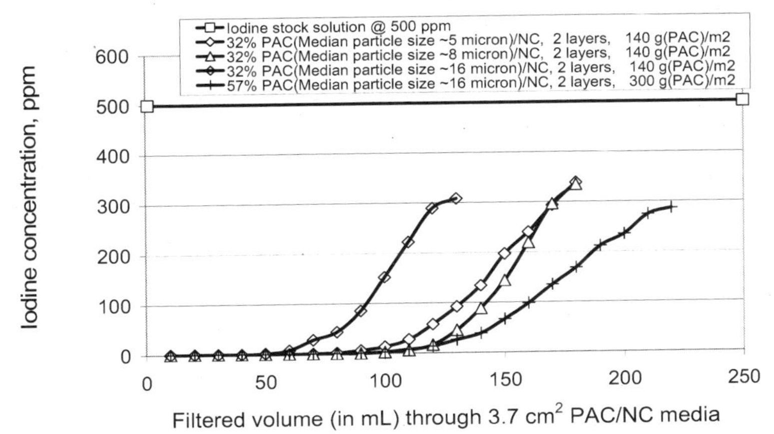

Figure 7. Iodine adsorption by PAC/NC media

Chlorine adsorption

A similar series was done using chlorine. In this case we used sodium hypochlorite to produce 2 ppm of free chlorine. The flowrate was 16 ml/min through a PAC media (32% PAC) that was 25 mm in diameter. Chlorine input and output concentrations were measured by a Tracer (LaMotte, Inc) with a detection limit of 0.01 ppm. The test results with PAC were compared to the three different manufacturers that were tested with iodine (see Figure 5). The results showed superior retention of chlorine by PAC media as compared to the other samples (see Figure 8). The adsorption capacity for the PAC was calculated as 350 mg Cl_2/g PAC to full saturation at input concentration of 2 ppm and flow velocity 10 cm/min.

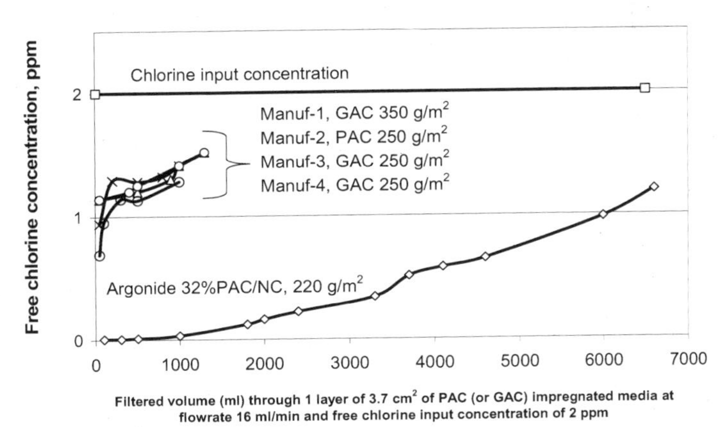

Figure 8. Chlorine adsorption by various media

Retention of Volatile Organic Compounds

Chloroform was selected for initial study because it is used as a simulant for volatile organic solvents when characterizing activated carbon beds for removal of trihalomethanes from water. The test method (EPA Method 502.2 (*13*)) chooses chloroform because its retention by activated carbon in water is rather low. In the first test, a 25 mm disc of 57% PAC (basis weight of media was 270 mg/m^2) was challenged by a solution containing 1300 ppb of chloroform at a flowrate of 15 ml/min. The filtrate contained 65 ppb of chloroform, or 95% removal. A second test involved four plies of the same media, under the same conditions. This filtrate contained 7.1 ppb of chloroform for a removal efficiency of 99.5%.

Retention of Particulates by PAC

Table I shows the retention of E coli and MS2 by 1 and 2 layers of the PAC media. Fifteen ml aliquots of concentrated bacteria and MS2 virus were passed through 25 mm diameter test samples at 40 ml/min. A single layer of the PAC media was capable of retaining >6 LRV of E. coli and approximately 3 LRV of MS2.

Two layers are necessary for retention of >4 LRV of virus at a flow velocity of 10 cm/min. However a pleated cartridge, offering a larger surface area is able to retain virus to greater than 4.5 at flow velocity 2.5 cm/min. For instance, a 2.5"diameter, 10" long PAC (median particle size ~8 μm) cartridge operating at flowrate of 1 GPM is able to retain >4.5 LRV.

Table I. Retention of E. coli and MS2 by 1 and 2 layers of the 32%PAC/NC media[a]

PAC	*Median particle size, μm*	*Thickness, mm (number of layers)*	*MS2, LRV[b]*	**E. coli, LRV[b]**
Calgon Carbon Corp., grade WPH	16.8	2.0(2)	5.1±0.1	>7
		1.0(2)	2.9±0.4	>6
Calgon Carbon Corp., grade 3164	8	2.0(2)	>5.3	>7
		1.0(2)	3.3±0.2	>6

Notes: a) at flow velocity 10 cm/min; b) logarithm retention value

A 25 mm disc of 57% PAC media was challenged with a stream containing 250 NTU (nephelometric turbidity units) of A2 fine test dust until the differential pressure reached 40 psi. The dirt holding capacity was 18 mg/cm^2 as compared to 17 mg/cm^2 for the nano alumina media without PAC. It had been anticipated that the considerable amount of PAC in the structure would have an adverse affect on the retention of additional particles. The turbidity of both filtrates was less than detectable throughout the whole experiment. In both cases, the filtration efficiency exceeded 99.99%. In addition to demonstrating retention of fine test dust, the low turbidity of the effluent suggested that shedding of activated carbon particles into the stream was very low.

Silt density index (SDI) is a sensitive method for determining the ability of a filter to remove colloidal particles (*14*). SDI (in min^{-1}) is used extensively as a criterion in minimizing fouling of reverse osmosis membranes. The lower the SDI value, the cleaner is the stream. Manufacturers of reverse osmosis membranes recommend that the stream be prefiltered so that it has an SDI factor less than 3.0. Typically '1 μm absolute' (*6*) filters have an SDI of about 4-5. Manufacturers of hollow fiber membrane filters claim SDI's in the range of 1.75 to 2.25. SDI measurements of effluents from the nano alumina media range from 0.5-1.0. The SDI of 32% PAC media was measured and found to be about 1.0. Turbidity as well as SDI tests have confirmed that the extent of shedding of PAC particles into effluent streams is minimal.

Environmental, Safety and Health Issues

There are significant concerns about the impact of nanomaterials on the environment and on safety and health. A review of the literature and accompanying experimental work was done relative to the EH&S of microcrystalline boehmite.

Fibrillated boehmite is used as a thickener in food products, as a food dye carrier (*15*) and as an analgesic in the treatment of peptic ulcers (*16, 17*). Microcrystalline boehmite is an important component of vaccines, and is classified an "adjuvant," for binding antigens. Alhydrogel, first described in 1926, remains the only immunologic adjuvant used in human vaccines licensed in the United States (*18*). It is used in diphtheria, tetanus, and pertussis vaccines as well as veterinary vaccines. Alhydrogel consists of primary particles of boehmite (AlOOH) having a fibril morphology and a high surface area, with many of the same characteristics as the boehmite fibers in NanoCeram®. Once the fibrillated boehmite is injected into the muscle and performs its function, it is solvated by citrate ion in the blood stream and carried to the kidney and eliminated (*19*). Approximately 2 billion doses of Alhydrogel-containing vaccines have been used in humans and countless billions more in animals.

In collaboration with Purdue University, we demonstrated (*20*) that nano alumina substrates can serve as scaffolding for the growth of osteoblast (human bone) cells. The study was aimed at developing bone growth cements. The nano alumina, blended into biodegradable (e.g.-lactic acid based) polymers were found to facilitate the proliferation of such cells. There was no toxic affect on osteoblast cells, and by inference nano alumina would have no affect on other mammalian cells.

We have extensive test data where we challenged nano alumina filters with various bacteria (E. coli, R. terrigina, B. globiggi, B. diminuta). We found that all such bacteria in the presence of a food source, can proliferate on the filter, and we therefore presume the filters are compatible with bacteria. We have also collected MS2 virus on NanoCeram and have been able to recover ~90% of viable phage. The toxicity of raw nano alumina fiber towards E. coli, Staphylococcus aureus, B. subtilis, B. pumilis, and Candida albicans has been measured. Filter discs with a diameter of 0.7 cm and nano alumina powder or nano alumina/microglass composite, were placed on the surface of dishes containing the cultures in 2-3 replications and wetted with 20 ml of sterile water. The dishes were incubated at 37°C for 24 hours and 13 samples were tested for the presence of a zone of inhibited growth of the test strain in the presence of the filter or powder. Controls used were granulated compositions including aluminum and titanium oxides supplemented with Fe^{+3} and carbon granules as carriers, that are known to have a toxic affect on test strains cells. No toxic affects were noted for any of the cultures with the exception of the controls.

Dinman (*21*) reviewed the data on occupational exposures to alumina and concluded that the full range of aluminas have little if any bioreactivity: "Until evidence is accrued to the contrary, it appears that considering either transitional aluminas, i.e., gamma, eta, chi or gelantinous, high surface area boehmite (γ-AlOOH) as a subspecies of special concern is unwarranted as regards occupational exposures." We conclude that there are no apparent deleterious EH&S affects of nano alumina.

Discussion and Applications

A non-woven electropositive filter media has been shown to be effective in filtering bacteria and virus. It is currently being manufactured and is commercialized in the form of media, as cartridges and as filter discs.

The component that contributes to the electropositive character of the filter is a nano alumina monohydrate fiber that is attached to a microglass fiber scaffolding. This study has demonstrated that non-fibrous fine and nano size particulate can be attached to the nano alumina by electrostatic forces. Once there they can provide adsorptive capability for chemical toxins that the original media is unable to retain.

A practical application is the integration of PAC into the structure. The resulting media has retention efficiencies that exceed the state of the art for both particulate removal and as compared to other activated carbon filters. This is particularly the case with thin beds operating at moderate to high flowrates (residence time 0.1-1 second).

PAC media has been produced in several manufacturing trials. Cartridges constructed of 32% PAC are being beta tested in a number of industrial and drinking water applications. The benefits include rapid adsorption kinetics as well as better retention of carbon particles than in a GAC bed and at moderate to high flowrates. One application is to use the PAC media downstream of a GAC bed to polish the stream of sorbates leaking from the GAC bed. Since the PAC media is also a highly efficient particulate filter it would retain carbon fines leaking from the GAC bed. Such fines would tend to have a higher concentration of contaminants sorbed on their surfaces, and intercepting them would minimize the contamination of the downstream water.

Drinking Water

A variety of filter devices are currently used for purifying water. Most of these devices employ a number of different types of filtration stages. For example, reverse osmosis (RO) is relied upon to produce high purity water but it requires both sediment and activated carbon prefilters to extend the life of the RO membrane. Nano alumina filters with 250 GPM flow capacity, have been qualified by Toyota and are being used in several plants as a prefilter for their industrial RO units. Such filters are in beta tests as prefilters for RO in desalinization plants. They would be as effective in protecting RO's used in drinking water systems. The ability to integrate PAC into the nano alumina structure would eliminate the need of a separate GAC prefilter for removing residual chlorine, which can damage the RO membrane.

The nano alumina filter system has been commercialized as a point of use (POU) and point of entry (POE). Such a system is far less complicated and less expensive than an RO based system. Moreover it is capable of delivering high rates of flow, eliminating the need for a storage tank that is used to collect RO purified water. One of the deficiencies of an RO system is if the tank becomes contaminated with bacteria, then they can proliferate causing a secondary contamination. A nano alumina or PAC media downstream of an RO is being considered by some users.

Both the nano alumina and the PAC media can filter virus even though the pore sizes are ~2 μm. This capability has been commercialized for sampling virus. A 2.5" diameter, 5" long nano alumina filter (no PAC) is used to collect and concentrate virus for the purpose of assay of pollution at beaches and upstream and downstream of a municipal plant. The virus are extracted using a solution containing beef extract (1.5% or 3%) and glycine (0.05 *M*) at pH 9.

Matrix Effects

The minor ingredients of water can play havoc with filtration systems. They include inorganic particles such as silica, live particulates such as protozoa (cysts) and bacteria (both alive and dead) and algae. The nano alumina and PAC media have a high capacity for inorganic and organic particulates that typically cause clogging of fibrous depth filters as well as fouling of membranes. Contamination of the water by natural organic matter (NOM), in the form of tannic, humic and fulvic acids, is very difficult to filter by either membranes or conventional non-woven depth filters. Test data (not shown) indicates that the humic acid adsorption capability is greatest for PAC media less so with nano alumina media and far less with membranes. Ultraporous (UP) membranes will pass humic acid particles, which have a unit cell size about that of a virus, and the limited amount of humic that is retained by UP membranes rapidly cause fouling.

The PAC media would also be useful as a POE for protecting the drinking water of commercial and military buildings against chemical and biological attack. And while no work has been done on capturing radiological contaminants from water, the filter will likely retain those radioisotopes that are present as particulates, such as metal oxides.

A project has been initiated for development of a portable water purifier for the third world that is capable of retaining biological and chemical toxins. It will use PAC media.

Air Filtration

The feasibility of low pressure drop HEPA based on nano alumina media and subsequently the development of a PAC media for air is under study (*22*). We recognized that the water filter, with its 2 μm pore size would have too high a pressure drop to serve as an air filter. The formulation was therefore modified using coarser microglass that resulted in pore sizes in the 20-40 μm range (*22*).

The first attempts were to match the ΔP of a commercial HEPA and evaluate filtration efficiency. Figure 9 shows the retention and capacity of 0.3 μm NaCl aerosol by 3 plies of such media as compared to a commercially available HEPA. Testing was done by Nelson Laboratories (Salt Lake City, UT). Also shown is the retention of a single ply of the media used as a prefilter for the HEPA. Note that both the 3 plies of HEPA and the combination of a single ply prefilter with HEPA resulted in an ULPA (Ultra low particulate air) filter.

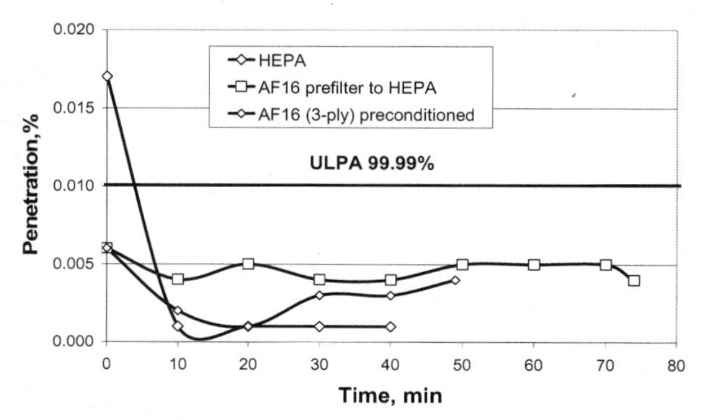

Figure 9. Retention of HEPA and test media

Figure 10 shows the air resistance of the respective samples shown in Figure 9. Note that the HEPA's pressure drop increased rapidly as it was loaded while that of the three plies of media, which initially had a higher pressure drop than the HEPA, did not rise as rapidly as the HEPA. When one ply of the media was used as a prefilter for the HEPA, the rate of rise of ΔP was even shallower. The data show that such a prefilter can extend the life of a HEPA by five to seven times, depending upon the terminal ΔP.

Efforts are now focused on reducing the initial pressure drop with a target of 50% reduction of ΔP. A secondary task is the incorporation of PAC into the media.

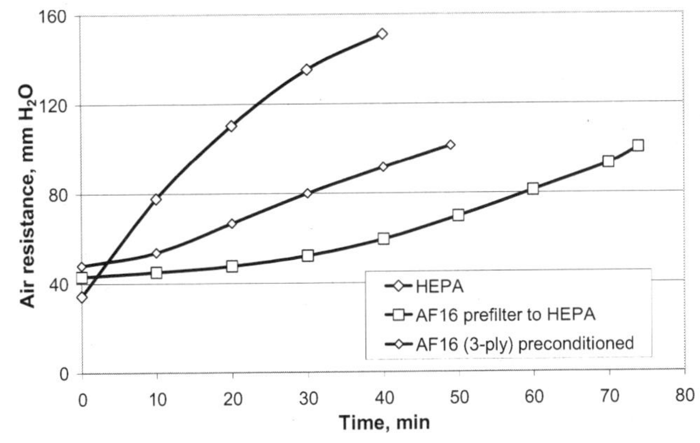

Figure 10. Build up of ΔP with increasing 0.3 μm NaCl loading

Conclusions

A high capacity electrostatic filter containing powdered activated carbon was developed. The filter is a homogeneous combination media capable of filtering bacteria and virus from water at high rates of flow and also having a high dirt holding capacity. At the same time it has a high dynamic efficiency for physically adsorbing soluble halogens and small organic molecules. An air filter version of the media is under development. The combination filter is suitable for removal of chemical and biological warfare agents from water. The combo filter has diverse application in purification of drinking water and for purifying industrial water and waste water.

References

1. Al-Shakhshir, R. ; Regnier, F. ; White, J. L. ; Hem, S. L. *Vaccine* **1995,** *13*, 41.
2. Tepper, F.; Kaledin, L. *High Performance Carbon Filter*, Amer. Filtration Soc. Annual Meeting, Orlando, March 27, 2007
3. Tepper, F.; Kaledin, L. *Phamaceutical Applications for a Combined Particulate and Activated Carbon Filter*, UltraPure Water Pharma Conference, New Brunswick, NJ, Aug. 2007.
4. Leiman, P. G.; Kostyuchenko, V. A.; Shneider, M. M.; Kurochkina, L. P.; Mesyanzhinov, V. V.; Rossmann, M. G. *J. Mol. Biol.* **2000**, *301*, 975.
5. Golmohammadi, R.; Valegård, K.; Fridborg, K.; Liljas, L. *J. Mol. Biol.* **1993**, *234*, 620.
6. Purchas, D. B; Sutherland, K. *Handbook of Filter Media;* Elsevier Science Ltd.: New York, 2002; p 18.
7. Tepper, F.; Kaledin, L. *BioProcess International* **2006, 4**, No. 6, 64.
8. *Forming Handsheets for Physical Tests of Pulp*; Standard TAPPI T-205, TAPPI press, Atlanta, GA; 2000.

9. Knappe D. R. U.; Li, K.; Quinlivan, P. A.; Wagner, T. B. *Effects of Activated Carbon Characteristics on Organic Contaminant Removal*; IWA Publishing: Denver, 2003; p 3.
10. Tom, G. M.; Haltquist, L. J. *Channelized sorbent media, and methods of making same,* US patent 6764755; 2004.
11. Liang, S. H.; Harrison, B. H.; Pagotto, J. G. *Analysis of TEDA on ASC-Whetlerite Charcoal* Defense Research Establishment, Ottawa. Available at http://handle.dtic.mil/100.2/ADA192035.
12. *Water Treatment. Principles and Design*; Crittenden, J. C.; Trussell, R. R.; Hand, D. W., Howe, K. J., Tchobanoglous, G., Eds.; J. Wiley & Sons, Inc., Hoboken, NJ, 2005; p 1296.
13. *Drinking water treatment units. Health effects.* Standard NSF/ANSI 53, American National Standards Institute; 2002.
14. *Standard Test Method for Silt Density Index (SDI) of Water;* Standard D4189-95; American Society for Testing and Materials, Philadelphia, PA; 2001.
15. Bugosh, J. *Fibrous Alumina Monohydrate and its Production*, U S Pat. 2915475 ; 1959.
16. Sepelyak, R. J.; Feldkamp, J. R; Regnier, F. E.; White, J. L.; Hem, S. L. *J Pharm Sci* **1984**, *73,* 1514.
17. Nishida, M.; Yoshimura, Y.; Kawada, J.; Ookubo, A.; Kagawa, T.; Ikawa, A.; Hashimura, Y.; Suzuki, T. *Biochem Intern.* **1990**, *22*, 913.
18. Vogel, F.R.; and Powell, M. F. *Pharm Biotechnol.* **1995**, *6*, 141.
19. Hem, Stanley L. Purdue Univ.; Private Communication, June 29, 2007.
20. Price, R.L.; L.G. Gutwein, L.G.; Kaledin, L.; Tepper, F.; Webster, T.J *J Biomed Mater Res.* **2003**, *A67*, 1284.
21. Dinman, B. A, in *The Aluminas and Health. Alumina Chemicals Science and Technology Handbook*, Hart, L. D. Ed.; American Ceramic Society: Washington, DC, 1990; pp 533-543.
22. Tepper, F.; Kaledin, L. *Electrostatic Air Filter*, US Patent 7,311,752 December 2007.

Chapter 22

Polymeric Membranes: Surface Modification by "Grafting to" Method and Fabrication of Multilayered Assemblies

Oleksandr Burtovyy[1], Viktor Klep[1], Tacibaht Turel[2], Yasser Gowayed[2], Igor Luzinov[1*]

[1]School of Material Science and Engineering, 161 Sirrine Hall, Clemson University, Clemson SC 29634
[2]Department of Polymer and Fiber Engineering, Auburn University

Surface modification of poly(ethylene terephthalate) (PET) membranes was carried out by grafting various polymers using the "grafting to" technique. To increase the initial surface reactivity of the membranes, air plasma treatment was applied for a short time. The reactive anchoring interface was created by deposition of several nanometers of poly(glycidyl methacrylate) (PGMA) onto the membrane boundary. Next, polymers containing reactive (amino and carboxy) groups were grafted to the macromolecular PGMA anchoring layer. Scanning electron microscopy (SEM) and a gas permeation test demonstrated that the polymers could be chemically attached to the PET membranes modified with PGMA. Grafted polymers uniformly covered the surface and did not block pores. We also demonstrated that the PET membranes modified with the PGMA anchoring layer can be successfully used to build membrane assemblies by incorporating silica or titanium oxide microparticles as spacers.

Intoduction

Polymeric membranes are vital elements for defense against chemical and biological weapons (*1,2*). It is also well known that the surface modification of the polymer membranes is on the same level of importance as characteristics of bulk material and process development (*3*). In recent years, various methods for changing the surface properties of membrane materials have been developed (*4-9*). For instance, for membranes with reactive sites on a surface, a dense polymer grafted layer (polymer brush) can be prepared via the polymerization of a monomer ("grafting from" approach) and/or the covalent reaction of a preformed polymer with functional end groups ("grafting to" approach). Plasma modification, γ-ray irradiation, corona discharge and UV photo induced graft polymerization are common techniques that have been used to modify different surfaces including the non-reactive ones.

Among the methods employed for surface modification, grafting polymerization is one of the most widely used techniques to durably modify the surface properties of membranes (*10-14*). The entire surface of a polymer membrane can be potentially covered with a layer of grafted polymer molecules. The "grafting from" method allows the preparation of a polymer brush layer of high grafting density through the polymerization of a monomer from the initiator immobilized on the membrane surface. The main disadvantages of the "grafting from" approach for industrial applications are long polymerization times, the use of toxic monomers (which may also swell/dissolve the membrane) and, in most cases, an inability to determine the exact macromolecular characteristics (e.g. molecular weight) of the grafted polymer.

The "grafting to" method is an alternative approach to creating a polymer brush on the surface. In this method, a preformed polymer is grafted to the polymer surface of interest. The polymer can be synthesized under the appropriate conditions for a particular application and characterized before modification. An additional advantage of using an already synthesized polymer for surface modification is that antimicrobial agents and hydrophobic, hydrophilic and charged groups can be also incorporated. The "grafting to" approach has been used almost exclusively to attach polymers to flat nonporous surfaces (*15-21*). To the best of our knowledge, studies describing the application of this method to the surface modification of polymer membranes are very limited (*22*).

To this end, the present work focuses on the attachment of hydrophobic and hydrophilic polymers with reactive groups to PET model films and track membranes using the "grafting to" approach. Prior to the grafting, a poly(glycidyl methacrylate) [PGMA] anchoring layer was deposited on the PET film/membrane. PGMA, an epoxy containing polymer, is used for the initial surface modification and generation of the primary reactive layer on a substrate surface (*23,24*). As a result of the surface activation, polymers possessing different functional groups (carboxy, anhydride, amino and hydroxyl) can be grafted to the surface modified with the anchoring layer. Since epoxy groups are highly active in various chemical reactions, the approach becomes virtually universal towards both surface and (end-)functionalized (macro)molecules being used for the grafted layer formation. Water contact angle measurements, ellipsometry, SEM, and atomic force microscopy (AFM) were used to characterize the grafted polymer films. The permeability, wettability and morphology of membranes grafted with polymers were investigated. We also demonstrated that the PET membranes modified with the PGMA anchoring layer can be successfully used to build membrane assemblies by incorporating silica or titanium oxide microparticles as spacers.

We expect that the proposed approach for the surface modification of polymeric membranes and the generation of the multilayered membrane assemblies can be straightforwardly employed as an efficient platform to fabricate breathable protective materials. The platform is highly tunable and upgradeable, since various parameters can be varied at will. First of all, membranes of different natures with different pore sizes can be employed. Second of all, various pre-modified (re)active/hydrophilic/hydrophobic membranes can be assembled together in a number of sequences. An additional advantage is the possibility of loading intermembrane space with functional micro- and nanoparticles, such as catalysts and/or adsorbents. Finally, in the assembly, protective elements are prefabricated and located at different levels and, thus, the compatibility issue can be resolved and multi-functionality can be achieved.

Experimental

Materials

PET membranes (Poretics, polyester, pore size 0.2 μm, membrane thickness 10 μm) were purchased from Osmonics, Inc. Diethylether (99%), 2,2'-azobisisobutyronitrile (AIBN), copper (I) bromide (CuBr) (98%), acetone (min. 99.5%), methanol (99%), ethanol (99%), methyl ethyl ketone (MEK) (99%) and polyethylenimine (hygroscopic) were purchased from VWR International and used as received. 4,4-Dinonyl-2,2'-dipyridine (DNBP) (97%) and ethyl-2-bromoisobutyrate (98%) were purchased from Sigma-Aldrich. Tetrahydrofuran (THF) was supplied by the EMD Co. Poly(acrylic acid) (PAA) (25% aqueous solution) and 2,3,4,5,6-pentafluorostyrene (PFS) were obtained from Polyscience Inc. and Oakwood Products Inc., respectively. Glycidyl methacrylate (min 95%) was purchased from TCI, America.

Grafting of the polymers to a model flat PET substrate

We initially studied the polymer grafting by employing model PET substrates. Namely, thin polymer films from PET materials were deposited on silicon wafers. The silicon wafer located under the film was capable of reflecting light and, therefore, ellipsometry could be used to monitor the polymer grafting. The deposition of a model PET film on a silicon wafer and its further modification via air plasma treatment and the grafting of a PGMA anchoring layer is described elsewhere (*24*). In brief, highly polished single-crystal silicon wafers of {100} orientation (Semiconductor Processing Co.) were first cleaned in a hot piranha solution and then rinsed several times with high purity water. PET (Wellman) was dip-coated on the silicon substrate from a (~ 1.5 wt/vol %) hexafluoro-2-propanol solution. The samples were annealed at 140°C for 3 hours. The thickness of the dip coated PET layer, as measured by ellipsometry (InOMTech, Inc), was 33$\pm$ 3nm. The PET film was plasma treated for 40 seconds using a Harrick Scientific Corporation (Model PDC-32G) plasma cleaner.

Glycidyl methacrylate was radically polymerized to give PGMA, M_n=250,000 g/mol. The polymerization, initiated by AIBN, took place in MEK at 60°C. The polymer obtained was re-precipitated several times from the MEK solution using diethylether. The PGMA layer was deposited on the plasma treated PET film by dip coating from 0.1 wt/vol % MEK solution and annealed at 120°C for 2 hours. The thickness of the PGMA layer (measured by ellipsometry) was 2.5 $\pm$ 0.5 nm.

The synthesis of carboxy-terminated PPFS (number average molecular weight of 93,250 g/mol) by atom transfer radical polymerization has recently been described (*24*). PPFS films were dip-coated on the PET substrate modified with a PGMA anchoring layer from 5.0 wt/vol % diethyl ketone (DEK) solution. The thickness of the dip coated films was 75 $\pm$ 15.0 nm (as measured by ellipsometry). The samples were annealed for 10 minutes in a temperature range from 100 to 160°C (*24*). After the annealing procedure, they were washed 3-4 times with THF to remove any unattached polymer chains.

Polyacrylic acid (PAA) (M_n=100,000 g/mol) and polyethylenimine (PEI) (M_n=25,000 g/mol) were deposited on the PET film modified with a PGMA layer using the same technique as for PPFS. The solution concentration (1 wt/vol %) and solvent (methanol) were the same for both polymers. The thickness of the dip coated films was 30 $\pm$ 2 nm. Typical grafting times were 10 minutes and 1 hour. The temperature of grafting varied from 40°C to 120°C. After the thermal treatment, the samples were rinsed with methanol several times.

Procedures for grafting to the membrane

To remove any residual substances from the surface, the PET track membranes were washed with acetone for 30 min and dried in a vacuum oven at room temperature for 2 hours. The membranes were exposed to plasma at a low setting (680V DC, 10mA DC, 6.8 W) for 40 seconds. After the plasma treatment, the membranes were taken out of the chamber and washed with ethanol. Then the samples were dried to a constant weight and kept under a nitrogen atmosphere prior to further modification. The plasma treated PET membrane was dipped into 1 wt/vol % PGMA solution in MEK and vacuumed several times to force the solution inside the pores. Next, the membrane was withdrawn from the polymer solution at a constant speed. The sample was annealed in a vacuum oven at 120°C for 2 hours. After the annealing, the membrane was washed several times with MEK and dried under the vacuum to a constant weight. The same procedure was used to graft PPFS, PAA

(M_n=100,000 g/mol) and polyethylenimine (M_n=25,000 g/mol) to the membranes covered with the PGMA anchoring layer. The concentration of the solution for all three polymers was 1 wt/vol %. PPFS was dissolved in THF and both PAA and PEI were dissolved in methanol. The annealing temperature and time for PPFS was 160°C for 15 minutes, for PAA – 100°C for 40 minutes and for PEI – 80°C for 1 hour.

Preparation of multilayered membrane assemblies with incorporated microparticles

PGMA coated membranes were dip-coated in 1% SiO_2 or TiO_2 particles suspension in methanol. After the particle deposition, two membranes (one was PGMA modified and the other was decorated with particles) were pressed together for 30 min at 140°C under 39.0 x 105 Pa using a top rubbery plate made of extreme-temperature silicon foam rubber (Aldrich).

Characterizations

Static water contact angle measurements of the model and membrane surfaces were carried out using a contact angle goniometer (Kruss, Model DSA 10). The static time before the angle measurements was 60 seconds and the calculation was made using a tangent method. Atomic force microscopy (AFM) studies were performed using a Dimension 3100 (Digital Instruments, Veeco, Inc.) microscope operated in tapping mode. Silicon tips with a spring constant of ~ 50 N/m were used to obtain the morphology of the films and membranes in air at ambient conditions. Root mean square roughness (RMS) of the samples was evaluated from the recorded AFM images.

Membranes were sputtered with platinum and examined using FESEM-Hitachi 4800 scanning electron microscopy (SEM). A CSI-135 permeability gas cell was used to measure the gas transmission rate (GTR) of air through the unmodified and modified membranes. To measure the adhesion between the layers of the membrane assembly, a 180° peel-off test was carried out using Instron-5582.

Results and Discussion

Grafting to the model PET substrate

We initially studied the polymer grafting employing model substrates, nanothick PET films deposited on a silicon wafer, to determine the optimal conditions for the surface modification of the membranes. Figure 1 shows the morphology of the PET films after annealing (140°C for 3 hours). The film uniformly covers the wafer at both micro- and nano-levels (Figure 1a). After the annealing procedure (Figure 1), a crystalline structure was formed and RMS roughness increased from 0.3 nm to 1.3 nm (10 x 10 μm^2 area).

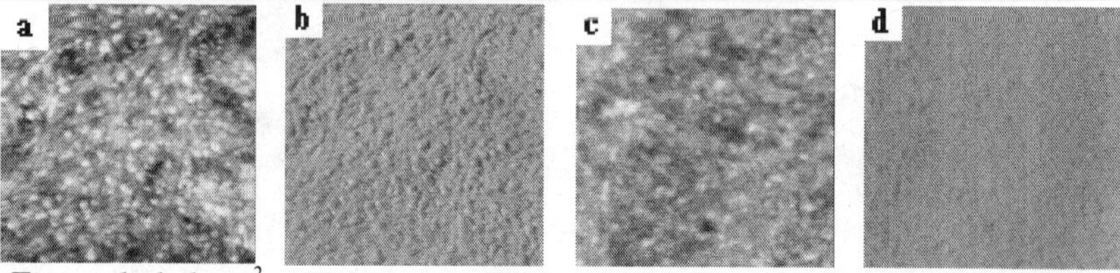

Figure 1. 1x1 μm^2 AFM topography (a, c) and phase (b, d) images of the PET films. (a) and (b) annealed films; (c) and (d) covered with PGMA anchoring layer and annealed at 120°C for 2 h. Vertical scale is 20 nm for topography and 25° for phase images

To make possible further surface modification with the PGMA anchoring layer, containing epoxy reactive groups, the PET films were treated with air plasma for 40 seconds. It was determined previously that this is an optimal time for this procedure (*24*). To remove any unattached materials, the samples were washed with ethanol. Ellipsometry measurement showed a decrease of PET film thickness by 3 ± 0.5 nm as a result of the plasma treatment. The water contact angle, measured for the treated PET films, was around 40° (for the untreated ones it was 65°). The wettability result indicated the formation of oxygen containing functional groups on the polymer surface. The roughness of the samples increased slightly after the treatment (RMS=1.4 nm), showing that the etching rate of the crystalline part of the PET film was lower than that of the amorphous part. The plasma action did not destroy the films and was uniform over the whole treated area.

The primary surface modification was carried out by dip-coating the PGMA layer on the plasma treated PET surface from 0.1 wt/vol % MEK solution. Next, the samples were annealed in a vacuum oven at 120°C for 2 hours. The thickness of the PGMA layer (as measured by ellipsometry) was 2.5 ± 0.5 nm. The epoxy groups of the polymer chemically anchor the PGMA to a surface. The glycidyl methacrylate units located in the "loops" and "tails" sections of the attached macromolecules are not connected to a substrate boundary. These free groups can serve as reactive sites for the subsequent attachment of molecules with functional groups (carboxy, anhydride, amino and hydroxyl), which exhibit an affinity for the epoxy-modified surface. The PGMA layer uniformly covered the PET substrate following the crystalline structures of the PET film at the nano- and micro-level. AFM RMS (10 x 10 μm^2 area) roughness decreased slightly to 1.1 nm. It was estimated that the PGMA layer had a surface concentration of active epoxy groups not less than 2 groups/nm^2 or 2 x 10^6 epoxy groups/μm^2 (*24*).

In fact, the PET film modified with the epoxy functionalities was active toward the grafting of reactive polymers. Initially, the anchoring of carboxy terminated PPFS was successfully conducted. Details of the PPFS grafting to the PGMA modified PET film are reported in our prior publication (*24*). It was determined that the thickness of the grafted layer increased with temperature and time of grafting. Rapid (5-10 min) and significant change in the surface properties of PET can be realized at grafting temperatures above 140°C. In these conditions, we observed a significant level of grafting and, consequently, high hydrophobicity (water-contact angle 100°C) with the modified PET film. Next, we used the grafting approach to attach PAA and PEI to the PET film surface via the PGMA anchoring layer. The AFM images (Figure 2) clearly show that the PAA and PEI macromolecule chains uniformly cover the substrate and the value of the AFM roughness (1.2 $\pm$ 0.5 nm) is close to the roughness of PET films after PGMA deposition.

In Figure 3, the thickness of the grafted PAA and PEI layer for different grafting times is plotted versus the temperature of grafting. In general, the rate of grafting increases with temperature. The increase can be connected to the rate of coupling reaction between the functionalities and the diffusion of the polymer reactive carboxylic or amine groups to the epoxy modified surface. The water contact angle measurements allowed us to estimate how well the grafted layers screen the underlying surface. Figure 3b shows that the contact angle of the PEI layer is almost independent of grafting amount, and is very close to the contact angle measured for the PEI free standing film (57°). Also, for all conditions, it is less than the contact angle of the PGMA layer on the PET film (60$\pm$2). The secondary amino groups of PEI are highly reactive and, therefore, a short time and low temperature are required to reach the suitable grafting density to fully screen the surface covered with the PGMA layer. A further increase of temperature and time does not change the thickness and contact angle of the PEI grafted layer significantly. However, at temperatures close to 80-90°C, we observed an increase of the grafted thickness. The observed phenomena may be connected with the higher mobility of the PET and PGMA polymer chains at this temperature, resulting in a higher probability of the reaction between the amino groups of the PEI and the epoxy rings of the PGMA macromolecules. (It is necessary to highlight that the glass transition temperature (T_g) for PGMA (*25*) and PET (*26*) is between 60 and 75°C, while T_g for PEI is around –13°C (*27*)).

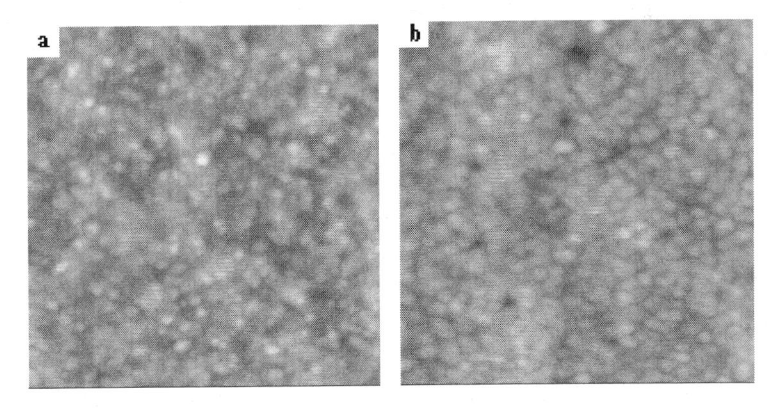

Figure 2. 1x1 μm² AFM topography images of PAA (a) and PEI (b) grafted to PET film. Grafting conditions: temperature – 120°C and time – 1 hour. Vertical scale is 25 nm.

For the grafting of PAA macromolecules, there is a quasi linear dependence between the thickness of the layer grafted and temperature. We suggest that the difference between the grafting of PEI and PAA chains can be explained by the different reactivity of amino and carboxy groups towards the epoxy functionality of PGMA. Specifically, the amino groups are more reactive. Also, T_g for PAA is around 100-109°C (*28,29*). This is a possible reason for the absence of the grafting acceleration at 80-90 °C, which was found for PEI macromolecules with lower T_g.

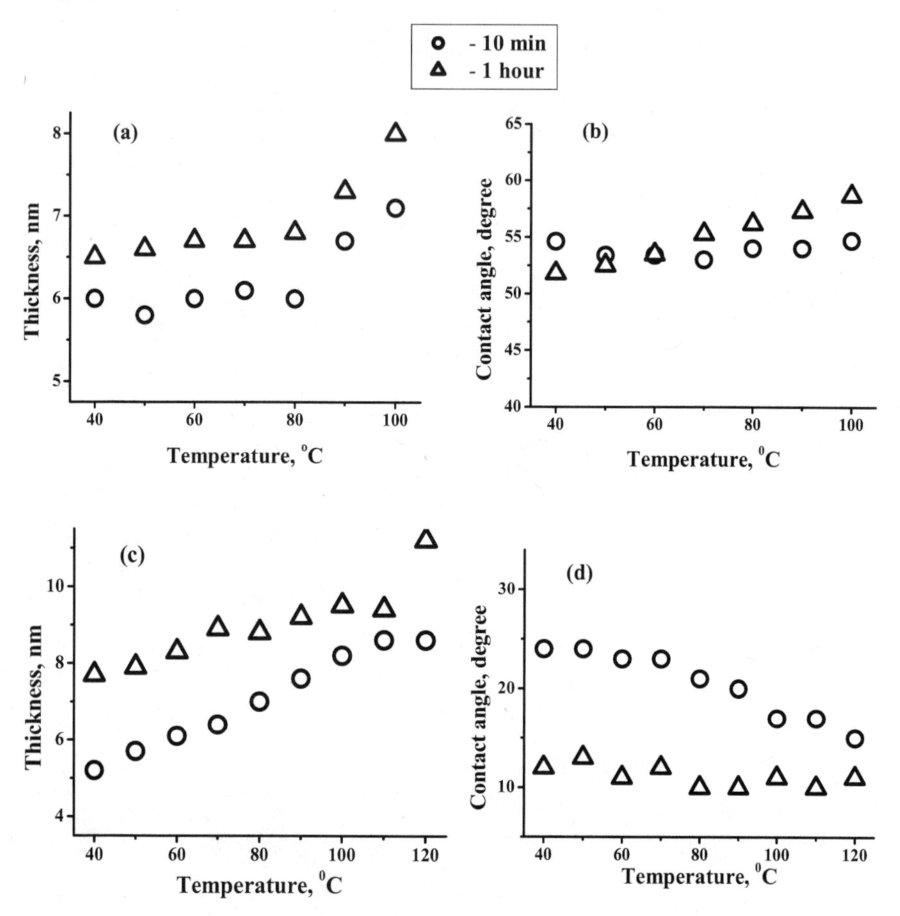

Figure 3. Thickness (a, c) and water-contact angle (b, d) for the grafted (to PET/PGMA substrate) PEI (a, b) and PAA (c, d) layer as a function of grafting temperature. The samples were washed with methanol before the measurements. Time of the grafting: 10 min and 1 hour.

It appeared that to reach a sufficient hydrophilicity, the PAA thickness should be not less than 8 nm (**Figures 3c** and **d**). This can be achieved by conducting the grafting for a longer time (1 hour) or at a higher temperature (120°C).

Grafting to the PET membrane

The method of surface grafting that was developed employing the model PET films was applied to the surface modification of the PET membranes. The membranes were first cleaned in acetone and then treated with plasma at the same condition as the films, where only 2-3 nm of PET material was etched. Figure 4 shows AFM images of the unmodified and plasma treated membrane. It is evident that morphology and pore size did not change significantly as a result of the plasma treatment.

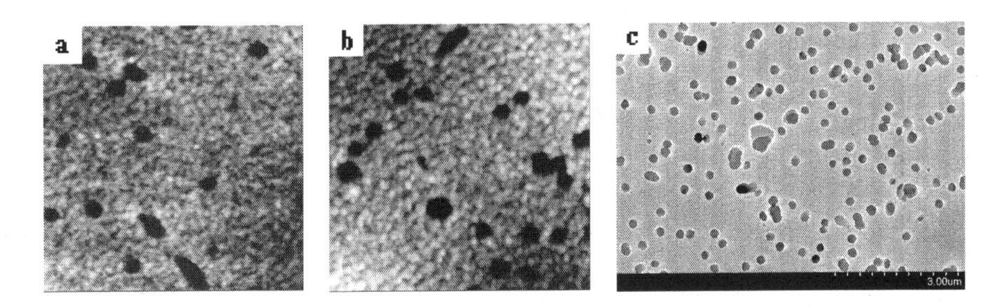

Figure 4. AFM topographical (a) and SEM (c) images of unmodified and AFM (b) of plasma treated PET membranes. Size of the AFM images is 2x2 μm^2. Vertical scale is 25 nm.

The plasma treated PET membrane was dipped into 1 wt/vol % PGMA solution in MEK and vacuumed several times to force the polymer solution inside the pores. Then, the membrane was withdrawn from the solution at constant speed using the dip-coater. The sample was annealed at 120°C for 2 hours, washed several times with MEK, and dried to a constant weight at ambient conditions. In our preliminary experiments, it was found that it is necessary to use a relatively concentrated solution of PGMA (1 wt %) to avoid dewetting the PGMA film on the membrane surface. Such grafting conditions allow for the anchoring of 10-15 nm of the PGMA reactive layer, as was determined employing the model PET film. SEM and AFM images (Figure 5) clearly demonstrate that the PGMA layer obtained is continuous and covers the substrate uniformly. Most importantly, the pores of the membrane were not visually blocked after the deposition of the reactive coating. Comparing the AFM images taken after the plasma treatment (Figure 4b) and the PGMA deposition (Figure 5a), a smoothing of the membrane surface and a slight decrease of the pore size due to the anchoring of the PGMA reactive layer can be observed. The water contact angle on the membrane increased from 34° to 57° (the measured contact angle of the PGMA flat film is $60 \pm 2°$). The water contact angle for the unmodified PET membrane is 50°.

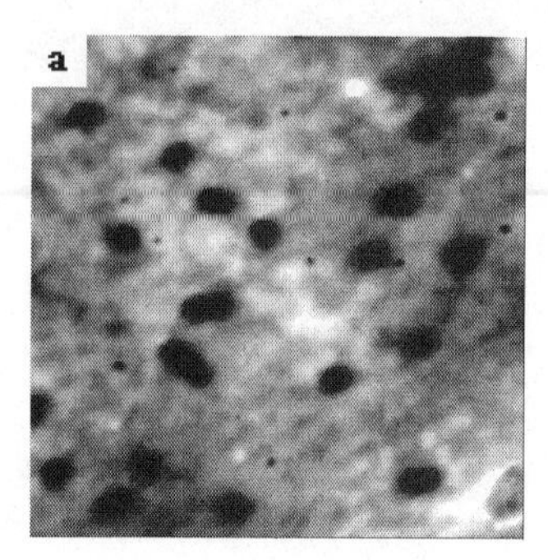

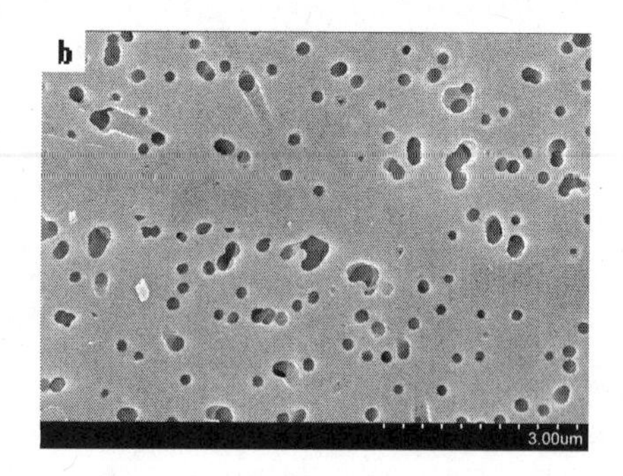

Figure 5. AFM (2x2 μm²) topographical (a) and SEM (b) images of PET membrane modified with the PGMA reactive film. Vertical scale of the AFM image is 25 nm.

To prove the presence of the PGMA layer inside the pores, we performed a cross-section analysis of the membranes. After modification, the samples were immersed in liquid nitrogen and broken after 1 minute. Figure 6 shows the cross-section of the PET membranes after plasma treatment and PGMA deposition. From the images with higher magnification, it is clearly visible that after PGMA modification we have a 15-30 nm PGMA layer located inside the pores, which are not blocked (Figure 6b). Thus, the AFM/SEM images and contact angle measurements indicate an attachment of the PGMA macromolecule chains to the outside and internal surfaces of the plasma treated PET membrane.

The membranes covered with the PGMA reactive film were further modified by the grafting of PAA, PPFS, and PEI layers. The major experimental procedures (besides time and temperature) were the same as for the deposition of PGMA. After the anchoring of PPFS (at 160^{0}C for 15min), the water contact angle of the modified membrane increased to 140°, indicating the effective grafting of the hydrophobic macromolecules. In fact, the PPFS layer uniformly covered the surface and pores of the PET membrane (Figure 7). The pores were not blocked, and we created a thin hydrophobic polymer layer and significantly changed the surface properties of the PET membrane. From the results obtained on the PET model substrate, it was expected that at the grafting conditions employed, the thickness of the PPFS layer should be around 12 nm (*24*).

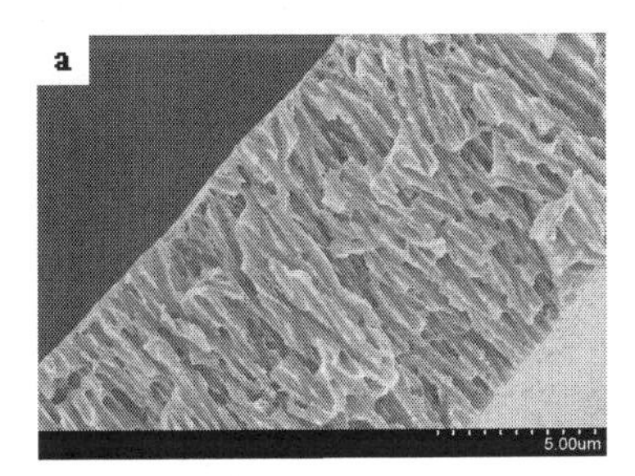

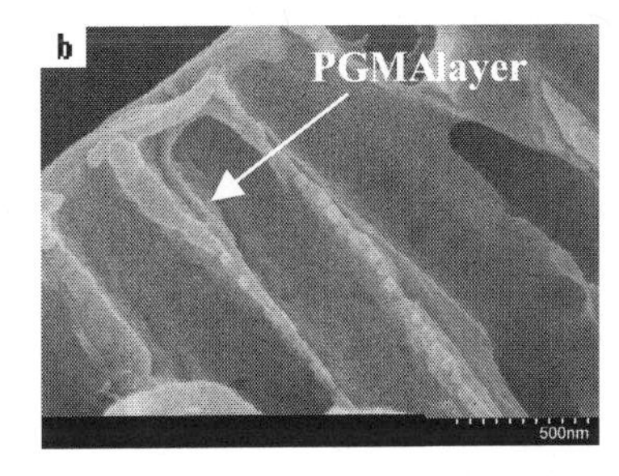

Figure 6. SEM cross-section images of PGMA modified (a, b) PET membrane.

From the SEM cross-section analysis, where it is impossible to differentiate between the PGMA and PPFS layers, the total thickness of the deposited layer (PGMA+PPFS) can be estimated to be 40 $\pm$ 5 nm. If we estimated that a PGMA layer with a thickness of 15-30 nm was deposited, the thickness of the grafted PPFS layer is around 15- 20 nm. We suggested that the discrepancy between the total thickness of the polymer layers on the model PET substrate and the PET membrane can be explained by the different levels of crystallinity and surface chemistry of the PET constituting the model films and membrane.

Figure 7. SEM (a) and SEM cross-section (b,c) images of PPFS modified PET membrane.

Next, we grafted a highly hydrophilic and reactive polymer (PAA), and a moderately hydrophilic polymer that contains reactive amine groups (PEI), to the PET membranes modified with the PGMA anchoring layer. The grafting of PAA was conducted at 100^0C for 40 min and the anchoring of PEI was carried out at 80^0C for 60 min. Figures 8 and 9 clearly indicate that the surface of the PGMA modified PET membrane is uniformly covered with the PAA and PEI thin films and the pores are not blocked. The total thickness of the grafted layers was estimated from SEM cross-section images (Figures 8 and 9), and was larger than the one obtained for the model substrates.

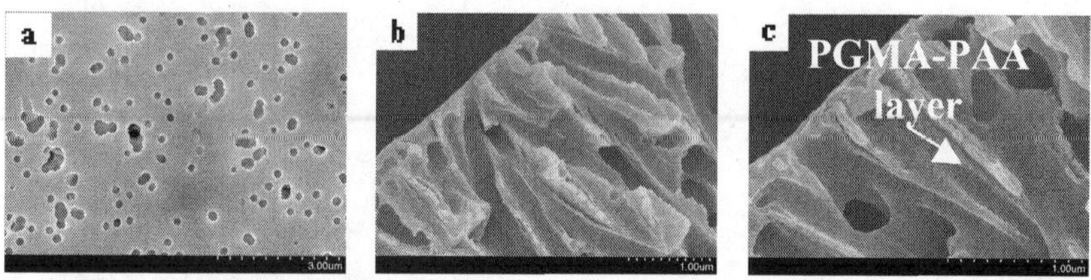

Figure 8. SEM (a) and SEM cross-section (b, c) images of PAA modified PET membrane

The thickness was determined to be 20-30 nm and 15-20 nm for the PEI and PAA layers, respectively. After the grafting, the water contact angle was measured and found to be 14° for the PAA modified membrane and 56° for the PEI coated membrane. In both cases, the results of the contact angle measurement accord well with the values recorded for the model PET substrates.

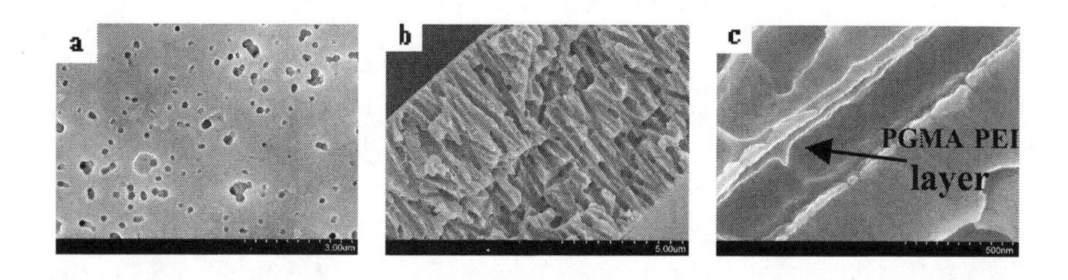

Figure 9. SEM (a) and SEM cross-section (b, c) images of PEI modified PET membrane.

An air permeability test for the unmodified and modified membranes was carried out to determine their breathability. The results in Figure 10 indicate a decrease of GTR values for the modified membrane as compared to the original (untreated) ones. The testing was performed at two different pressures and the same tendency was observed. In general, the PGMA modified membrane had a lower GTR value than the untreated one because of a decrease in the pore diameter. Modification with grafted polymers (PPFS, PEI, and PAA) further decreased the pore size and consequently the permeability. In essence, the permeability test revealed that the grafted polymers were present on the membrane surface and did not block the pores.

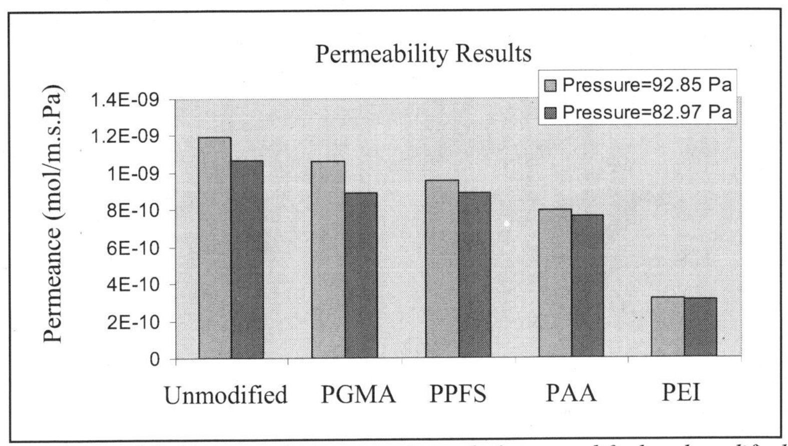

Figure 10. Air transmission rate through the unmodified and modified membranes.

Multilayered membrane assemblies

The possibility of generating multilayered membrane assemblies employing PET membranes modified with PGMA anchoring layers was examined. The multilayered assemblies can be straightforwardly employed as an efficient tunable and upgradeable platform to fabricate breathable protective materials, where membranes of different natures with different pore sizes can be combined. An additional advantage of this is the possibility to load intermembrane space with functional micro- and nanoparticles, such as catalysts and/or adsorbents.

To begin with, the appropriate conditions (pressure, time and temperature) had to be determined for the successful bonding of two thin PET membranes with a thickness on the order of 10-20 microns. To this end, PET membranes were first cleaned with acetone in an ultrasonic bath for 30 minutes, then dried at ambient conditions and pressed together at elevated temperatures using a laboratory press. Attempts to press the membranes employing traditional metal plates resulted in uneven contact between the membranes. Next, two plates made of extreme-temperature silicon foam rubber with plain backs were used instead of the metal plates. The application of the top rubbery plate significantly improved the quality of the two-layered assembly. In a typical experiment, two membranes were pressed together at 140°C for 30 min. AFM images (Figure 11) of the tested samples show that under pressures above 39.2x105 Pa, the integrity of the membranes was compromised. Below this pressure, however, the shape of the pores and the morphology of the membrane surface remained intact.

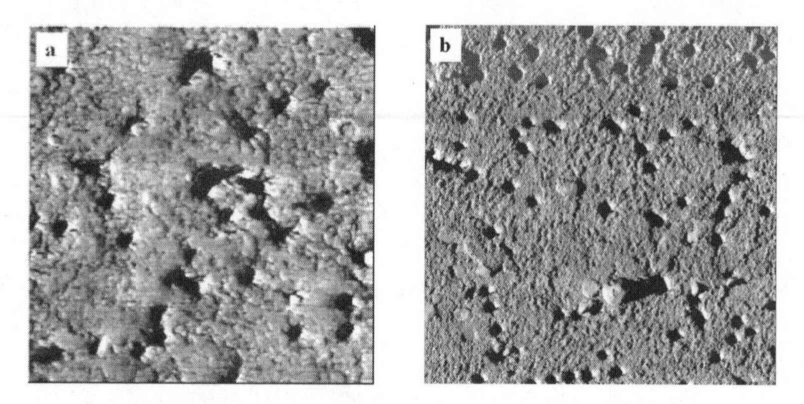

Figure 11. AFM images of the unmodified tested PET membrane (a, 3x3 μm²) above 39.2x10⁵ Pa (b, 5x5 μm²) below 39.2x10⁵ Pa

The peel-off test (Figure 12a) shows that the contact between the two unmodified PET membranes under three different pressures (4.9 x10^5, 19.6 x10^5 and 39.2x10^5 Pa) is weak, but is the same along the tested area. Analogous experiments were conducted for two PET membranes modified with a PGMA anchoring layer. The results demonstrated (Figure 12a) that the strength of adhesion between the membranes modified with the anchoring epoxy layer was much higher due to the reaction of one layer of PGMA with another. AFM imaging of the delaminated assembly (Figure 12b) showed that the pores were not destroyed. In addition, the transfer of material from one membrane to another signified good adhesion between the layers. The surface morphology resulting from the three different pressures applied during the fabrication of the assembly (4.9 x10^5, 19.6 x10^5 and 39.2x10^5 Pa) was observed to be very similar.

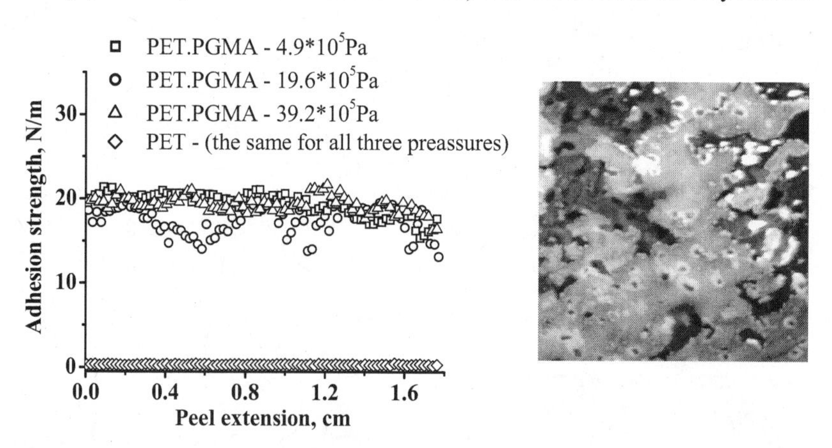

Figure 12. (a) Results of peel off tests for two-layer assembly made of unmodified membranes and membranes modified with PGMA layer. (b) AFM image (5x5 μm²) of surface of PET membrane covered with PGMA after the peel off test. Vertical scale 30 nm.

To create an assembly with microparticles as spacers, unmodified silica (~ 1 μm) or TiO_2 (1-2 microns) particles were deposited on the PET membrane modified with PGMA from suspension in methanol. The SEM images (not shown) demonstrated that the monolayer of the particles covered the membrane surface. The two-layered assemblies of the membranes modified with PGMA and silica or TiO_2 microparticles were prepared by pressing them together at 140°C for 30 min under 39.0 x 10^5 Pa. The SEM cross-section images (Figure 13) show that there is a gap between the membranes filled with the microparticles, which are not pressed into the membranes. The peel-off test indicated that incorporated microparticles increased adhesion between the layers, and revealed that adhesion between the membranes was larger than the cohesion of the PET material of the membranes. The increase of adhesion could be attributed to the better reactivity and stronger interaction of silicon oxide or titanium oxide and PGMA as compared with PGMA-PGMA interaction. In addition, higher values of pressure developed at the contact area between the particles and the membrane during the assembly preparation.

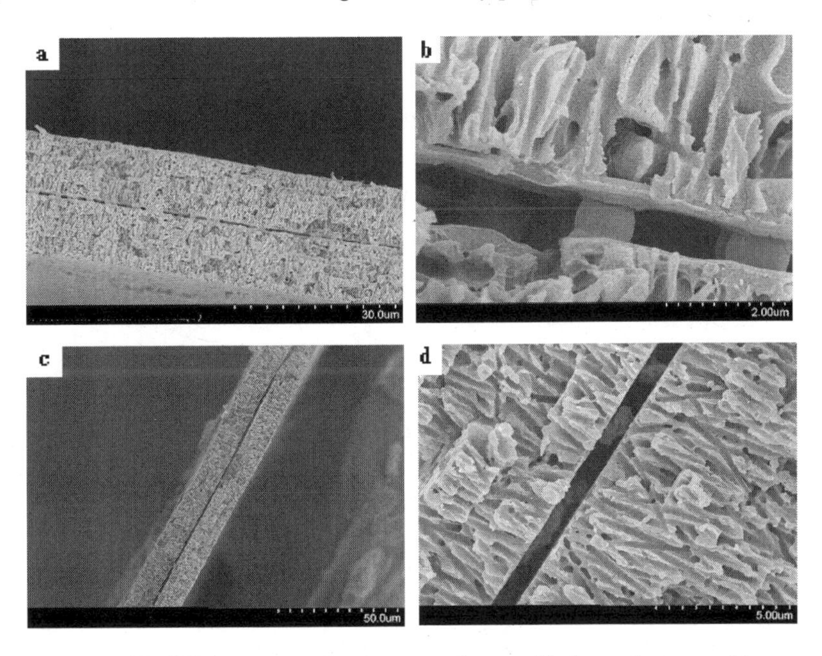

Figure 13. SEM cross-section images of assembled membranes with incorporated microparticles. (a, b) – SiO_2 and (c, d) – TiO_2.

To evaluate the importance of the modification of the membranes with the reactive PGMA layer for the generation of robust membrane assembly, we prepared the microparticle/membrane assembly according to the procedure described above employing the unmodified PET membranes. It was found that, without the PGMA modification, the adhesion between the membranes in the assembly was very weak.

Conclusions

The "grafting to" approach was successfully used for the attachment of hydrophobic and hydrophilic polymers to the PET substrate and PET membranes via a PGMA anchoring layer. To achieve a significant level of grafting and simultaneously change surface properties, a high temperature and a short time or a low temperature and a long time had to be used. AFM and SEM images clearly revealed that the nanolayer of grafted polymers is uniform and does not block the pores. A wettability test confirmed that the grafting of hydrophobic and hydrophilic polymers changed surface properties and created new surface functionality in the PET membrane. SEM cross-section analysis and the peel-off test showed that PGMA modified PET membranes can be used to create membrane assemblies by incorporating silicon or titanium oxide microparticles. The particles were not meshed into the membranes and increased adhesion strength (adhesion was higher than the cohesion of PET material).

Acknowledgments

This work has been supported in part by the Homeland Security Advanced Research Projects Agency, the Department of Commerce via the National Textile Center, and the ERC Program of National Science Foundation under Award Number EEC-9731680.

Reference

1. Truong Q.; Rivin D. NATICK/TR-96/023L (U.S. Army Natick Research, Development, and Engineering Center, Natick, MA, 1996).
2. Koros M.; Fleming G. *J. Membr. Sci.* **1993,** *83,* 1.
3. Matsuura, T. Synthetic Membranes and Membrane Separation Processes; CRC Press: Boca Raton, FL, 1993.
4. Carroll, T.; Booker, N.A.; Meier-Haack, J. *J. Membr. Sci.* **2002**, *203,* 3.
5. Xu, Z.K.; Wang, J.L.; Shen, L.Q; Meng, D.F.; Xu, Y.Y. *J. Membr. Sci.* **2002**, *196,* 221.
6. Ma, H; Kakim, L.F; Bowman, C.N.; Davis, R.H. *J. Membr. Sci.* **2001**, *189,* 255.
7. Pieracci, J; Crivello, J.V.; Belfort, G. *Chem. Mater.* **2002**, *14,* 256.
8. Piletsky, S.; Matuschewski, H.; Schedler, U.; Wilpert, A.; Piletska, E.; Thirle, T.A.; Ulbricht, M. *Macromolecules* **2000**, *33,* 3092.
9. Wavhal, D.S.; Fisher, E.R. *Langmuir* **2003**, *19,* 79.
10. Bergbreiter, D. E.; Bandella, A. *J. Am. Chem. Soc.* **1995**, *117,* 10589.
11. Ulbricht, M.; Yang, H. *Chem. Mater* **2005**, *17,* 2622.
12. Xu, Z.; Wang, J.; Shen, L.; Men, D.; Xu, Y. *J. Membr. Sci.* **2002**, *196,* 221.

13. Singh, N.; Husson, S. M.; Zdyrko, B.; Luzinov, I. *Journal of Membrane Science* **2005,** *262*, 81.
14. Chen, J.-P.; Chiang, Y.-P. *J. Membr. Sci.* **2006**, *270,* 212.
15. Michielsen, S. *J. Appl. Polym. Sci.* **1999**, *73,* 129.
16. Iyer, K. S.; Zdyrko, B.; Malz, H.; Pionteck, J.; Luzinov, I. *Macromolecules* **2003**, *36,* 6519.
17. Zdyrko, B.; Varshney, S. K.; Luzinov, I. *Langmuir* **2004**, *20*, 6727.
18. Ionov, L.; Zdyrko, B.; Sidorenko, A.; Minko, S.; Klep, V.; Luzinov, I.; Stamm, M. *Macromol. Rapid Commun.* **2004**, *25,* 360.
19. Ionov, L.; Sidorenko, A.; Stamm, M.; Minko, S.; Zdyrko, B.; Klep, V.; Luzinov, I. *Macromolecules* **2004**, *37,* 7421.
20. Ionov, L.; Houbenov, N.; Sidorenko, A.; Stamm, M.; Luzinov, I.; Minko, S. *Langmuir* **2004**, *20,* 9916.
21. Hongjie Z.; Yoshihiro I. *Langmuir* **2001**, *17,* 8336.
22. Burtovyy, O.; Klep, V.; Turel, T.; Gowayed, Y.; Luzinov, I. *Polymer Preprints* **2007**, *48,* 725.
23. Luzinov I., Iyer K. L. S., Klep V., Zdyrko B. US patent 7,026,014 B2, Apr. 11, **2006**.
24. Burtovyy, O.; Klep, V.; Chen, H.-C.; Hu, R.-K.; Lin, C.-C.; Luzinov, I. *J. Macromol. Sci. Part B: Physics* **2007**, *46,* 137.
25. Tsyalkovsky, V.; Klep, V.; Ramaratnam, K.; Lupitskyy, R.; Minko, S.; Luzinov, I. *Chemistry of Materials* **2007,** ACS Article ASAP.
26. Escala, A.; Stein, R. Advance in Chem. Series **1979**, *176,* 455.
27. Guo, O.; *Macromol. Rapid Commun.* **1995**, *16,* 785.
28. Mark, J.E.; “Polymer Data Handbook”. *Ox. Univ. Press*, N.Y. **1999**, p. 559.
29. Greenberg, A.; Kusy, R. *J. Appl. Polym. Sci.* **1980**, *25,* 1785.

Chapter 23

Polymer-Polymer Nanocomposite Membranes as Breathable Barriers with Electro-Sensitive Permeability

Hong Chen, Aflal M. Rahmathullah, Giuseppe R. Palmese, and Yossef A. Elabd

Department of Chemical and Biological Engineering, Drexel University, Philadelphia, PA 19104

In this study, a new class of membranes has been developed, which is based on a polymer-polymer nanocomposite that consists of nanodomains of hydrophilic polyelectrolyte gels within a hydrophobic polymer host matrix. The hydrophobic host matrix provides a mechanically strong, durable, flexible barrier, while the polyelectrolyte provides a highly water permeable (breathable) membrane. A co-continuous morphology ensures transport across the membrane in the ionic phase. Two membrane concepts have been developed: (1) nanopore-filled membranes and (2) nanofiber encapsulated membranes. Nanopore-filled membranes entail the synthesis of nanoporous polymers that are subsequently filled (via graft polymerization) with polyelectrolyte gels, while nanofiber encapsulated membranes consist of fabricating polyelectrolyte nanofiber meshes (via electrospinning) that are subsequently encapsulated in a durable barrier material. A key feature in these membranes is responsiveness, where the nanoscale co-continuous phase of ionic gel can contract and expand in response to electrical stimuli providing "tunable" permeability valves – maximum water permeability in non-threat situations and reduced agent permeability in situations when a chemical threat is imminent. This responsive behavior (electro-sensitive permeability) was demonstrated in the nanopore-filled membranes, where the membranes were nearly impermeable to dimethyl methylphosphonate (DMMP; a simulant of the nerve agent Sarin) with the application of a low voltage and selectively permeable with the removal of voltage. In addition to responsive behavior, nanopore-filled membranes showed intriguing properties, such as a 10 times higher selectivity (water/agent permeability) compared to current standard materials used in chemical protective clothing with a high water vapor transmission rate (breathable) and mechanical strength similar to the matrix membrane in both dry and hydrated state, unlike the bulk ionic gel, which was many orders of magnitude lower in strength in the hydrated state. Also, successful synthesis of nanofiber encapsulated membranes was demonstrated.

Overview

Expectations for protective clothing for the military include maximum survivability and combat effectiveness for the soldier against extreme weather conditions, radioactive contamination, and chemical and biological warfare agents. Specifically, chemical and biological defense requires protective clothing that is both breathable (to provide comfort and reduce heat stress) and an absolute barrier to harmful agents (i.e., highly selective membranes). In addition to the military, garments for homeland defense should also be both breathable and provide protection from chemical warfare agents, such as Sarin, Soman, Tabun, and mustard gas from inhalation and adsorption through the skin (*1*). Current protective clothing contain activated carban absorbents, which results in limited protection time, breathability, and impose extra weight to the garment. Therefore, a lightweight and breathable fabric, which is permeable to water vapor, resistant to chemicals and highly reactive with nerve agents and other deadly biological agent, is desirable.

Recently, protective clothing developed for other applications, such as clean-up of chemical spills, has been suggested for homeland defense. These garments based on non-permeable and chemical resistant materials provide good barrier properties for chemicals and harmful agents (*2*), however, are also impermeable to water vapor. Recently, electrospun nanofibers have been investigated for theirapplication to protective clothing due to their ultrahigh surface areas. For example, Gibson et al.(*3-5*) demonstrated that electrospun elastomer fiber mats are efficient in trapping aerosol particles between 1-5 μm, while maintaining high water vapor transmission rates. However, porous electrospun fiber mats are still not ideal to be used alone as barriers for gaseous toxins.

For dense polymeric materials, a major setback in materials development is that highly selective materials are usually neither robust nor flexible. For example, ionic polymer gels or polyelectrolytes have high water permeabilities, high selectivities for water over organics, and can contract and expand under an electrical stimulus (i.e., responsive). However, polyelectrolytes are not mechanically stable, particularly in a hydrated state. To overcome this shortcoming, while maintaining high perm-selecitivity, a new membrane concept was adopted, where composite materials containing polyelectrolyte nanodomains within a hydrophobic thermoplastic polymer. The hydrophobic host matrix provides mechanical strength, durability, flexibility, and barrier properties, while the polyelectrolyte polymer provides high water vapor permeability and perm-selectivity. In addition, the polyelectrolytes can contract and expand within the pores in response to electrical stimuli providing "tunable" permeability valves. In this study, two strategies were employed: filling nanoporous membranes with polyelectrolytes and encapsulated electrospun nanofiber mats within a polymer matrix, where a co-continuous morphology ensures transport across the membrane through ionic phase. In both strategies, the size and volume fraction of polyelectroyte nanodomains can be tailored easily for optimization.

Nanocomposite with Oriented Polyelectrolyte Nanodomains

Researchers have observed enhanced transpot properties in membranes with organized and oriented nanodomains (*6-9*). The oriented structures developed thusfar are mostly aligned along the plane of membrane, which is not desirable for many applications. For example, Cable et al. (*6*) stretched Nafion® to induce orientation of the ionic nanostructure and observed a 40% increase in proton conductivity when comparing measurements in and normal to the plane of the membrane (anisotropic conductivity). Maki-Ontto et al. (*7*) also reported anisotropic conductivity (an order of magnitude difference) in proton conductive block copolymers that were sheared to induce an oriented lamellar nanostructure. Anisotropic conductivity was also observed in proton conductive triblock copolymers of sulfonated poly(styrene-b-isobutylene-b-styrene) by Elabd and coworkers (*8,9*). A lamellar morphology with a preferred orientation in the plane of the membrane was confirmed by small-angle X-ray scattering and the proton conductivity measured in plane was over an order of magnitude higher than when measured normal to the plane.

Efforts have been made to induce oriented cylindrical nanodomains in diblock copolymers normal to the plane of the film with the use of mechanical sheer, electric field, magnetic field, and other surface treatments (*10-12*). The diameter of these domains are ~10 nm, but the orientation or length of the nanodomains thus far has been limited to ~10 nm (i.e., small aspect ratios). To induce orientation of nanodomains normal to the plane of the membrane over a larger length scale (~1 μm) is difficult.

Track-etched polymer membranes, which have straight cylindrical pores that are oriented normal to the plane of the membrane, provide a promising platform to design membranes with nano-domains with high aspect ratios oriented in the desired direction. Presently, track-etched membranes are commercially available in polyester and polycarbonate with various thicknesses (6 μm or above), pore sizes (10 nm to microns), and porosities (~0.05% to

~20%). Filling the pores of a track-etched membrane with a polyelectrolyte results in a new class of polymer composite membrane, where the filler can enhance or regulate transport in the desired direction and the track-etched membrane can provide mechanical stability and durability. Compared to dense membranes, these composite membranes can be tailored by the type of filler, pore size and porosity of the membrane to provide tunable transport properties.

Synthesis and Characterization of Nanocomposite

For a typical synthesis, polyester track-etched (PETE) membranes were pretreated using a dielectric barrier discharge (DBD) plasma configuration, consisting of two electrodes (19×19 cm^2) separated by 0.4 cm, at atmospheric conditions. Details of this plasma apparatus and treatment conditions are described elsewhere (*13,14*). After the plasma treatment, the membranes were transferred to a N_2-purged reaction flask charged with monomer, 2-acrylamido-2-methyl-1-propanesulfonic acid (AMPS) or acrylic acid (AA). After a 4-hour reaction at 50°C, the membranes were removed from the reaction system and then thoroughly washed with deionized water. Graft polymerization with 2 wt% (of monomer) *N,N'*-methylenebisacrylamide (BisA) crosslinker was conducted using the same procedure. The membranes were immersed in deionized water for 2 days after the reaction and the excess gel surrounding the membranes were gently removed with tweezers. The membranes with various pore sizes (50 to 500 nm) and porosities (Table 1) were used to study the effect on transport properties.

Table 1. Properties of Polyester Track-Etched (PETE) Membranes

Sample	Thickness (μm)	Pore Size (nm)	Porosity (%)
PETE50-2	12	50	1.57
PETE100-3	12	100	3.14
PETE100-6	12	100	6.28
PETE500-20	12	500	19.63

Oxygen plasma treatment adopted in this synthesis functionalizes the polymer surface with oxygen-containing groups, such as hydroxyl, carbonyl, and hydroperoxide groups (*15*). The hydroperoxide groups decompose and form free radicals when exposed to heat, which initiates graft polymerization on the surface. In addition, the oxygen-containing groups introduce hydrophilic groups to the hydrophobic track-etched membrane. This facilitates the filling of pores with an aqueous reaction solution. The grafting yield is tunable by reaction time, monomer concentration, and crosslinker concentration. Under selected reaction conditions, up to 15 wt% of linear PAMPS and 30 wt% of crosslinked PAMPS were grafted.

Figure 1 shows micrographs of track-etched membranes before and after grafting with PAMPS or PAA gel. The original membranes have a flat surface and uniform-cylindrical pores randomly oriented normal to the plane of the membrane (Figure 1a1 and 1a2). The membranes grafted with linear polyelectrolyte are characterized by partially closed pores and smoother pore surfaces (linear grafting domains, data not shown). In comparison, the grafting with crosslinker results in the top surface of membrane completely covering with a layer of gel (Figure 1b1) and the pores fully filled with crosslinked gel rods or fibrils (Figure 1b2 and 1b3). The micrographs also demonstrate that a broad range of domains sizes (50 and 2000 nm) can be filled by this grafting technique, where polyelectrolyte nanodomains with aspect ratios as high as 240 was achieved. Similar graft patterns are obtained with PETE and PCTE membranes. As revealed by the micrographs, excluding the changes in surface morphologies, the bulk, as well as, the dimensions of track-etched membranes have not been altered by the grafting reaction, confirming that grafting is limited to the membrane surface.

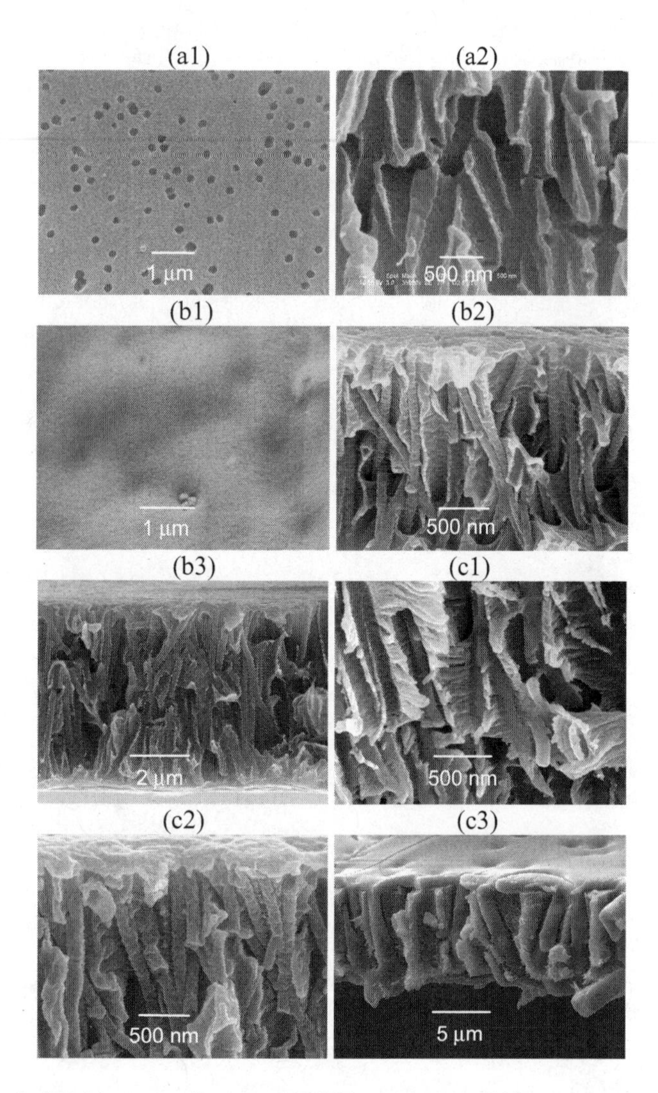

Figure 1. SEM images of original PETE membrane (100 nm pore size): (a1) surface view, (a2) cross-section view; PETE composite membranes with (b1-b2) 100 nm crosslinked PAMPS domains, (b2) 100 nm crosslinked PAA domains; and polycarbonate track-etched (PCTE) membrane composites with (c1) 50 nm crosslinked PAMPS domains,(c2) 100 nm crosslinked PAMPS domains, (c3) 2000 nm crosslinked PAMPS domains.

The surface grafting was further confirmed by ATR-FTIR with the appearance of new bands at 1055 cm^{-1}, 1658 cm^{-1}, and a broad band between 3150-3650 cm^{-1} (data not shown), which represent $-SO_3H$, C=O, and N-H groups in PAMPS, respectively. To study the distribution of PAMPS on the pore walls, X-ray maps were collected on the cross-section of grafted membranes, where the yellow color represents the sulfur atoms in PAMPS (Figure 2). While no sulfur was observed in the original PCTE membrane, clear patterns of sulfur across thecross-section were obtained in the grafted membranes. These patterns suggest that PAMPS was uniformly grafted on the pore walls.

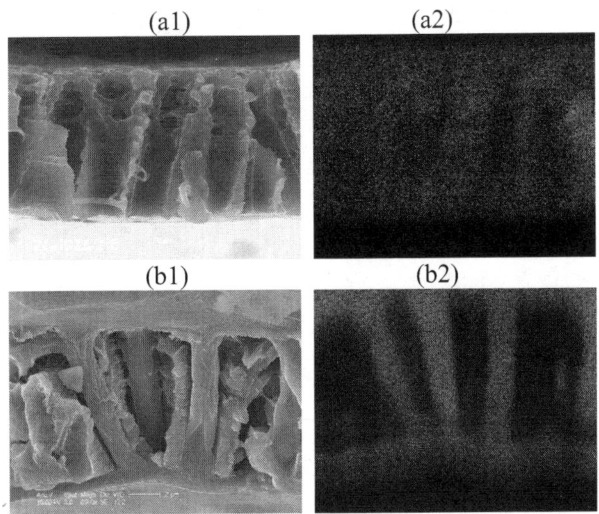

Figure 2. SEM images (1) and X-ray maps (2) of the cross-section of (a) PCTE (2000 nm pore size) with linear PAMPS and (b) crosslinked PAMPS domains. Yellow color in the X-ray maps represents sulfur K_{α} in PAMPS.

Water and DMMP Vapor Transport

Vapor transport properties of the composite membranes were investigated with pure water and dimethyl methylphosponate (DMMP), which is a simulant of sarin gas, based on the ASTM E 96-95 (Standard Test Methods for Vapor Transmission of Materials) procedure (*16,17*). Selectivity was defined as the effective permeability of water vapor over DMMP vapor. Figure 3 shows the water vapor transmission rate and selectivity of selected PETE membranes filled with linear and crosslinked PAMPS domains compared to a dense polyester film (control) and Tychem-LV (DuPont). Both the dense polyester film and Tychem-LV result in a low water vapor transmission rate and selectivity (i.e., nonbreathable barrier). The selectivity increased to ~3 by including 500 nm linear PAMPS domains in the membrane, and further increased to ~7.5 and 10.2 by reducing the tubular domain size to 100 nm and 50 nm, respectively. The composite membrane containing 50 nm crosslinked PAMPS domains has a selectivity of 21 – an order of magnitude higher than materials currently used in chemical protective clothing. Increased selectivity compared to the dense film is due to the incorporation of the highly selective polyelectrolyte, which has a higher affinity to water compared to organics.

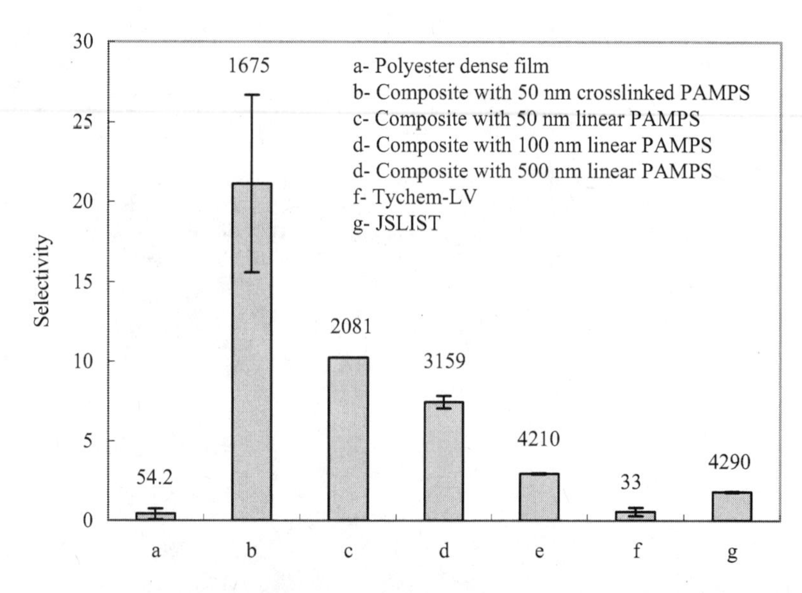

Figure 3. Selectivity and water vapor transport property of selected PAMPS-PETE composites and reference barrier materials. Numbers above bars correspond to water VTR (g m^{-2} day^{-1}).

The effect of polyelectrolyte domain size and volume fraction on vapor transport properties is more clearly shown in Table 2. Due to a porous structure, PETE membranes with selected pore sizes and porosities exhibit high permeability to both water and DMMP vapor. No obvious effect of the pore structures was observed on transport properties. Filling the pores with linear PAMPS did not change the transport properties of PETE membrane with 500 nm pore sizes. A distinct reduction on DMMP vapor permeability was observed when the domain size of the linear PAMPS decreased to 50 nm, which increased the selectivity by nearly 5 times compared to the original membrane. Table 2 also shows that by increasing the volume fraction of linear PAMPS (in 100 nm pores), the composite was less permeable to DMMP vapor with a higher selectivity.

Table 2. Water and DMMP Transport Properties of PETE and Nanocomposites

Sample	Water Vapor		DMMP Vapor		Selectivity
	VTR [a]	$P_{eff}(10^{-4})$ [b]	*VTR* [a]	$P_{eff}(10^{-4})$ [b]	
PETE500-1E8	4163	13.5	248	4.39	3.1
PETE100-4E8	4237	16.0	225	4.65	3.4
PETE100-8E8	4197	15.9	279	5.78	2.7
PETE50-8E8	3763	15.8	298	6.88	2.3
PETE500-1E8-tubular PAMPS	4210	14.8	260	5.00	3.0
PETE100-4E8-tubular PAMPS	2142	7.78	114	2.26	3.4
PETE100-8E8-tubular PAMPS	3159	10.2	77.4	1.37	7.5
PETE50-8E8-tubular PAMPS	2081	7.58	38.5	0.74	10.2
PETE50-8E8-fibril PAMPS	1675	6.13	13.8	0.29	21.1
PETE50-8E8-fibril PAA	915	11.5	26.9	1.87	6.1

[a] $g \cdot m^{-2} \cdot day^{-1}$
[b] $g \cdot mmHg^{-1} \cdot m^{-1} \cdot day^{-1}$

Completely filling the 50 nm pores of PETE with crosslinked PAMPS drastically decreases DMMP vapor permeability by ~20 times while maintaining a high water vapor transport rate. Consequently, a selectivity of ~ 21 was observed. In comparison, a lower selectivity was observed with composites containing a weak polyelectrolyte, polyacrylic acid (Table 2). With selectivities above 20 and water vapor transmission rate higher than 1000 $g \cdot m^{-2} \cdot day^{-1}$, the acceptable breathable region (*17*), these composite membranes with oriented PAMPS nanodomains provide a promising avenue for the development of light-weight breathable barriers for chemical and biological defense.

Electro-Sensitive Transport Properties

Polyelectrolyte hydrogels exhibit reversible contraction and expansion (actuation) in response to an "on-off" direct current electric field (*18,19*). This electrosensitivity results from the migration of hydrated ions and water in the network to the electrode bearing opposite charge opposite in sign to that borne by the polymer network. To investigate the actuation behavior of the polyelectrolyte nanocomposite, the transport properties of DMMP liquid were measured using a side-by-side glass diffusion cell equipped with a thermal jacket (*20,21*). Membranes were prehydrated in ultra-pure deionized, reverse osmosis (RO) water (resistivity ~ 16 MΩ cm) for at least 48 h. The membrane, which was sandwiched by two pieces of porous carbon cloth (electrodes), was clamped between the two compartments. In a typical experiment, the donor compartment was charged with 10 vol% DMMP aqueous solution, while the receptor compartment was filled with water. The electrodes were connected to a DC power supply using platinum wire. The concentration of DMMP that permeates through the membrane was measured continuously as a function of time on the receptor side with a real-time in-line Fourier transform infrared, attenuated total reflectance (FTIR-ATR) (Nicolet 6700 Series, Specac Inc.) spectrometer for detection. In all experiments, both the side-by-side diffusion cell and ATR cell were temperature controlled (35°C) with the same circulating water bath (Neslab RTE10, Thermo Electron Co.).

Figure 4a shows the down-stream DMMP concentration after permeation trhough the nanocomposite membranes with and without electric actuation. When an electric field (10 V) was applied, the membrane was nearly impermeable, but permeable when the electric field was removed. The electric field was applied again after ~3 h and the same electrosensitive barrier properties were observed (reversible process). Without applied voltage, the nanocomposite membrane exhibited a permeability of 2.84×10^{-8} cm^2/s (Figure 4b). This was similar to the permeability in the region where no voltage was applied in Figure 4a and was an order of magnitude lower than that of the original (unfilled) PETE membrane (2.45×10^{-7} cm^2/s; Figure 4c).

The break through time (~13 min) after the removal of applied voltage in Figure 4a was similar to the break through time in Figure 4b, where no voltage was applied. In comparison, no actuation behavior was observed with the original PETE membrane, where the electric signal had no impact on the permeability of DMMP (Figure 4c).

Mechanical Properites

Tensile tests show that the pure PAMPS gel membrane is more rigid and brittle than track-etched matrix in the dry state (Table 3). As expected, the hydrophobic track-etched membranes showed no obvious changes in mechanical properties after being exposed to water. However, the polyelectrolyte gel, PAMPS, decreased in strength by many orders of magnitude after hydration. By grafting polyelectrolyte gel within the pores of the matrix, the resulting composite showed similar mechanical behavior in the dry state compared to the matrix membrane. In the hydrated state, the composites mechanical properties reduced slightly, but were still on the same order of magnitude as the matrix.

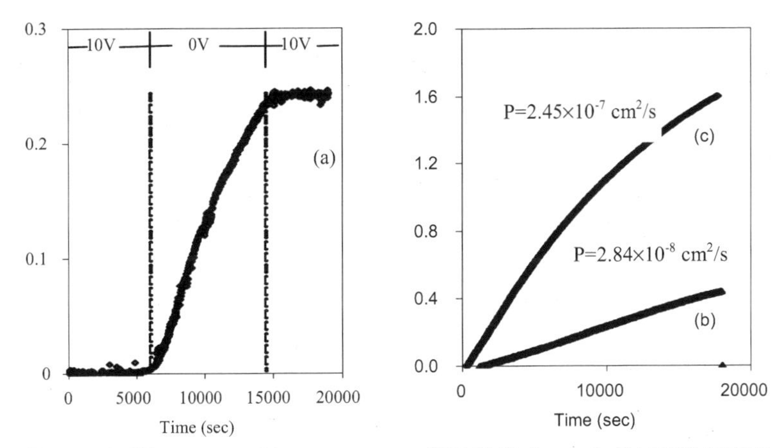

Figure 4. Electro-sensitive transport of DMMP through PAMPS-PETE composite (a) with and (b) without acutation and (c) through the original (unfilled) PETE membrane (50 nm pore diameter).

Table 3. Mechanical Properties of Track-Etched Membranes and Nanocomposites with Oriented Polyelectrolyte Nanodomains

Sample	Breaking Strength (MPa)		Breaking Elongation (%)		Modulus (MPa)	
	Dry	Hydrated	Dry	Hydrated	Dry	Hydrated
PAMPS gel	72 [a]	0.012 [b]	3.2 [a]	40 [b]	3108 [a]	0.03 [b]
PETE50	87±9	77±9	21±3	19±1	762±203	788±77
Composite with fibril PAMPS	67±19	21±5	16 ±4	9.8±0.4	467±120	121±4
Composite with fibril PAA	77±8	23±1	15±1	14±5	636±111	164±8
PCTE50	48 ± 4	86 ± 26	57 ± 5	22 ± 7	1228 ± 64	1304 ± 264
Composite with tubular PAMPS	24 ± 13	19 ± 1	4 ± 1	5 ± 1	1087 ± 7	704 ± 417
Composite with fibril PAMPS	30 ± 8	16 ± 3	4 ± 1	5 ± 1	1482 ± 265	503 ± 124

[a] containing 5 wt% BisA crosslinker. [b] 98% swelling (*22*). [c] general purpose grade (*23*).

Nanocomposites with Nanofiber Encapsulation

Nanofiber fabrics prepared by electrospinning possess ultra-high surface area, high porosity, but very small pore size (*21,24*). These characteristics provide efficient trapping and good resistance to the penetration of harmful chemical agents in aerosol form, meanwhile providing minimal resistance to moisture vapor diffusion (*3-5*). However, the porous structure also limits their application in barriers for gaseous toxins. In this work, a nanocomposite of polyelectrolyte nanofiber mats encapsulated in a hydrophobic barrier material was explored. While both the polyelectrolyte fiber and matrix material provide barrier properties to gaseous toxins, the polyelectrolyte fibers also function as the diffusion channels for moisture vapors and provide breathability.

Besides the desired transport properties, an additional merit of fiber reinforcement is expected from the fiber encapsulation composite structure. Kim and Reneker (*25*) investigated the reinforcing effect of electrospun nanofibers of polybenzoimidazole (PBI) in an epoxy matrix. Multiple plies of PBI fabrics were impregnated with the epoxy resin and cured at elevated temperature under vacuum. Improvement in Young's modulus and fracture toughness was observed. Bergshoef and Vancso (*26*) fabricated a nanocomposite of electrospun Nylon-4,6 fabrics and epoxy resin by the same method. The stiffness and strength of this composite was significantly higher than the unreinforced matrix.

Preparation of Electrospun Nanofiber Mat and Encapsulation

Two types of polyelectrolyte nanofiber mats were prepared in this study. One was polyacrylic acid (M_v = 450,000) fibers, spun from 5 wt% of aqueous solution. The other one was polysulfone fibers, spun from 25 wt% of DMF solution. The polysulfone fiber was then coated with a layer of polyelectrolyte, PAMPS, by a surface grafting technique (PAMPS-*g*-PS) similar to the procedure described previously. Nucrel® membranes wereprepared by heat pressing Nucrel® 535 (DuPont) pellets at 125°C. For fiber encapsulation, the fiber mats with measured weight and thickness were placed between two preweighed Nucrel® membranes and then pressed at 125°C. The resulting composite membranes were flexible and robust with a homogeneous transparency.

The nanofiber encapsulated composites were freeze fractured in liquid nitrogen and their cross sections are coated with platinum and obversed by SEM. As shown in Figure 5a, a pure Nucrel® membrane exhibits a dense and homogeneous structure, while a porous structure was observed with the polyelectrolyte fiber mats (Figure 5b1, 5c1, and 5c2). After encapsulation, Figure 5b2 and 5b3 show that the pores of PAA nanofiber mat were fully filled with Nucrel® matrix. In comparison, a loose encapsulation structure was observed with the polysulfone-Nucrel® composite, in which polysulfone nanofiber were completely separate from the matrix (Figure 5c3). However, the micrograph reveals that grafting the fiber with linear PAMPS improved the fiber-matrix interface significantly (Figure 5c4).

Water and DMMP Vapor Transport

The vapor transport properties of nanofiber encapsulated composite membranes are listed in Table 4. Nucrel® membrane, which is a random copolymer of polyethylene and ~10 % polymethacrylic acid, has a low water vapor transmission rate and poor selectivity. Higher selectivity and high water vapor transmission was observed for both PAA and PAMPS-*g*-PS polyelectrolyte nanofiber mats. However, these fiber mats also exhibit a high permeability to DMMP due to the highly porous structure. Despite of expected higher selectivity, encapsulation of the fiber mats with a Nucrel® matrix results in vapor transport properties similar to that of the dense Nucrel® membrane. With a close observation on the cross-section of the composites, a dense layer of Nucrel® was found on both sides of membrane. These Nucrel® dense surface layers results in resistance to vapor transpor to the polyelectrolyte nanofibers. To ensure exposure of polyelectrolyte nanofibers to membrane surfaces, the encapsulation technique will need to be modified to minimize surface resistance. This work is currently ongoing.

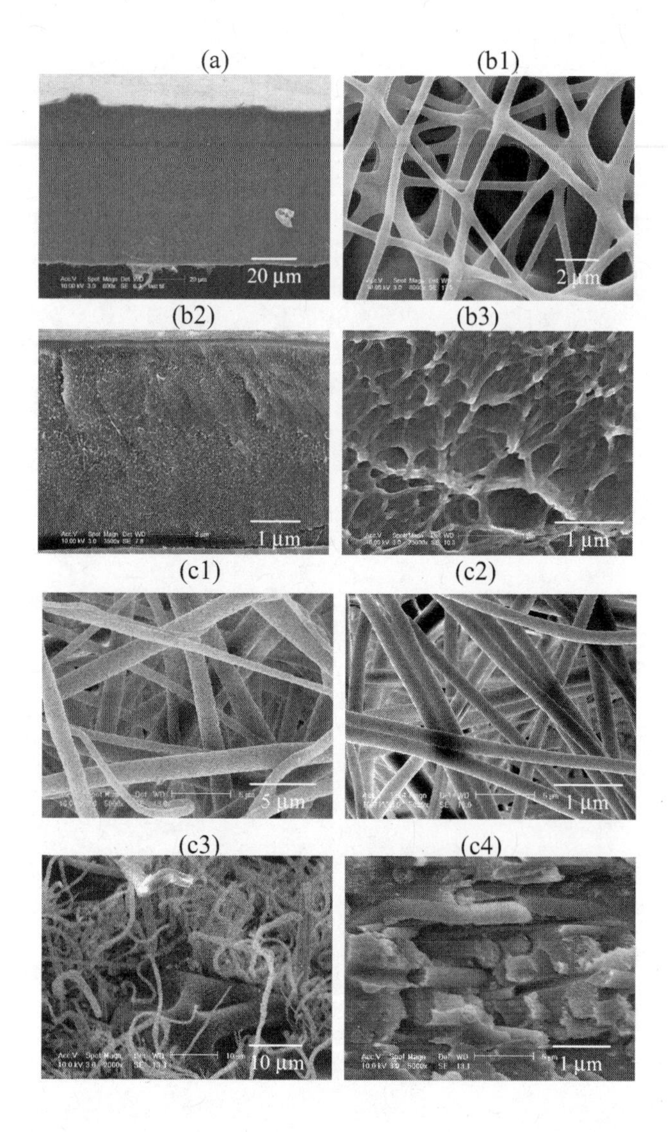

Figure 5. SEM image of (a) Nucrel® membrane (b1) PAA electrospun nanofiber mat (b2-b3) PAA nanofiber encapsulated Nucrel® membrane (c1) polysulfone (PS) electrospun nanofiber mat (c2) PAMPS-g-PS electrospun nanofiber mat (c3) PS nanofiber encapsulated Nucrel® membrane (c4) PAMPS-g-PS nanofiber encapsulated Nucrel® membrane.

Table 4. Vapor Transport Properties of Nucrel®, Nanofiber Mats, and Nanofiber-Encapsulated Composites

Sample	Water Vapor		DMMP Vapor		Selectivity
	VTR [a]	$P_{eff}(10^{-4})$ [b]	*VTR* [a]	$P_{eff}(10^{-4})$ [b]	
Nucrel membrane	27.1	0.61	81.4	10.1	0.1
PAA nanofiber mat	4203	136	242	43.1	3.1
PAMPS-*g*-PS nanofiber mat	4549	206	249	61.5	3.3
PAA-Nucrel® composite	28.6	1.47	57.9	16.3	0.1
PAMPS-*g*-PS-Nucrel® composite	44.7	1.90	49.2	11.5	0.2

[a] $g \cdot m^{-2} \cdot day^{-1}$

[b] $g \cdot mmHg^{-1} \cdot m^{-1} \cdot day^{-1}$

Conclusions

In this study, polymer-polymer nanocomposites membranes were synthesized using two strategies: filling (via graft polymerization) nanoporous polymer membranes with polyelectrolyte gels and encapsulating electrospun polyelectrolyte nanofiber meshes within polymer membranes. Membranes that were highly permeable to water vapor, highly selective (water/agent permeability), responsive to an electrical stimulus, and mechanically robust in hydrated conditions was demonstrated. In addition to protective clothing, these research results have a broad impact to a variety of applications, including electrochemical devices (e.g., fuel cells, batteries, actuators, sensors) and other perm-selective membrane applications (e.g., water purification, pervaporation, stimuli-responsive materials).

Acknowledgements

The authors acknowledge the financial support of the U.S. Army Research Office through Grant W911NF-05-1-0036.

References

1. Lee, B.T.; Wang, T.W.; Wilusz, E. *Polym. Eng. Sci.* **2003**, *28*, 209.
2. *Chemical Protective Clothing Performance Index*, 2nd ed., Frsberg, K.; Keith, L.H.; Wiley 1999.
3. Gibson, P.; Schreuder-Gibson, H.; Rivin, D. *Coll. Surf. A: Physic. Eng. Asp.* **2001**, *187–188*, 469.
4. Schreuder-Gibson, H.L.; Gibson, P. *Int. Nonwoven J.* **2002**, *11*, 21-26.
5. Schreuder-Gibson, H.L.; Gibson, P.; Senecal, K.; Sennett, M.; Walker, J.; Yeomans, W. *J. Adv. Mater.* **2002**, *34*, 44-55.
6. Cable, K.M.; Mauritz, K.A.; Moore, R.B. *Chem. Mat.* **1995**, *7*, 1601.
7. Mäki-Ontto, R.; de Moel, K.; Polushkin, E.; van Ekenstein, G.A.; ten Brinke, G.; Ikkala, O. *Adv. Mater.* **2002**, *14*, 357.
8. Elabd, Y.A., Napadensky, E.; Walker, C.W.; Winey, K.I. *Macromolecules* **2006,** *39,* 399.
9. Elabd, Y.A.; Beyer, F.L.; Walker, C.W. *J. Membrane Sci.* **2004**, *231*, 181.
10. Zhu, Z.Q.; Kim, D.H.; Wu, X.D.; Boosahda, L.; Stone, D.; LaRose, L.; Russell, T.P. *Adv. Mat.* **2002**, *14*, 1373-1376.
11. Thurn-Albrecht, T.; Schotter, J.; Kastle, G.A.; Emley, N.; Shibauchi, T.; Krusin-Elbaum, L.; Guarini, K.; Black, C.T.; Tuominen, M.T.; Russell, T.P. *Science* **2000**, *290*, 2126-2129.
12. Xu, C.; Fu, X.F.; Fryd, M.; Xu, S.; Wayland, B.B.; Winey, K.I.; Composto, R.J. *Nano Letters* **2006**, *6*, 282.
13. Robinette, E.J.; Toughening Vinyl Ester Matrix Composites, PhD Dissertation, Drexel University, Philadelphia, PA **2005**.
14. Chen, H.; Palmese, G.R.; Elabd, Y.A. *Chem. Mater.* **2006**, **18**, 4875-4881.
15. Denes, F.S.; Manolache, S. *Prog. Polym. Sci.* **2004**, *29*, 815.
16. ASTM E 96-95. Standard Test Methods for Vapor Transmission of Materials. *Annu. Book of ATM Stand.* **2002**.
17. Napadensky, E.; Sloan, J.M.; Elabd, Y.A. *ACS PMSE Preprints* **2004**, *91*, 752-753.
18. Osada, Y.; Takeuchi, Y. *J. Polym. Sci.: Polym. Lett.* **1981**, *19*, 303.
19. Tanaka, T.; Nishio, I.; Sun, S.T.; Ueno-Nishio. S. *Science* **1982**, *218*, 467.
20. Elabd, Y.A.; Napadensky, E.; Sloan, J.M.; Crawford, D.M.; Walker, C.W. *J. Mem. Sci.* **2003**, *217*, 227.
21. Chen, H.; Palmese, G.R.; Elabd, Y.A. *Macromolecules* **2007,** *40,* 781.
22. Siddhanta, S.K.; Gangopadhyay, R. *Polymer* **2005**, *46*, 2993.
23. *Polymer Handbook*, 4th ed.; Brandrup, J.; Immergut, E.H.; Grulke, E.A., Eds.; Wiley: 1999.
24. Formhals, A. U.S. Pat., 1,975,504, **1934**.
25. Kim, J.S.; Reneker, D.H. *Polym. Compos.* **1999**, *20*, 124-131.
26. Bergshoef, M.M.; Vancso, G.J. *Adv. Mater.* **1999**, *11*, 1362-1365.

Chapter 24

Scanning Atmospheric Plasma Processes for Surface Decontamination and Superhydrophobic Deposition

Seong H. Kim,[1*] Jeong-Hoon Kim,[2] and Bang-Kwon Kang[3]

1. Department of Chemical Engineering, The Pennsylvania State University, University Park, PA 16802.
2. APPLASMA, 111 Gibbons Street, State College, PA 16801.
3. APPLASMA Co, Ltd. 211 Sanhakwon, Ajou University, Suwon, Korea.

A scanning atmospheric radio-frequency (rf) plasma source was developed for surface decontamination and protective coating deposition. This plasma source was a dielectric barrier discharge system that utilizes a cylindrical electrode and a dielectric barrier surrounding the electrode. The gaseous species present in the plasma were analyzed with in-situ spectroscopic technique. Since the plasma generation region was open to ambient air, nitrogen, oxygen and water molecules in the air were readily excited. This system was used for decontamination of organophosphorus nerve agents and deposition of hydrophobic and superhydrophobic coatings on various substrates including ceramics, metals, fabrics, and nanofibers.

Introduction

There is a great need for plasma-based surface cleaning processes for pesticide decontamination, chemical and biological warfare agent decontamination, biomaterial sterilization, polymer film stripping, etc (*1-3*). The use of a plasma for surface cleaning and modification provides many advantages over conventional solution-based processes which may raise material compatibility issues or generate toxic liquid wastes (*4*). In the case of polymeric materials, the penetration of solvent into the bulk can alter the bulk properties of the polymer. The plasma-based surface modification approach can circumvent these problems since the reactive gaseous species modify only the surface and do not alter the bulk chemistry. In addition, the use of a high energy plasma process allows the generation of reactive species (such as excited atoms and radicals as well as electrons and ions) that are much more potent than chemical reagents in solution (*5-8*). Among various plasma types, an atmospheric non-thermal plasma is of specific interest since it can be used to scan over large surface area samples without enclosing the samples in a vacuum chamber. So, it can be operated in an in-line mode, not in a batch mode, and scaled up for applications to large substrate surfaces or continuous processing. We have developed a new scanning atmospheric rf plasma generation source that can operate at low voltage and low power and generated the glow plasma without arcing over a large area (*9*). This paper reviews the characterization of the scanning rf plasma with optical emission spectroscopy and transmission Fourier-transform infrared spectroscopy and its uses for decontamination of organophosphorus (OP) nerve agents on model sample surfaces and deposition of hydrophobic and superhydrophobic coatings as a protective layer.

Experimental

The schematic of the scanning atmospheric rf plasma system is shown in Figure 1. A stainless steel cylindrical electrode (1.5 cm dia., 12 cm long) is shielded inside a quartz tube (inner dia. = 1.5 cm, wall thickness = ~2 mm) which separates the electrode from the grounded cover. The plasma was generated with Ar or He gas (flow rate = 4 liter per minute) by applying a 13.56 MHz rf electric field across the center electrode and the grounded cover. Typical operating rf power was 150 ~ 400 W. Various process gases can be fed into the plasma depending on the application purpose. The plasma generated in the open space between the dielectric barrier and the ground electrode landed on the substrate surface when the substrate was brought ~3 mm from the plasma head. The effective plasma area was ~0.9 cm wide and 10 cm long. Since the plasma was operated in a glow discharge mode, it could be directly applied to metallic substrates as well as non-conducting substrates without arc or streamer damages. The plasma exposure to the sample was made by automated scanning of the sample under the fixed plasma source. The typical scan speed was varied within the 3~10 mm/sec range.

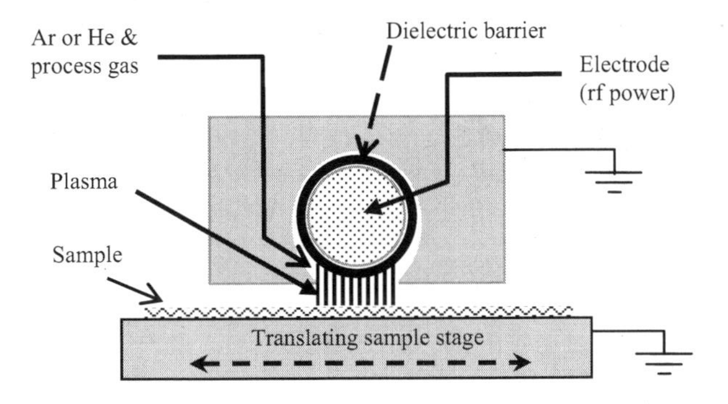

Figure 1. Schematic of the scanning atmospheric rf plasma system.

The optical emission from the excited gaseous species in the Ar and Ar/O_2 plasma was analyzed using a Varian Cary Eclipse spectrophotometer with a 1 nm wavelength resolution. A clean blank substrate was placed ~5 mm below the plasma head and the optical emission was analyzed from the side of the plasma generated between the head and the substrate. The center of the plasma was positioned at the focal point of the spectrometer. The gas phase species produced in the plasma used for the hydrophobic and superhydrophobic coating deposition were analyzed with transmission FTIR spectroscopy (Thermo Nicholet Nexus 670) with a 1 cm^{-1} spectral resolution. The IR beam was passed through the plasma zone between the plasma head and the clean blank substrate. The vibrational spectrum of the process gas was collected in atmospheric flow conditions before turning on the plasma and used as a background.

In the case of OP decontamination, parathion and paraoxon were spin-coated on glass or gold film surfaces from 0.023 mM solution in ethanol. The loadings of parathion and paraoxon were determined with a quartz crystal microbalance (QCM). The deposited amount of parathion and paraoxon film was ~8.5 $\mu g/cm^2$ and ~5.0 $\mu g/cm^2$, respectively. These OP films were then treated with an O_2/Ar plasma. The treated samples were analyzed with polarization-modulcation reflection absorption infrared spectroscopy (PM-RAIRS; Thermo Nicholet Nexus 670 with a Hindz polarization modulation unit and a GWC demodulator) and tested for residual toxicity.

For deposition of hydrophobic coatings, a mixture of CH_4/He gases was used in the plasma. Superhydrophobic coatings were created with a CF_4/H_2/He plasma. The deposited coatings were analyzed with PM-RAIRS, x-ray photoelectron spectroscopy, scanning electron microscopy (SEM, JEOL 6700 F - FESEM), atomicforce microscopy (AFM, Molecular Imaging SPM microscope with RHK controllers), ellipsometry, and water contact angle measurements.

Results and Discussion

Organophosphorus Compound Decontamination Using O_2/Ar Plasma

The plasma-generated gaseous species are typically in electronically excited states. Figure 2a shows a typical optical emission spectrum of the Ar plasma produced in air (*10*). The characteristic emission peaks from the excited Ar molecules are clearly seen in the region from 696 nm to 812 nm. The peaks at 337 nm and 674 nm originate from the excited N_2 molecule (*11*). The peak at 309 nm is due to the OH radical produced by dissociation of water vapor in the plasma (*11*). The peaks at 777 nm and 845 nm are due to the excited atomic oxygen.

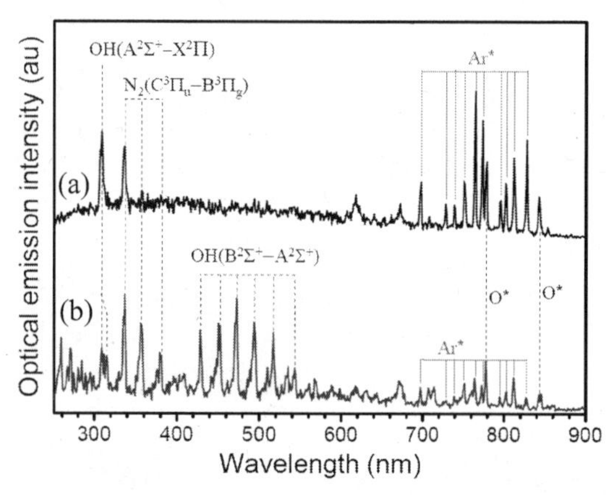

Figure 2. Optical emission spectra of (a) Ar and (b) O_2/Ar plasma generated in ambient air (rf power = 200 W).(10)

When oxygen is added into the plasma (Figure 2b), the ratio of the atomic oxygen peaks to the excited Ar peaks increases significantly. This is accompanied by the increase of the peaks corresponding to the electronic transitions involving highly excited OH and N_2 species (*11*). Although the optical emission at ~322 nm from the excited ozone molecules is not detected, it is very likely that the atmospheric rf plasma also produces ozone. It is well known that the atomic oxygen and other electronically-excited OH, N_2, and Ar species undergo a three body reaction with molecular oxygen to produce ozone (*12,13*). These highly excited OH radicals, atomic oxygen, and ozone species are very efficient for oxidative modification of surface chemistry.

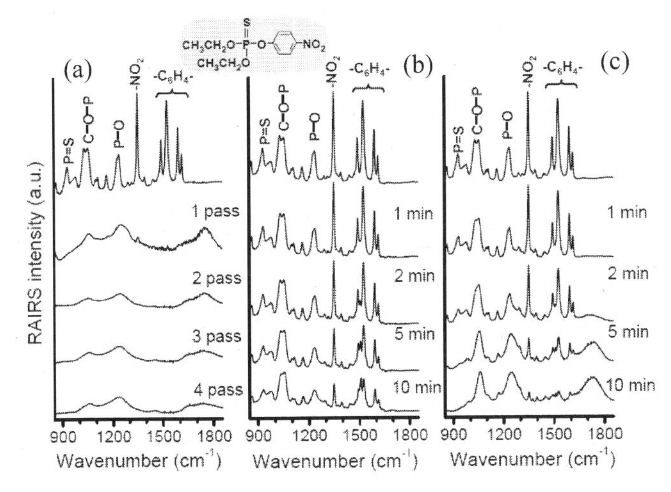

Figure 3. PM-RAIRS spectra of parathion films after being treated with (a) O_2/Ar plasma (rf power = 200 W), (b) UV, and (c) UV/ozone.(10)

The decomposition products of parathion and paraoxon after the atmospheric rf plasma treatment were analyzed with PM-RAIRS. In order to find whether photolysis or oxidation is responsible for the decomposition reaction, the plasma-induced decomposition products were compared with the surface species remaining on the surface after treatments with UV (200W Hg lamp) and ozone (a commercial UV/ozone cleaner). These results are shown in Figure 3. Upon a single pass of the Ar/O_2 plasma (rf power = 200 W, scan rate = 3 mm/sec) over the OP film, the nitrobenzyl and alkyl peaks disappear almost completely and broad peaks corresponding to amorphous phosphate glass (1055 and 1245 cm^{-1}) (14-16), water adsorbed in the glassy film (1650 cm^{-1}), and carbonyl species (1750 cm^{-1}) appear. The carbonyl peak eventually decreases as the plasma exposure time is increased. The changes in the paraoxon spectrum upon O_2/Ar plasma exposure are basically the same as the parathion case. The spectral changes of parathion and paraoxon films upon the oxygen plasma exposure are not due to the irradiation of high-flux UV photons from the plasma source. The PM-RAIRS data for the OP films irradiated with UV (Figure 3b) show the monotonic and slow decrease of the nitrobenzyl peaks; but the phosphorus ether peaks are not changed much. The plasma-induced reaction products appear to be similar to the ozonolysis product (Figures 3c). The decrease of the nitrobenzyl characteristic peaks is accompanied by the appearance of the broad peaks at 1055, 1245, and 1750 cm^{-1}. From the degree of peak intensity decrease as a function of process time, it is obvious that the atmospheric plasma treatments are several orders of magnitude faster than the UV and UV/ozone treatments.

The efficacy of the OP decomposition and the nontoxicity of the decomposition reaction product were tested by culturing *Drosophila* in a confined space containing the substrates which were initially coated with OP films and treated with the O_2/Ar plasma, UV, and UV/ozone.(*10*) The UV and

UV/ozone treated parathion and paraoxon samples (treatment time = 5 min) were still toxic enough to kill all *Drosophila* in less than 12 hours in the culture bottle. In contrast, the *Drosophila* in the bottle containing the plasma-treated sample (treatment time = 10 sec) lived their full life cycle and laid eggs. Even the newly hatched larvae were able to develop into adult flies. These culture test results qualitatively demonstrate the efficacy of the atmospheric rf plasma process for the complete decomposition and detoxification of OP nerve agents.

Hydrophobic Coating Deposition Using CH_4/He Plasma

CH_4 is used as a process gas and He as a plasma generation gas to deposit hydrophobic coatings on various surfaces (*17*). The transmission FTIR analysis of the gas phase region of the atmospheric CH_4/He plasma showed the consumption of the CH_4 process gas (negative peaks at 3017.6 cm^{-1} and 1306 cm^{-1}) and water (small negative peaks near 1600 cm^{-1} and 3700 cm^{-1}) in the plasma and the positive absorbance is due to production of new species in the plasma (*18*). The changes in rotational fine structures and width of these vibration peaks indicated that CH_4 molecules are highly excited in the plasma. Although hydrocarbon radicals and ions are expected to be the primary dissociation products of CH_4 in the plasma, they are not directly observed in the FTIR spectrum of the scanning atmospheric rf plasma. Instead, the products of their reactions with other gas-phase species are observed in FTIR. These include C_2H_2, C_2H_4, and C_2H_6 (formed by reactions between hydrocarbon radicals) as well as CO (oxidation by atomic oxygen and/or OH radical) (*18*). These are the consequence of the short mean free path of molecules (10^{-9} ~ 10^{-8} m) at the atmospheric pressure. Since typical molecular speed at room temperature is about 500 m/s, the collision frequency in the gas phase is about ~10^{12} per second.

The activation and dissociation of CH_4 near the substrate surface lead to deposition of hydrophobic coatings on the surface. The hydrophobic coating layer thickness can be easily controlled with the number of passes under the scanning atmospheric CH_4/He plasma. The deposition rate is ~14 nm for the first pass and then slightly decreases in subsequent passes. (Figure 4). The deposited film surface is very smooth. The coating thickness is very uniform over the entire treated area.

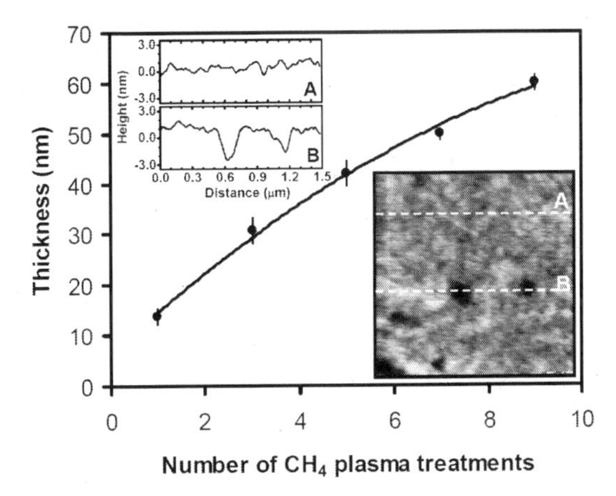

Figure 4. Thickness of hydrophobic coatings deposited with scanning atmospheric CH_4/He plasma (rf power = 250 W). The insets show the AFM image (1.5μm×1.5μm) and line profiles of the 11-pass deposited film.(17)

The chemical composition of the deposited hydrophobic coatings is determined using XPS. Figure 5 shows the XPS analysis results for the coating deposited on a clean gold film substrate. The C1s peak is the only main peak detected in XPS. The elemental analysis shows that the oxygen concentration is less than 2% regardless of substrates. The incorporation of this small amount of oxygenated species is inevitable because the plasma is generated in atmospheric air. The exact nature of the carbonaceous species is determined with infrared vibration spectroscopy. The C-H vibration peak positions in the PM-RAIRS spectra (Figure 5) are very close to those of polypropylene IR spectrum (*19*). This indicates that the deposited film is composed mostly of CH_2 polymeric backbones with CH_3 side groups. The peaks at 1708 cm^{-1} and 1660 cm^{-1} are due to carbonyl containing groups which are inevitable because the plasma is generated in atmospheric air.

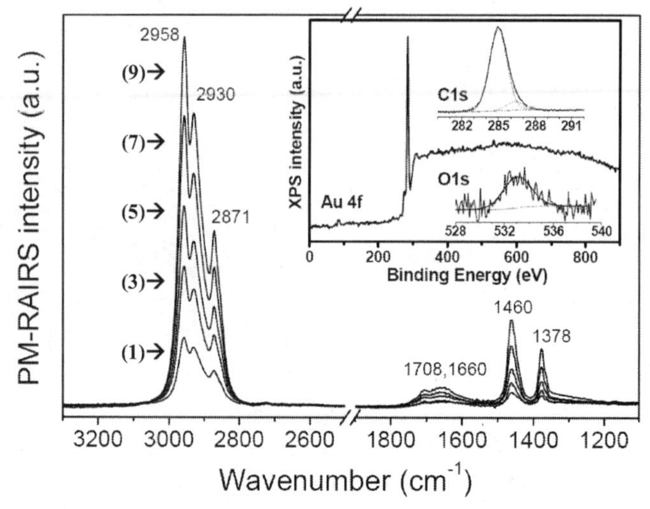

Figure 5. PM-RAIRS and XPS (inset) spectra of the hydrophobic coating deposted with scanning atmospheric CH_4/He plasma on a gold film. The number of plasma scanning is indicated in the figure.(17)

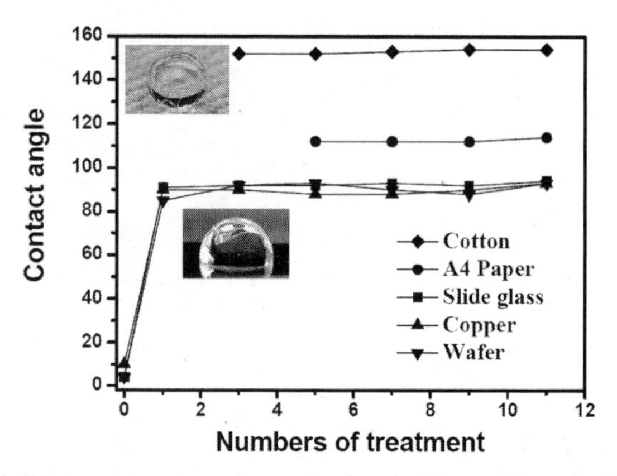

Figure 6. Water contact angles on hydrophobic coating produced with scanning atmospheric CH_4/He plasma (rf power = 250 W).(17)

All coatings deposited by the atmospheric CH_4-He plasma process show good hydrophobicity regardless of the substrate materials. Water contact angles on these coatings are plotted in Figure 6 as a function of the plasma deposition cycles. On flat surfaces such as Cu foil, Si wafer, and glass slide, the contact angle increases to ~90° after a single treatment with the CH_4 plasma. This value is very close to the water contact angle of polypropylene (93°). The contact angle remains unchanged upon further deposition. The paper and cotton substrates require at least 3 deposition cycles to be hydrophobic. Until then, the water droplet is completely absorbed into the substrate. This probably arises from lower deposition yields on side walls present in rough surfaces, so repeated treatments are required for conformal coating. After multiple treatments with the scanning atmospheric CH/He plasma, the contact angles on paper and cotton substrates show 112.5±1° and 150±1°, respectively. The high contact angle on these substrates is due to the surface roughness effect that amplifies the hydrophobicity (*20,21*). In general, the water contact angle becomes larger on rough surfaces than on flat surfaces. On the treated cotton sample, the water droplet is easily rolled off with slight tilting of the substrate, showing a self-cleaning capability. It should be noted that the untreated side of the cotton sample still absorbs water readily. These properties will make the plasma-treated cotton fabric a good candidate for self-cleaning garments.

Superhydrophobic Coating Deposition Using $CF_4/H_2/He$ Plasma

A mixture of CF_4 and H_2 is used in the He plasma to deposit superhydrophobic coatings on various surfaces (*9*). The transmission FTIR analysis of the gas-phase region of the atmospheric $CF_4/H_2/He$ plasma allowed to monitor the consumption of the CF_4 and the gas-phase reaction products. The consumption of CF_4 in the plasma was the highest when the CF_4 and H_2 ratio was 2:1, which is the optimum gas composition ratio for the superhydrophobic coating depositon (*18*). The CF_4 consumption in the atmospheric rf plasma was accompanied by the production of HF (sharp peaks between 3593 cm^{-1} and 3920 cm^{-1}), C_2F_6 (peaks at 1250 cm^{-1} and 1114 cm^{-1}), CF_3H (a weak peak at 1153 cm^{-1}), and COF_2 (a sharp peak at 774 cm^{-1}) (*18*).

The $CF_4/H_2/He$ plasma deposited film shows excellent superhydrophobic properties with a water contact angle of >170° on all surfaces tested (silicon wafer, smooth gold film, paper, and cotton). Figure 7 shows selected time-sequence images taken with a high speed camera. Upon landing on the superhydrophobic coating surface (Figures 7a and 7c), water droplet advances (panels 3 and 4) and then recedes (panels 5 and 6) with contact angle greater close to 170°. At this high contact angle, the three-phase (solid-liquid-air) contact line circumference is so small that the momentum loss due to liquid surface tension at the contact line is negligible and the droplet bounces back off the surface almost elastically (panels 7 and 8). On hydrophilic surfaces (Figure 7b), the water droplet spreads with a very low contact angle and never recedes back. On untreated cotton surface (Figure 7d), the water droplet advances with a contact angle greater than the hydrophobic flat surfaces, but does not recede. The water is eventually absorbed into the cotton.

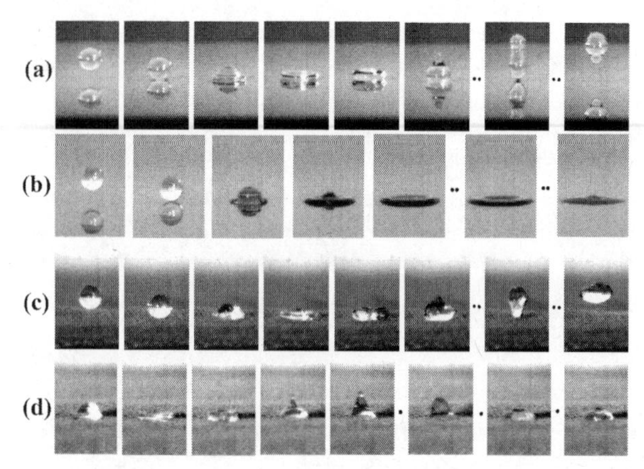

Figure 7. Time-sequence images of water droplet falling on (a) plasma-treated superhydrophobic gold film, (b) clean Si wafer, (c) plasma-treated superhydrophobic cotton and (d) untreated cotton. All panels are at 2 msec interval, except the ones separated with dots.(9)

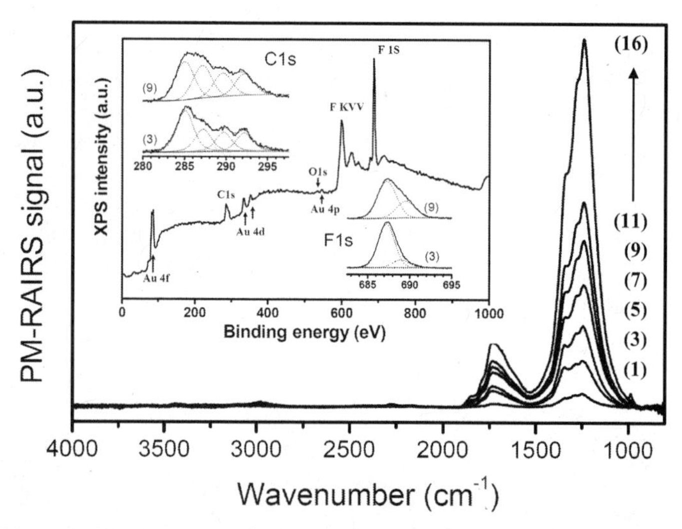

Figure 8. PM-RAIRS and XPS (inset) of superhydrophobic coatings deposited with scanning $CF_4/H_2/He$ plasma (rf power = 300 W). The number of plasma treatments is indicated in the figure.(9)

The chemical functional groups in the superhydrophobic coatings deposited via the scanning atmospheric $CF_4/H_2/He$ plasma process are identified with PM-RAIRS (Figure 8). The observed vibrational peaks are the C-F stretching vibrations at 1245 cm^{-1}, 1285 cm^{-1}, and 1348 cm^{-1} and the C=O stretching vibration at 1730 cm^{-1}. The fact that C-F and C=O vibrations of the coating are broad and not well resolved indicates that the coated material is amorphous and highly cross-linked.

The chemical composition of the superhydrophobic coating is obtained from XPS analysis. Figure 8 shows a survey XPS spectrum of a 3-pass coating on a gold film. The main components in the coating are carbon and fluorine. There are trace amounts of nitrogen and oxygen detected. These species are incorporated because the plasma is generated in air. The high-resolution C1s region shows 4 different carbon species (*22*). [C-C species at 285 eV, C-F and C=O species at 287.2 eV, CCF_2 at 289.8 eV, and CF_2–CF_2 species species at 292.4 eV]. The atomic composition of the superhydrophobic coating is represented as CF_x($x<1$).

There are several features in XPS data implying that the CF_x superhydrophobic coating is composed of particulates. The Au substrate peak is still detectable even after 12 passes of the plasma deposition process. This indicates that there are still small amounts of bare gold surface regions that are not covered with the CF_x coating. In high resolution spectra, the gold peaks are not shifted at all due to charging, while the C1s and F1s peaks are significantly shifted and become broader as the number of plasma deposition increases. These are due to differential charging of the sample in XPS measurements. The differential charging is observed for samples consisting of non-conducting domains with varying thickness and incomplete coverage on a conducting substrate.

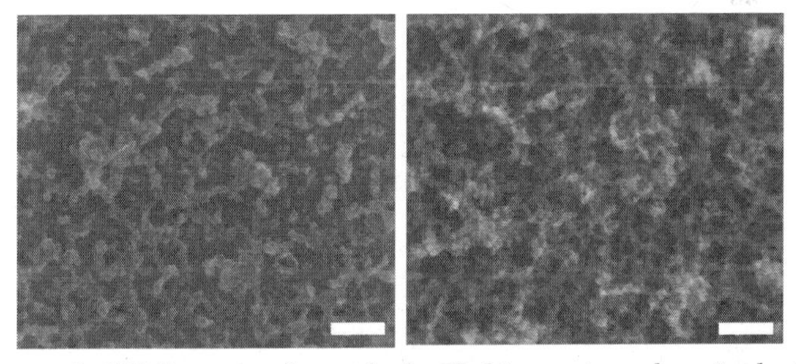

Figure 9. SEM images of superhydrophobic coatings deposited with scanning atmospheric $CF_4/H_2/He$ plasma (rf power = 300 W) after (a) 3 passes and (b) 10 passes. Scale bar = 200 nm.

The particulate nature of the plasma-deposited superhydrophobic coating is confirmed with high resolution SEM analysis. These images clearly show nanoparticulates of CF_x deposits and some flat regions of the Au substrate. Particulate formation is known to occur at relatively high density of reactive radicals and ions in the plasma (23,*24*). Under these conditions diffusion path lengths of reactive species are very short and gas-phase reactions take place. Upon reaching a critical number density, rapid condensation of the gas-phase species is triggered, resulting in particulate deposition onto the substrate (*25*). In AFM analysis of these coatings, it was found that the root-mean square roughness of the CF_x coatings increases from 7.2 nm for 3 passes to 11.8 nm for 7 passes, and 16.8 nm for 11 passes (*9*). These nano-scale topographic features gives the surface roughness required to produce the superhydrophobicty (*20,21*).

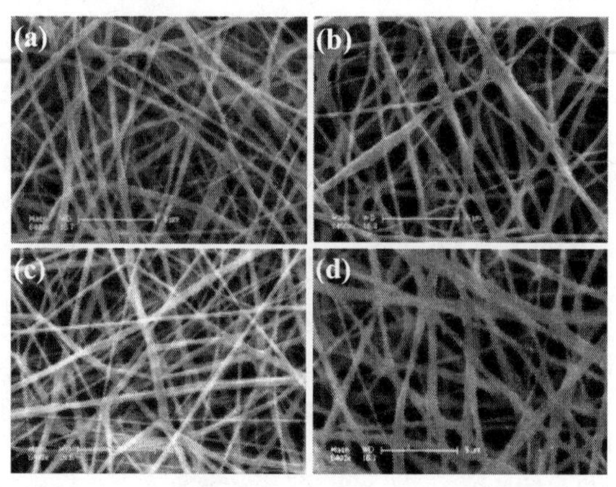

Figure 10. SEM images of (a) prestine, (b) 100W O_2/Ar plasma treated, (c) 200 W CH_4/He plasma treated, and (d) 250W CF_4/H_2/He plasma treated nanofibers of poly[bis(2,2,2-trifluoroethoxy) phosphazene].(26)

The scanning atmospheric rf plasma process is very powerful to change the surface chemistry of the substrate or deposited coating layers on the substrate, but gently enough that polymeric nanofibers can be treated without any morphological changes. This was demonstrated by treating electrospun non-woven nanofiber mats of poly[bis(2,2,2-trifluoroethoxy)phosphazene] with the scanning atmospheric rf plasma and comparing their fiber morphologies and surface chemistry (*26*). Figure 10 exhibits the SEM images of the poly[bis(2,2,2-trifluoroethoxy)phosphazene] naonfibers before and after the plasma treatment. There is no discernable changes in the fiber diameter distribution or any hint of fiber melting. By plasma treatments, the surface chemistry of these nanofibers can be tuned from highly hydrophilic (water contact angle = 5°) to superhydrophobic (water contact angle = 153°).

Conclusions

The scanning atmospheric rf glow-discharge plasma system is developed and demonstrated for surface decontamination and protective coating deposition. The plasma source consisted of a cylindrical electrode with a tubular dielectric barrier to separate the electrode and the grounded cover. The plasma is generated at the bottom open space of the grounded cover and used to treat the sample surface mounted on a moving stage. Since the plasma is generated in an open space in ambient air, nitrogen, oxygen and water vapor are also excited inside the plasma. The oxygen plasma is very efficient to decompose organophosphorous nerve agents. When CH_4 is fed into the plasma, hydrophobic hydrocarbon coatings can be deposited on various substrate surfaces. Without any pre-treatment of the substrate surface, superhydrophobic coatings can also be produced even on optically flat substrate surfaces as well as textured surfaces. A promissing aspect of the scanning atmospheric plasma process is that it can be applied to any surface regardless of the surface chemistry of the substrate – metals, metal oxides, Kimwipe papers, cottons, nanofibers, etc. Because of this versatility of applicable substrates, this plasma process can be utilized to a wide range of applications such as decontaminate of delicate substrates, self-cleaning fabrics, planar microfluidics, corrosion and adhesion protections, etc. The system does not require any vacuum lines and is operated in an in-line mode, not in a batch mode. So it can easily be scaled up for applications to large substrate surfaces or continuous processing.

Acknowledgements

This work was partially supported by the 3M Nontenured Faculty Grant.

References

1. Chu, P. K.; Chen, J. Y.; Wang, L. P.; Huang, N. *Mater. Sci. Eng.* **2002**, *R36*, 143.
2. Moon, S. Y.; Choe, W.; Kang, B. K. *Appl. Phys. Lett.* **2004**, *84*, 188.
3. Zhu, W.-C.; Wang, B.-R.; Yao, Z.-X.; Pu, Y.-K. *J. Phys. D: Appl. Phys.* **2005**, *38*, 1396.
4. Yang, Y.-C.; Baker, J. A.; Ward, J. R. *Chem. Rev.* **1992**, *92*, 1729.
5. Herrmann, H. W.; Selwyn, G. S.; Henins, I.; Park, J.; Jeffery, M.; Williams, J. M. *IEEE Trans. Plasma Sci.* **2002**, *30*, 1460.
6. Rahul, R.; Stan, O.; Rahman, A.; Littlefield, E.; Hoshimiya, K.; Yalin, A. P.; Sharma, A.; Pruden, A.; Moore, C. A.; Yu. Z.; Collins, G. J. *J. Phys. D. Appl. Phys.* **2005**, *38*, 17501.
7. Birmingham, J. G. *IEEE Trans. Plasma Sci.* **2004**, *32*, 1526.
8. Laroussi, M. *Plasma Processes & Polymers* **2005**, *2*, 391.
9. Kim, S. H.; Kim, J.-H.; Kang, B.-K.; Uhm, H. S. *Langmuir* **2005**, *21*, 12213.
10. Kim, S. H.; Kim, J.-H.; Kang, B.-K. *Langmuir* **2007** *23*, 8074.
11. Pearse, R. W. B.; Gaydon, A. G. The Identification of Molecular Spectra; 4th ed. John Wiley & Sons, Inc.: New York, 1976.
12. Eliasson, B.; Hirth, M.; Kogelschatz, U. *J. Phys. D: Appl. Phys.* **1987**, *20*, 1421.
13. Ahn, H.-S.; Hayashi, N.; Ihara, S.; Yamabe, C. *Jpn. J. Appl. Phys.* **2003**, *42*, 6578.
14. Masui, T.; Hirai, H.; Imanaka, N.; Adachi, G. *Phys. Stat. Sol. A* **2003**, *198*, 364.
15. Panda, R. N.; Hsieh, M. F.; Chung, R. J.; Chin, T. S. *J. Phys. Chem. Solids* **2003**, *64*, 193.
16. Tkalcec, E.; Sauer, M.; Nonninger, R.; Schmidt, H. *J. Mater. Sci.* **2001**, *36*, 5253.
17. Kim, J.-H.; Liu, G.; Kim, S. H. *J. Mater. Chem.* **2006**, *16*, 977.

18. Kim, S. H.; Kim, J.-H.; Kang, B.-K. *J. Vac. Sci. Technol.* **2008**, *26*, 123.
19. Hopkins, J;. Badyal, J. P. S. *Langmuir* **1996**, *12*, 4205.
20. Wenzel, R. N. *Ind. Eng. Chem.* **1936**, *28*, 988.
21. Cassie, A. B. D.; Baxter, S. *Tran. Faraday Soc.* **1944**, *40*, 546.
22. Beamson, G.; Briggs, D. High Resolution XPS of Organic Polymers: The Scienta ESCA300 Database. John Wiley & Sons (New York: 1992).
23. Teare, D. O. H.; Spanos, C. G.; Ridley, P.; Kinmond, E. J.; Roucoules, V.; Badyal, J. P. S. *Chem. Mater.* **2002**, *14*, 4566.
24. Takahashi, K.; Tachibana, K. *J. Vac. Sci. Technol. A* **2001**, *19*, 2055.
25. Hollenstein, C. *Plasma Phys. Control Fusion* **2000**, *42*, R93.
26. Allcock, H. R.; Steely, L. B.; Kim, S. H.; Kim, J. H.; Kang, B.-K. *Langmuir* **2007**, *23*, 8103.

Indexes

Author Index

Subject Index

E

F

G

P

Q

R

S

T

U

V

W